NCERT WORKBOOK
BIOLOGY

Volume 1
CLASS
XI

AF657585

NCERT WORKBOOK BIOLOGY

Volume 1
CLASS
XI

FOR SURE SHOT SUCCESS IN
NEET

Authors
Sanubia Saleem
Kavita Thareja
K. Anita

arihant
ARIHANT PRAKASHAN (Series), MEERUT

arihant

卐 **Administrative & Production Offices**

Regd. Office
'Ramchhaya' 4577/15, Agarwal Road, Darya Ganj, New Delhi -110002
Tele: 011- 47630600, 43518550

Head Office
Kalindi, TP Nagar, Meerut (UP) - 250002, Tel: 0121-7156203, 7156204

卐 **Sales & Support Offices**
Agra, Ahmedabad, Bengaluru, Bareilly, Chennai, Delhi, Guwahati, Hyderabad, Jaipur, Jhansi, Kolkata, Lucknow, Nagpur & Pune.

卐 **ISBN** 978-93-25294-01-1

PO No : TXT-XX-XXXXXXX-X-XX

Published by Arihant Publications (India) Ltd.

For further information about the books published by Arihant, log on to www.arihantbooks.com or e-mail at info@arihantbooks.com

Follow us on

PREFACE

NCERT books are the most valuable assets for NEET aspirants. More than 90% questions asked in NEET 2019 and 2020 were based on concepts of NCERT. This simply means conquering NEET requires a firm grip over NCERT concepts. However, it is not so easy as NCERT is one of those theoretical books in which each line has many interpretations.

So far multiple resources are there in the market to help in gripping the NCERT concepts. But all of them are MCQs based practice books. Practice with all such resources makes the learning monotonous and boring.

"NCERT Workbook Biology" is altogether different from all such resources. This innovative book is written and developed keeping in mind the concept of 'Fun with learning'. The speciality of the book is, it makes you apply the conceptual knowledge in answering different types of questions which just require a word or maximum a sentence as their answer. In this way you can revise & strengthen your concept in an interesting manner.

A variable typology of questions given in each chapter, like;

- True-false
- Word correction in sentences
- Sequence arrangement and completion
- Small matchings
- Analogy completion
- Figure labelling and figure identification, etc.
- Fill in the blanks
- Sequential term arrangement
- Odd one out type
- Flow chart completion

Such a variety in question types reduces monotonicity of subject and increases concept understanding and figure visualisation.

This matchless fun filled practice book will help NEET aspirants in gripping the NCERT concepts to their maximum. We hope all of you will enjoy the learning along with practice while exploring this unmatched resource and secure sure shot success in NEET.

Authors

CONTENTS

CHAPTER ONE

THE LIVING WORLD

Topic 1 : What is Living?

1. How many (in number) of the following are the defining characteristics of living organisms?

   Growth, Reproduction, Metabolism, Consciousness, Cellular organisation

   ____

2. Growth is a common feature exhibited by both living and non-living.

   True ☐ False ☐

3. Multicellular and unicellular organisms grow by ______________ (cell division/cellular organisation).

4. Non-living organisms perform growth by ______________

5. In ______________ the growth by cell division occurs continuously throughout their lifespan and in ______________, the growth is seen only up to a certain age.

6. Write a common mode of reproduction in the following.

   Fungi, Filamentous algae, Protonema of mosses

   ____

7. For unicellular organisms like bacteria, unicellular algae or *Amoeba,* ______________ is synonymous with ______________ .

8. Out of the given organisms, how many of them do not reproduce?

   Mules, Worker bees, Infertile human couple, Horse

   ____

9. What are the twin characteristics of growth?

____

10. Match the following columns.

	Column I (Organisms)		Column II (Modes of reproduction)
A.	Fungi	1.	Asexual spores
B.	*Amoeba*	2.	Binary fission
C.	*Hydra* and yeast	3.	True regeneration
D.	*Planaria*	4.	Budding

A. ______________ B. ______________

C. ______________ D. ______________

11. The sum total of all chemical reactions occurring in our body is called ______________

12. Metabolic reactions can be demonstrated outside the body in cell-free system.

True ☐ False ☐

13. Isolated metabolic reactions *in vitro* are not living things, but surely living reactions.

True ☐ False ☐

14. ______________ affects reproduction in seasonal breeders in both plant and animals.

15. Arrange the following organisational complexity of living beings from simple to complex form.

Organ system, Tissue, Cellular level, Molecular lever, Atomic level, Individual, Organ

____

16. Human being is the only organism who is aware of himself, i.e. has metabolism.
(*Rewrite the statement by changing the underlined word.*)

____

17. All living organisms present, past and future are linked to one another by the sharing of the common genetic material, but to varying degrees. (Name the genetic material)

____

Topic 2 : Diversity in the Living World

18. Each different kind of plant, animal or organism that you see, represents a ______________ (genus/species).

19. Who is known as 'The Darwin of the 20th century'? ______________

20. There is a need to standardise the naming of living organisms such that a particular organism is known by the same name all over the world. This process is called

____

21. In a given habitat, we have 20 plant species and 20 animal species. Should we call it as diversity or biodiversity? ______________

22. Choose the odd one out from the following.

Characterisation, Identification, Classification, Nomenclature, Systematics

____

23. Give full forms of the following abbreviation.
   (i) ICBN _______________
   (ii) ICZN _______________
   (iii) ICVN _______________

24. Nomenclature is possible only when the organism is _______________ correctly.

25. The process by which anything grouped into convenient categories based on some easily observable character is called _______________

26. Binomial nomenclature was given by _______________

27. Scientific names are printed in _______________ and are derived from _______________ .

28. What are the two components in a scientific name?

____

29. The first word in a scientific name denoting genus starts with a _______________ letter, while the second word denoting species starts with a _______________ letter.

30. What does Linn., represent in the biological name *Mangifera indica* Linn.,?

____

31. In *Mangifera indica*, the word *Mangifera* is a _______________

32. Choose the odd one out from the following scientific names.

Panthera tigris. Mangifera indica, Panthera leo, Columba LIVEA

____

33. The scientific term for different categories like plants and mammals is _______________

34. _______________ (Systematics/Taxonomy) takes into account evolutionary relationships between organisms.

Topic 3 : Taxonomic Categories

35. A taxonomic group of any ranking is _______________ (taxa/taxon).

36. The taxonomic categories are distinct biological _______________ and not merely _______________ aggregates.

37. Ascending and descending arrangement of taxonomic categories is known as <u>classification</u>.
(*Rewrite the statement by changing the underlined word.)*

____

38. Species is considered as the _______________ (smallest/largest) taxon of taxonomy.

39. Individual of which taxa interbreed freely?________________

40. In *Solanum tuberosum,* first and second words stand for ________________, respectively.

41. Family comprises of related species which has more characters in common in comparison to species of other genera.

True ☐ False ☐

42. As we go higher from species to kingdom then what will happen to the number of common characteristics? ________________

43. Write the missing taxonomic categories.

Kingdom ← Phylum ← ________ ← Order ← ________ ← Genus ← Species

44. The suffix used for plant family is ________ and for animal family is ________.

45. If two plants belong to the same division but in different orders, they may belong to the same. (Family/Class) ________________

46. How many of the given plants belong to genus–*Solanum*?

Potato, Brinjal, *Datura,* Tulip

____

47. 'Order' related to 'Brinjal' is ________________

48. In case of plants, classes with a few similar characters are assigned to a higher category called ________________ (phylum/division).

49. Suffix used for plant order is________________

50. Given below are certain biological terms/names. Write a single collecting term for each group.

(i) *Mangifera, Solanum, Panthera, Homo* ________________
(ii) Class, Order, Family, Species ________________
(iii) Primata, Diptera, Sapindales, Poales________________
(iv) Poaceae, Convolvulaceae, Anacardiaceae. ________________

51. Recognise the following flowchart and write the missing taxonomic hierarchy.

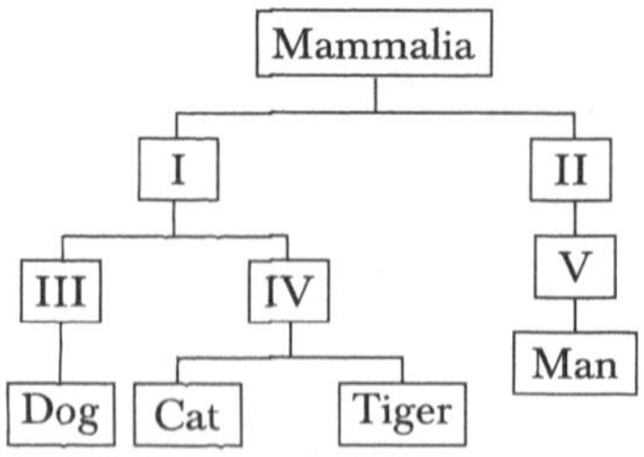

I. ________________ II. ________________

III. ________________ IV. ________________

V. ________________

52. Give the name of the category given in bracket on the basis of the adjoining information.

(i) *Felis* and *Panthera* (Family)____________________

(ii) Solanaceae and Convolvulaceae (Order) ____________________

53. Find the odd one out for housefly classification.

Musca, Chordata, Muscidae, Diptera, Insecta

____

54. Arrange the following taxonomic categories in ascending order.

I. Poaceae
II. Poales
III. Monocotyledonae
IV. *Triticum aestivum*
V. Angiospermae

____

55. Write the next higher taxonomic hierarchy for the following taxonomic groups.

(i) Pisces, Amphibia, Aves, Mammalia ____________________

(ii) Liliales, Poales, Commelinales ____________________

56. Give the scientific name of lion, leopard and tiger.

____

57. Write one similarity and one difference between cats and tigers on the basis of classification. ____________________

58. Match the following columns.

	Column I (Taxonomic categories)		Column II (Examples)
A.	Family	1.	*tuberosum*
B.	Kingdom	2.	Polymoniales
C.	Order	3.	Plantae
D.	Species	4.	Solanaceae

A. ______________ B. ______________

C. ______________ D. ______________

59. Consider the categories given in the box and group together those which belong to the similar hierarchy.

Cicer, *Petunia*, Asteraceae, Primata, Rodentia, Solanaceae

____

60. Name of an organism according to binomial nomenclature includes genus and species, likewise a new category added in trinomial nomenclature is ____________________

61. Complete the following analogy.

Taxonomic category : Taxonomical aid :: Kingdom : ____________

62. Names in which specific epithet and infra specific epithet are identical. ____________

63. Name the type specimen which is a duplicate of the holotype.____________

64. A specimen is called ____________ when no other type specimen exists.

65. *Rattus rattus* and *Homo sapiens sapiens* are ____________ and ____________, respectively.

66. Match the following columns.

	Column I (Types of taxonomy)		**Column II** (Features)
A.	Numerical taxonomy	1.	Nuclear and chromosomal studies
B.	Classical taxonomy	2.	Number of shared characters
C.	Practical taxonomy	3.	Utility of organism
D.	Karyotaxonomy	4.	α-taxonomy

A. ____________ B. ____________

C. ____________ D. ____________

67. Taxonomy based on identification of evolutionary units within species by determining their genetical inter-relationship is called ____________

68. Give a common term for the following words.

Holotype, Syntype, Epitype, Paratype

____

Topic 4 : Taxonomical Aids

69. A storehouse of collected plant specimens that are dried pressed and preserved on sheets is ____________

70. Herbaria serve as quick referral systems in taxonomic studies.

True ☐ False ☐

71. The label of a herbarium sheet does not carry information on ____________ (date of collection/height of the plant).

72. Plant species in botanical gardens are labelled indicating only its common name.

True ☐ False ☐

73. Largest herbarium in India is National Botanical Garden.
(*Rewrite the statement by changing the underlined word.*)

____

74. Match the following columns.

	Column I (Botanical gardens)		Column II (Locations)
A.	Indian Botanical Garden	1.	Kew (England)
B.	National Botanical Research Institute	2.	Howrah (India)
C.	Royal Botanical Garden	3.	Lucknow (India)
D.	Lloyed Botanical Garden	4.	Darjeeling (India)

A. ______________________ B. ______________________

C. ______________________ D. ______________________

75. Study the given flow diagram and write the missing word *A*.

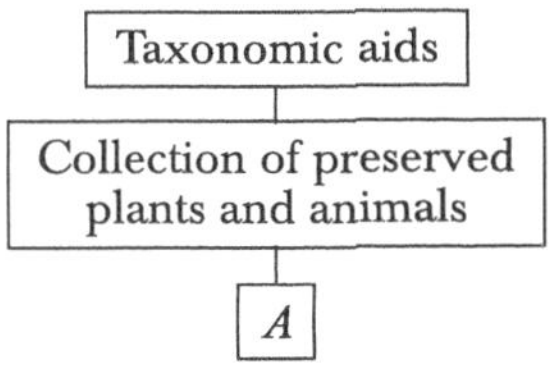

A. ____

76. Write the correct specimen name for the procedures done in museum for their preservation.

(i) Collecting, killing and pinning in boxes.______________________

(ii) Bodies stuffed with cotton. ______________________

77. Where would you keep the following for taxonomic study?

(i) Dinosaur skeleton ______________________

(ii) Living *Aloe vera* plant ______________________

(ii) *Cycas* cones ______________________

78. Wild animals are kept in protected environments under human care in

____

79. What is the purpose of zoological parks?

____

80. Name the scientist who introduced the concept of 'key'.______________________

81. In keys, a pair of contrasting characters is called ______________________

82. Couplet represents the choice made between two similar options during study.

True ☐ False ☐

83. Each statement in the couplet of key is called ______________________

84. Write the scientific terms for the statements given below.

(i) An actual account of habitat and distribution of plants of a given area.____________

(ii) A book useful in providing information for identification of names of species found in an area.____________

(iii) A book containing information of any one taxon. ____________

85. Taxonomic studies of various species of plants and animals are useful in agriculture, forestry, industry and in general for knowing our bioresources and their diversity.

True ☐ False ☐

86. A list of all species organised in some way (particularly alphabetised) is called

____

87. Match the following columns.

Column I	Column II
A. Robert Morison	1. *Systema Naturae*
B. Kew	2. Scientific names
C. ICBN	3. First monograph of a plant taxon
D. Carolus Linnaeus	4. Largest botanical garden

A. ____________ B. ____________

C. ____________ D. ____________

Answers

1. 3 (Metabolism, cellular organisation and consciousness).
2. True
3. Cell division
4. Accumulation of material on the surface
5. Plant, animal
6. Fragmentation
7. Reproduction, growth
8. 3 (Mules, sterile worker bees as well as an infertile human couple do not reproduce).
9. Increase in number of individuals and increase in mass.
10. A-1, B-2, C-4, D-3
11. Metabolism
12. True
13. True
14. Photoperiod
15. Atomic level → Molecular level → Cellular level → Tissue → Organ → Organ system → Individual
16. Self consciousness
17. DNA or RNA
18. Species
19. Ernst Mayr
20. Nomenclature
21. Biodiversity
22. Systematics is odd one out. Rest all are basic processes in taxonomy while systematics is the branch of biology that deals with comparative and evolutionary relationship between organisms.
23. (i) **ICBN**–International Code of Botanical Nomenclature
(ii) **ICZN**–International Code of Zoological Nomenclature
(iii) **ICVN**–International Code of Viral Nomenclature
24. Identified
25. Classification
26. Carolus Linnaeus
27. Italics, Latin
28. Generic name and specific epithet
29. Capital, small
30. Name of the author who first described the species.
31. Genus
32. *Columba LIVEA* is odd one out, as it is incorrectly written scientific name.
33. Taxa
34. Systematics
35. Taxon
36. Entities, morphological
37. Hierarchy
38. Smallest
39. Species
40. Genus and species
41. False. Genus comprises of related species which has more characters in common in comparison to species of other genera.
42. It will decreases.
43. Family, class
44. Acae, idae
45. Class
46. 2 (Potato, brinjal)
47. Polymoniales
48. Division
49. Ales
50. (i) Genus, (ii) Taxonomic categories, (iii) Order, (iv) Family
51. I. Carnivora, II. Primata, III. Canidae, IV. Felidae, V. Hominidae
52. (i) Felidae
(ii) Polymoniales
53. Chordata is odd one out. Housefly belongs to phylum–Arthropoda.
54. IV → I → II → III → V
55. (i) Chordata (Phylum),
(ii) Monocotyledonae (Class)
56. Lion – *Panthera leo*, Leopard – *Panthera pardus*, Tiger– *Panthera tigris*
57. They belong to same family – Felidae and different genus – *Felis* (cat) and *Panthera* (tiger)
58. A-4, B-3, C-2, D-1
59. *Cicer* and *Petunia* (Genus), Asteraceae and Solanaceae (Family), Primata and Rodentia (Order).
60. Subspecies
61. Key or Herbarium
62. Autonyms
63. Isotype
64. Neotype
65. Tautonym and autonym
66. A-2, B-4, C-3, D-1
67. Experimental taxonomy
68. Nomenclature type
69. Herbarium
70. True
71. Height of the plant
72. False. Plant species in botanical gardens are labelled indicating its botanical/scientific name and its family.
73. Indian Botanical Garden, Howrah.
74. A-2, B-3, C-1, D-4
75. *A*–Museum
76. (i) Insects, (ii) Large animals like birds and mammals
77. (i) Museum, (ii) Botanical garden, (iii) Herbarium in boxes
78. Zoological parks
79. To learn about food, habits and behaviour of wild animals.
80. John Ray
81. Couplet
82. False. Couplet represents the choice made between two opposite options during study.
83. Lead
84. (i) Flora, (ii) Manual, (iii) Monograph
85. True
86. Catalogue
87. A-3, B-4, C-2, D-1

CHAPTER TWO

BIOLOGICAL CLASSIFICATION

Topic 1 : Introduction and Kingdom–Monera

1. Biological classification was first proposed by ____________ who used simple morphological characters to classify plants and animals.

2. Aristotle divided animals into two groups on the basis of the presence and absence of ____________ (WBCs/RBCs/Lymphocytes).

3. Two kingdom classification system was given by ____________

4. The demerit of two kingdom system is that it does not distinguish between the eukaryotes and prokaryotes, unicellular and multicellular organisms and photosynthetic and non-photosynthetic organisms.

   True ☐ False ☐

5. Haeckel gave three kingdom classification which include kingdom ____________, ____________ and ____________.

6. Complete the following analogy given below.

   Copeland : Four kingdoms : : ____________ : Five kingdoms.

7. In two kingdom classification, *Euglena* was placed in kingdom ____________.

8. Out of the following how many criteria were used for five kingdom classification?

Cell structure, Body organisation, Mode of nutrition, Reproduction, Phylogenetic relationship, Nuclear membrane composition.

   ____

9. Study the following table carefully and write the correct words for *A*, *B*, *C* and *D*.

Characters	Monera	Protista	Fungi	Plantae	Animalia
Cell type	*A*	Eukaryotic	Eukaryotic	Eukaryotic	Eukaryotic
Cell wall	*B*	Present in some	Present	Present	Absent
Nuclear membrane	Absent	Present	Present	Present	*C*
Body organisation	Cellular	Cellular	*D*	Tissue/Organ	Tissue/Organ/ Organ system

A. ________________ B. ________________

C. ________________ D.________________

10. In the given diagram, *A*, *B* and *C* are autotrophs, heterotrophs and both (autotrophs and heterotrophs), respectively.

Write down how many of the kingdoms out of the five kingdoms belong to *A*, *B* and *C*?

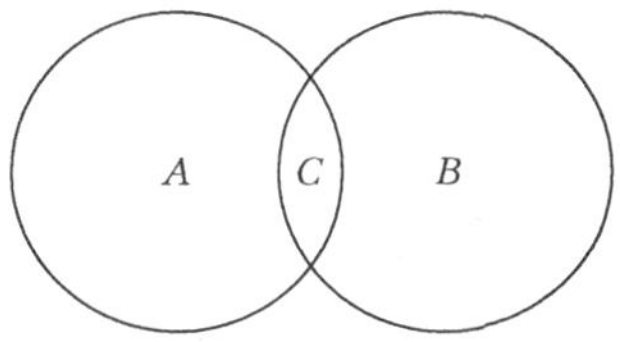

A. ________________________________

B. ________________________________

C. ________________________________

11. Match the following columns.

	Column I (Modes of nutrition)		Column II (Examples)
A.	Chemosynthetic autotrophs	1.	*Xanthomonas*
B.	Photosynthetic autotrophs	2.	*Beggiatoa*
C.	Parasitic organisms	3.	*Chlorobium*
D.	Holozoic	4.	*Amoeba*

A. ________________ B. ________________

C. ________________ D.________________

12. Earlier classification systems included bacteria, Blue-Green Algae (BGA), fungi, mosses, ferns, gymnosperms under plants. Name the character which unified all these organisms in one kingdom. ________________________________

13. The three domain classification divides the kingdom ______________ into two domains.

14. The sole members of kingdom–Monera are ______________ and ________________ .

15. Write the names of different bacteria according to shapes given below.

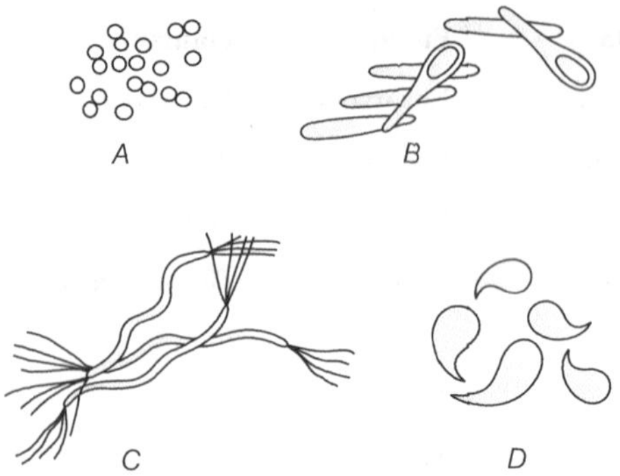

A. ____________________

B. ____________________

C. ____________________

D. ____________________

16. Write true and false for the given statements with respect to bacteria.

(i) They are the most abundant microorganism found almost everywhere.

True ☐ False ☐

(ii) They cannot survive in extreme habitats like hotsprings, desert, snow and deep sea.

True ☐ False ☐

(iii) Many of them live in or on other organisms as parasites.

True ☐ False ☐

17. Bacterial structure is very ____________, but they are very __________ in behaviour.

18. According to the mode of nutrition,

(i) the bacteria which depend on other organisms or on dead organic matter for food are ____________________.

(ii) the bacteria which synthesise their own food from inorganic substrate are ____________________.

19. Name the two types of autrophic nutrition performed by bacteria.

____

20. Match the following columns.

	Column I (Types of Archaebacteria)		Column II (Habitats)
A.	Halophiles	1.	Hotsprings and extreme acidity
B.	Thermoacidophiles	2.	Extreme salty areas
C.	Methanogens	3.	Marshy areas

A. ______________ B. ______________ C. ______________.

21. Archaebacterial cell wall is made up of ____________________ whereas other bacterial cell wall is made up of ______________.

22. Read the following characters.

(i) Found in the gut of ruminants.

(ii) Commercially used for methane (biogas) production.

Above characters are related to group of organisms collectively called as ________________.

23. Name the bacterial group which is also known as 'True bacteria'. ____________________ .

24. Identify the following organism *'X'* and its structures *A* and *B*.

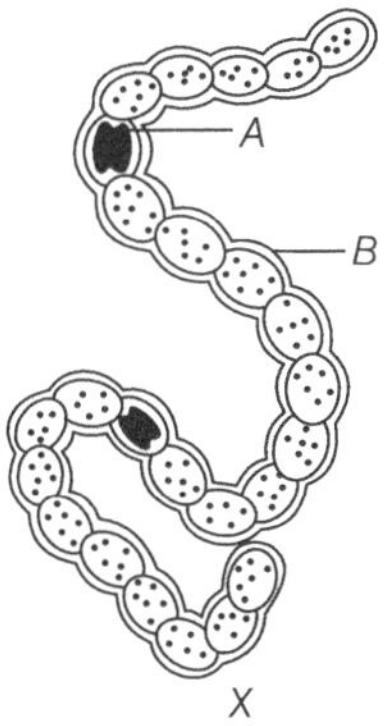

A. ____________________________

B. ____________________________

X. ____________________________

25. The cyanobacteria are also referred to as __________________________________.

26. Choose the odd one out with respect to eubacteria.

Spirulina, Methanococcus, Nostoc, Anabaena

____

27. Cyanobacteria have chlorophyll ___________ (*a*/*b*/*c*) similar to green plants.

28. Given below are characteristics of eubacteria, write whether these are photosynthetic autotrophic bacteria or chemosynthetic autotrophic bacteria.

(i) Oxidise various nitrates, nitrites, ammonia and use the released energy for their ATP production.____

(ii) Form blooms in polluted water bodies. ________________________________

(iii) Play a great role in recycling of nutrients like nitrogen, phosphorus, iron and sulphur.

____

29. Specialised cells present in cyanobacteria for fixing atmospheric nitrogen is known as

______________________ .

30. Choose the odd one out with respect to disease caused by bacteria.

Cholera, Typhoid, Citrus canker, Influenza

____

31. Blue-green algae are unicellular, colonial or filamentous, freshwater or marine or terrestrial algae.

True ☐ False ☐

32. Identify the below 'X' process in bacteria and write its parts A, B and C.

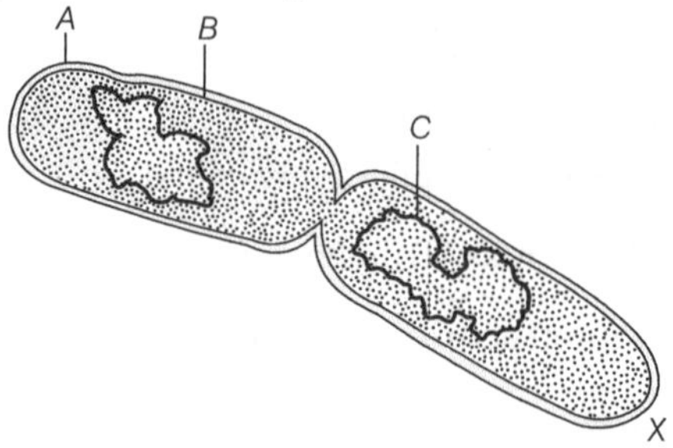

A. ______

B. ______

C. ______

X. ______

33. Given below are some economic importance of heterotrophic bacteria. How many of them are correct?

Antibiotic production, Curd making, Fodder crop, Decomposition, Diatomaceous earth, Nitrogen-fixation.

______

34. Sometimes under unfavourable conditions, bacteria produce fragmentation. (*Rewrite the statement by changing the underlined word.*)

______

35. Sexual reproduction in bacteria occurs by transfer of DNA from one to another by three methods, i.e. ______, ______ and ______.

36. Name the organism based on the following characters.

(i) Lack cell wall (ii) Anaerobic

(iii) Smallest cell (0.1-0.15 μm in length)

______

37. Write the full form of following.

(i) MLOs ______

(ii) PPLOs ______

38. Name one disease caused by mycoplasma in plants and animals each.

______

Topic 2 : Kingdom–Protista

39. In which kingdom, chrysophytes, dinoflagellates, euglenoids, slime moulds and protozoans are included?

______

40. How many of the given features are correct with respect to members of kingdom-Protista?

Single-celled, Eukaryotic, Membrane bound organelles, Lacks nucleus, Terrestrial, Connecting links between plants, animals and fungi

______

41. Complete the following analogy given below with respect to Protista.

Asexual reproduction : Sexual reproduction : : ______________ : __________________ .

42. Write the common biological term.

Diatoms, Desmids, Golden algae

____

43. The cell wall is composed of two thin overlapping shells which fit together like a soap case in ________________________ .

44. Siliceous frustules of diatoms being indestructible piled up at the bottom of ocean and formed a thick bed over billions of years. Such a thick bed is known as

__________________________ .

45. Which of the following is not the economic importance of 'Diatomaceous earth' or 'Diatomite'.

Polishing, Filtering oil and syrups, Biogas production, Making sound and fireproof rooms, Insulation material in refrigerators, Source of sodium silicate.

____

46. Dinoflagellates are mostly marine water ________________________________ (saprophytic/photosynthetic) organisms .

47. ________________ are the chief producers in the oceans.

48. Which group of organisms is represented by the given figure?

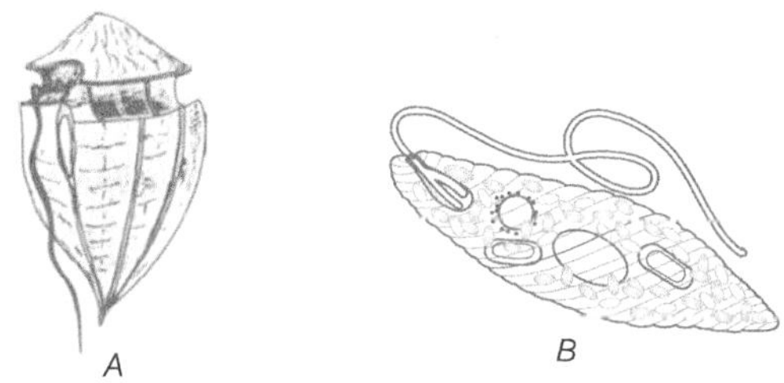

A. ________________________________

D. ________________________________

49. Which protist group possesses two flagella, one lies longitudinal and other transversely?

____

50. Name a colourless dinoflagellate famous for bioluminescence. ______________________

51. Choose the odd one out.

Ceratium, Gonyaulax, Noctiluca, Euglena

____

52. Identify the dinoflagellate based on the given features.

(i) Form 'red tides' in warm coastal water.

(ii) Release toxin called saxitoxin to kill prey.

(iii) Cell wall has stiff cellulosic plates on the outer surface.

____

53. Dinoflagellates appear yellow, green, brown, blue or red depending on the main pigments present in their cells.

True ☐ False ☐

54. Euglenoids are freshwater organisms having a protein rich layer called ________________, which makes their body flexible.

55. Write the number of flagella in euglenoids.____________________________

56. *Euglena* is called a plant due to the presence of ______________________ and an animal due to _______________ nutrition.

57. The slime moulds are _________________ (photosynthetic protists/saprophytic protists).

58. The multinucleate slimy mass of protoplasm forming the body of slime moulds is known as <u>myxamoeba</u>.
(*Rewrite the statement by changing the underlined word.*)

____

59. Plasmodial slime mould under unfavourable conditions differentiates and forms fruiting bodies, i.e. ________________ bearing spores. A spore possesses a cell wall made up of _______________. These spores are dispersed *via* ____________________.

60. The body of slime moulds moves along the decaying twigs and leaves engulfing organic material.

True ☐ False ☐

61. Name the protist group based on the given below characters.

(i) Heterotrophs (ii) Predators or Parasites

(iii) Believed to be primitive relative of animals.

____

62. Match the following columns.

Column I (Groups of protozoan)	**Column II** (Examples)
A. Amoeboid	1. *Paramoecium*
B. Flagellated	2. *Plasmodium*
C. Ciliated	3. *Entamoeba*
D. Sporozoans	4. *Trypanosoma*

A.______________ B. ______________ C. ______________ D.______________

63. ______________________ is a protozoan parasite of large intestine of human beings and causes the disease dysentery.

64. Identify the given figure of protozoan protist which causes kala-azar and sleeping sickness in human.

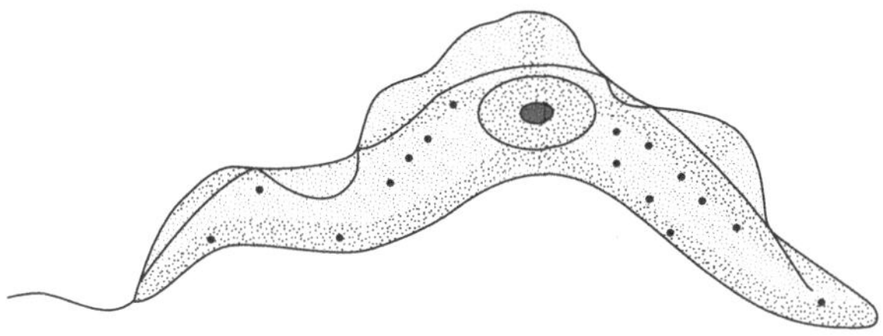

____

65. The sporozoans are all internal ____________________ that typically have an infective cyst stage in their life cycle. An example of sporozoan is the genus ______________, which causes malaria.

66. Identify the given figure *'X'* and write the name of structures *A*, *B*, *C* and *D*.

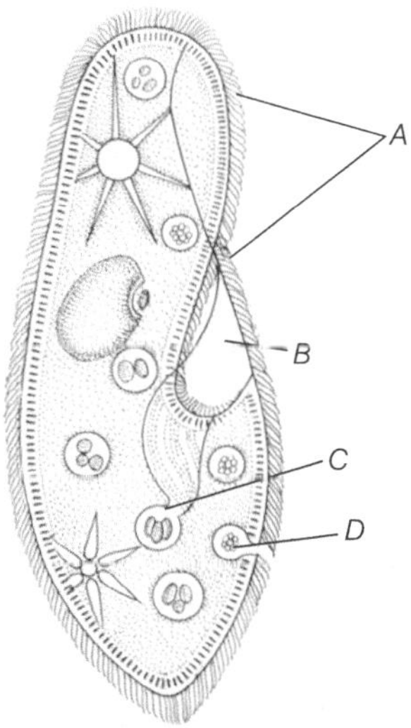

X

A. ______________________________

B. ______________________________

C. ______________________________

D. ______________________________

X. ______________________________

67. Write the names of the group of protozoan based on their characters.

(i) They move to capture their prey by forming pseudopodia and some marine forms have silica shells on their surface. ______________________________

(ii) Actively moving organisms having a cavity (gullet) that opens to the outside of the cell surtacc. ______________________________

Topic 3 : Kingdom-Fungi

68. The fungi constitute a unique kingdom of ______________ organisms.

69. Give one an example of each

(i) Edible fungi ____________

(ii) Poisonous fungi ____________

70. With the exception of mushrooms, all other fungi are filamentous. (*Rewrite the statement by changing the underlined word*).

____

71. The causal organism of wheat rust is ______________ (*Ustilago/Puccinia*).

72. Fungi prefer to grow in ________________ environment.

73. Fungal cell walls composed of ____________ and ____________ .

74. A unicellular fungi, ____________ is used to make bread and beer.

75. Write the scientific names of the following fungi.

(i) Bread mould ________________________________

(ii) Mushroom ________________________________

(iii) Smut ________________________________

(iv) Rust fungus________________________________

76. Categories the types of spores into their production type in fungi.

Zoospores, Aplanospores, Conidia, Ascospores, Basidiospores

(i) Exogenous ________________________ .

(ii) Endogenous ________________________ .

77. Coenocytic hyphae have septal walls in its filaments.

True ☐ False ☐

78. Except yeast, the bodies of fungi consist of long, slender, thread-like structure called ____________, which form a network called ____________ .

79. Name the class of fungi which shows all the three types of sexual reproduction.

Isogamous, Anisogamous, Oogamous

____

80. The lichens live as symbionts having ____________ and ____________ as symbiotic partners.

81. Write the biological term for the following sentences.

    (i) Fungi that absorb soluble organic matter from dead substrates. ____________

    (ii) Fungi that absorb nutrients directly from the cytoplasm of living host. ____________

    (iii) Association of fungi with roots of higher plants. ____________

82. Give an example of mycorrhizal fungi. ____________

83. Dikaryophase is a specific property of ____________ and ____________ classes of fungi.

84. Given below is the flowchart of the types of reproduction in fungi.

Reproduction

Vegetative means	Asexual spores	Sexual spores
(i) ________	(i) ________	(i) ________
(ii) ________	(ii) ________	(ii) ________
(iii) ________	(iii) ________	(iii) ________

Write the correct words in the blanks given.

85. Name the three steps of sexual cycle in fungi.

    ____

86. In fungi, karyogamy is the fusion of two ____________.

87. In fungi, the fusion of protoplasms between two motile or non-motile gametes is called <u>cytokinesis</u>.
    (*Rewrite the statement by changing the underlined word.*)

    ____

88. Given flowchart is the representation of life cycle of members of class–Ascomycetes and Basidiomycetes. Identify and write *A*, *B* and *C*.

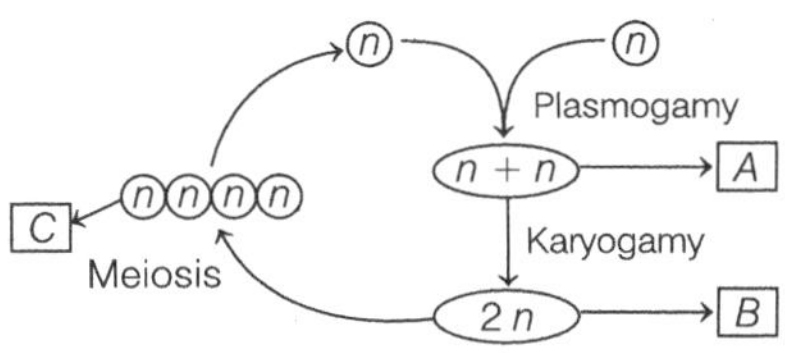

A. ____________

B. ____________

C. ____________

89. The morphology of mycelium, mode of spore formation and type of fruiting bodies forms the basis of classification of fungi.

    True ☐ False ☐

90. Name the type of fungi, based on its description.

Fungi
- True fungi, perithecium fruiting bodies. ____________
- Exogenous spores, asexual reproduction absent, non-motile spores. ____________

91. Members of ________________ are found in aquatic habitats and on decaying wood in moist and damp places or as obligate parasites.

92. Identify the fungi *'X'* and its parts *A, B, C* and *D* in the diagram given below.

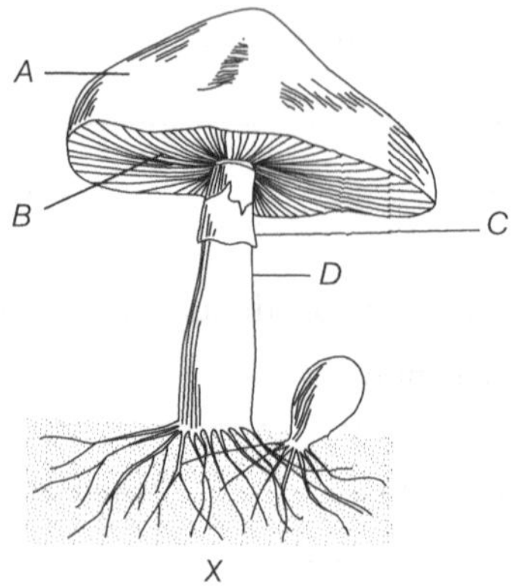

A. ______________________

B. ______________________

C. ______________________

D. ______________________

X. ______________________

93. Choose the odd one out.

Phycomycetes, Ascomycetes, Zygomycetes, Basidiomycetes, Deuteromycetes

____

94. Name the class of fungi on the basis of their characters.

(i) They are saprophytic, decomposers, parasitic or coprophilous (growing on dung).

____

(ii) They grow in soil, on logs and tree stumps and living in plant bodies as parasites.

____

95. Fill in the table for classification of fungi.

Fungi classes	Features	Members
(A)	Mycelium septate and branched	*Neurospora*
Phycomycetes	(B)	*Rhizopus*
Deuteromycetes	Mycelium septate and branched	(C)
(D)	Mycelium branched and septate	(E)

A. ____________ B. ____________ C. ____________

D. ____________ E. ____________

96. Identify the fungi shown below in figures *A, B* and *C*.

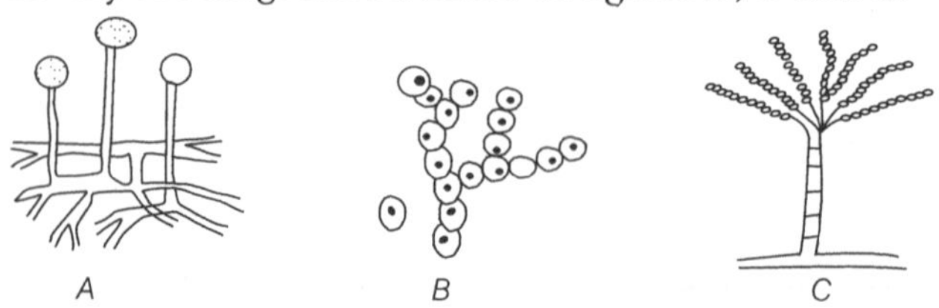

A. ______________________

B. ______________________

C. ______________________

97. Rewrite the statements by changing the underlined words.
    (i) The fruiting body of *Agaricus* is called mericarp. ______
    (ii) Coenocytic mycelium is uninucleate and septate. ______
    (iii) The parasitic fungi on mustard is *Mucor*. ______

98. Which fungus is used extensively in biochemical and genetic works? ______

99. The basidiospores are ______ produced on the ______ .

100. Name the common class of these fungi.

Mushroom, Bracket fungi, Smut fungi

______

101. Most of the members of the class Ascomycetes are decomposers of litter and help in mineral cycle.
(*Rewrite the statement by changing the underlined word.*)

______

102. The wonder drug, penicillin is extracted from which of the species of *Penicillium*?

______

103. In *Penicillium*, the asexual reproduction takes place by ______ spores.

104. Match the following columns.

	Column I		Column II
A.	Source of antibiotics	1.	*Penicillium, Streptomyces*
B.	Rust and smut diseases	2.	*Puccinia, Ustilago*
C.	Early blight of potato	3.	*Alternaria solani*
D.	Late blight of potato	4.	*Phytophthora infestans*

A. ______ B. ______ C. ______ D. ______

105. Which fungi is called '*Drosophila* of plant kingdom'? ______

106. In Ascomycetes, ascospores are produced in a sac called ______

107. Choose the odd one out for the edible form of Ascomycetes.

Morels, Puffballs, Truffles

______

108. Rewrite the statements by changing the underlined word.

(i) In Phycomycetes, asexual reproduction takes place by zoospores (motile) and conidiophores (non-motile).

___

(ii) In Ascomycetes, asexual spores are conidia and sexual spores are basidiospores.

___

(iii) In Deuteromycetes, sex organs are absent but plasmogamy is brought about by the fusion of two vegetative cells. ___

109. In Phycomycetes, zygospore is formed by the fusion of two gametes. These gametes are similar in morphology, i.e. ___ or dissimilar, i.e. ___.

110. In which class of fungi, asexual spores are generally not found, but vegetative reproduction by fragmentation is common?

___

111. How many basidiospores are produced from the basidium? ___

112. Match the following columns.

	Column I (Fungi classes)		Column II (Common names)
A.	Ascomycetes	1.	Club fungi
B.	Basidiomycetes	2.	Sac fungi
C.	Deuteromycetes	3.	Algae fungi
D.	Phycomycetes	4.	Fungi imperfecti

A.___ B. ___ C. ___ D.___

Topic 4 : Kingdom–Plantae and Animalia

113. In kingdom–Plantae, few members are partially heterotrophic like insectivorous plants or parasites.

With reference to the above statement give examples of insectivorous plant and the parasitic plant. ___

114. Give a common term.

Bryophytes, Pteridophytes, Gymnosperms, Angiosperms

___

115. Life cycle of plants have two distinct phases, the diploid ___ and the haploid ___ (sporophytic/gametophytic).

116. Out of the words/features given in the box, choose the correct for describing plants and animals.

Chlorophyll, Cellulose present, Glycogen present, Holozoic nutrition, Mobility, Alternation of generation

Kingdom–Plantae	Kingdom–Animalia
__________	__________
__________	__________
__________	__________

117. State true or false.

(i) Animals show definite growth pattern and grow into adults that have a definite shape and size.

True ☐ False ☐

(ii) All animals have elaborated sensory and neuromotor mechanisms.

True ☐ False ☐

(iii) Sexual reproduction is followed by embryological development in animals.

True ☐ False ☐

Topic 5 : Viruses, Viroids, Prions and Lichens

118. Name the acellular organisms not included in five kingdom classification of Whittaker.

____

119. Viruses are ____________ (non-cellular/cellular) organisms that are characterised by having an inner crystalline structure outside the living cell.

120. The genetic material of viruses consists of ____________

121. Match the following columns.

	Column I (Scientists)		Column II (Contributions)
A.	Pasteur	1.	*Contagium vivum fluidum*
B.	Beijerinek	2.	Crystallisation of viruses
C.	Stanley	3.	Virus means venom

A. ________ B. ________ C. ________

122. Given below is the diagram of a virus. Identify the virus '*X*' and write its parts *A* and *B*.

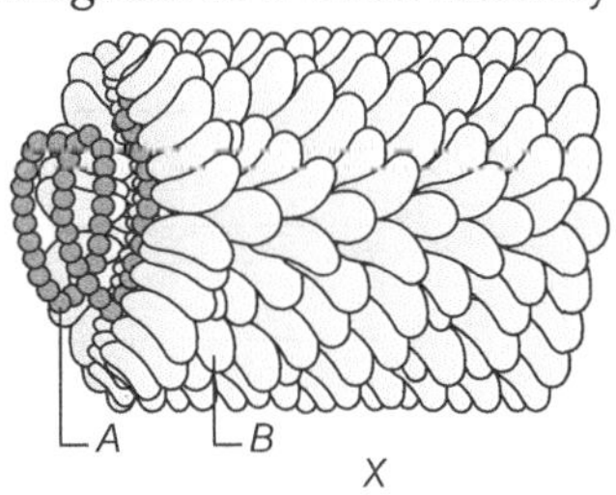

A. __________

B. __________

X. __________

123. The name 'virus' means ______________________

124. Name the scientists who said that 'viruses are smaller than bacteria'.

____

125. Viruses infect, by taking over the machinery of the host cell to replicate themselves and then kill the host cell.

True ☐ False ☐

126. Complete the following analogy given below.

Retrovirus : RNA containing virus : : ______________ : DNA containing virus.

127. No virus contains both RNA and DNA.

True ☐ False ☐

128. Write a common term for the diseases mentioned in the box given below.

Common cold (flu), Mumps, Herpes, Influenza

____

129. Name the scientific term for the statements given below.

(i) The protein coat of a virus. ______________________

(ii) The subunit of capsid. ______________________

130. Out of the following symptoms, how many are shown by plants infected with virus?

Mosaic formation, Leaf rolling and curling, Leaf spots Yellowing and vein clearing, Dwarfing, Stunted growth

____

131. Rewrite the statements by changing the underlined word.

(i) Virus is an infectious <u>lipoprotein particle</u>. ______________________

(ii) Bacteriophage is a <u>*ds* RNA virus</u>. ______________________

(iii) Virus that attacks bacteria is called <u>prions</u>. ______________________

132. Identify the virus '*X*' and name its structures marked as *A*, *B*, *C*, *D* and *E*.

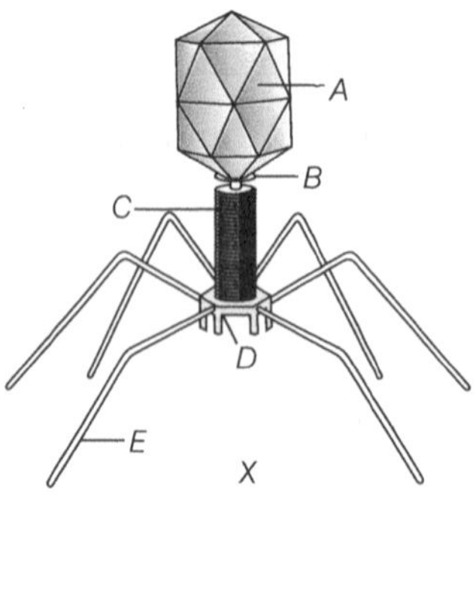

A. ______________________

B. ______________________

C. ______________________

D. ______________________

E. ______________________

X. ______________________

133. Complete the following analogy given below.

Tobacco mosaic disease : Plant viral disease : : ________ : Animal viral disease

134. The genetic material of virus is *ss*RNA, *ds*RNA, *ss*DNA and *ds*DNA. Out of these the genetic material of viral that infect.

(i) Plants ________________________________

(ii) Animals ________________________________

135. Name the microorganism discovered by TO Diener in 1971, which is smaller than viruses.

____

136. Viroids are infectious agents made up of free ________________ (DNA/RNA) that lacks ________________ (protein/carbohydrate) coat found in virus.

137. TO Diener discovered a popular disease in plants called ________________ caused by ________________ .

138. Certain infectious neurological diseases are transmitted by abnormally folded proteins called ________________ (virions/prions).

139. Match the following columns.

	Column I (Names of diseases)		Column II (Causal agents/Organisms)
A.	Little leaf of brinjal	1.	Virus
B.	Cr-Jakob disease in humans	2.	Viroids
C.	Citrus exocortis	3.	Prions
D.	Mosaic disease of okra	4.	Mycoplasma

A. ____________ B. ____________ C. ____________ D. ____________

140. The other name of Bovine Spongiform Encephalopathy (BSE) is

________________________________ .

141. The full form of CJD in humans is ________________________________

142. The algal and fungal components of a lichen are respectively called ________________ and ________________ .

143. Write one function of each

(i) Algal component of lichen.________________________________

(ii) Fungal component of lichen.________________________________

144. Lichens grow very well in ________________ (less/more) polluted areas.

145. Name the three types of lichens on the basis of their structure.

____

146. Write the correct term against each statement from the box given below.

Virion, Viroid, Prions, Virus, Lichen

(i) A completely assembled and infectious virus outside the host cell. ________

(ii) Proteinaceous infectious particles with small amount of nucleic acid. ________

(iii) Small infectious agent with low molecular weight of RNA. ________

Answers

1. Aristotle

2. Red Blood cells (RBCs)

3. Carolus Linnaeus

4. True

5. Protista, Plantae, Animalia

6. Whittaker

7. Animalia

8. 5(All of them except nuclear membrane composition).

9. *A*–Prokaryotic, *B*–Present (non-cellulosic made up of polysaccharide and amino acid), *C*–Present, *D*–Multicellular

10. *A*–Autotrophs–Plantae, *B*–Heterotrophs–Animals and fungi, *C*–Both autotrophs and heterotrophs–Monera and Protista

11. A-2, B–3, C–1, D–4

12. Present of cell wall in their cells.

13. Monera

14. Bacteria and mycoplasma

15. *A*–Cocci, *B*–Baccilli, *C*–Spirilla, *D*–Vibrio

16. (i) True
(ii) False. Bacteria can surive in extreme habitats like hotsprings, deserts, snow and deep oceans.
(iii) True

17. Simple, complex

18. (i) Heterotrophic,
(ii) Chemosynthetic autotrophic

19. Chemoautotrophs, photoautotrophs

20. A-2, B-1, C-3

21. Protein and non-cellulosic polysaccharide, peptidoglycan

22. Methanogens

23. Eubacteria

24. *X* – *Nostoc*; *A*–Heterocyst, *B*–Mucilaginous sheath.

25. Blue-green algae

26. *Methanococcus* is odd one out as it is a methanogen bacterium.

27. *a*

28. (i) Chemosynthetic autotrophic bacteria
(ii) Photosynthetic autotrophic bacteria
(iii) Chemosynthetic autotrophic bacteria

29. Heterocyst

30. Influenza is odd one out as it is a viral disease.

31. True

32. *X* – Binary fission, *A*–Cell wall, *B*–Cell membrane, *C*–DNA

33. 5(All of them are economic importance of heterotrophic bacteria except fodder crop.)

34. Spores by sporulation

35. Conjugation, transformation transduction.

36. *Mycoplasma*

37. (i) *Mycoplasma* Like Organisms
(ii) Pleuropneumonia Like Organisms

38. Little leaf disease of brinjal–Plant, Pleuropneumonia–Animal

39. Protista

40. 4 (Single-celled, eukaryotic, membrane bound organelles, connecting links between plants, animals and fungi)

41. Cell fusion, zygote formation

42. Chrysophytes

43. Diatoms

44. Diatomaceous earth

45. Biogas production is not done by diatomite.

46. Photosynthetic

47. Diatoms

48. *A*–Dinoflagellate, *B*–Euglenoid (*Euglena*)

49. Dinoflagellates

50. *Noctiluca*

51. *Euglena* is odd one out as it is euglenoid and others are dinoflagellates.

52. *Gonyaulax*

53. True

54. Pellicle

55. 2 (One long and one short)

56. Chloroplast, heterotrophic

57. Saprophytic protists

58. *Plasmodium*

59. Sporangia, cellulose, air currents

60. True

61. Protozoans

62. A-3, B-4, C-1, D-2

63. *Entamoeba histolytica*

64. *Trypanosoma*

65. Parasites, *Plasmodium*

66. *X–Paramoecium*, *A*-Cilia, *B*-Oral groove, *C*-Mouth pore, *D*-Anal pore

67. (i) Amoeboid protozoan,
(ii) Ciliated protozoan

68. Heterotrophic

69. (i) Common mushroom
(ii) Toad stools

70. Yeast

71. *Puccinia*

72. Warm and humid

73. Chitin and polysaccharides

74. Yeast

75. (i) *Rhizopus*
(ii) *Agaricus*
(iii) *Ustilago*
(iv) *Puccinia*

76. (i) Conidia, basidiospores
(ii) Zoospores, aplanospores, ascospores

77. False. Coenocytic hyphae have no septal walls in its filaments.

78. Hyphae, mycelium

79. Phycomycetes

80. Fungi, algae

81. (i) Saprophytes
(ii) Parasites
(iii) Mycorrhiza

82. *Glomus*

83. Ascomycetes, Basidiomycetes

84. Vegetative means-Fragmentation, Budding, Fission.
Asexual spores-Conidia, Zoospores, Sporangiospores.
Sexual spores-Oospore, Basidiospore, Ascospore.

85. Plasmogamy → Karyogamy → Meiosis in zygote for haploid spore formation.

86. Nuclei

87. Plasmogamy

88. *A*-Dikaryon, *B*-Synkaryon, *C*-Meiospores

89. True

90. Ascomycetes, Basidiomycetes

91. Phycomycetes

92. *X–Agaricus*, *A*–Pileus, *B*–Gills, *C*–Annulus, *D*–Stipe

93. Deuteromycetes is odd one as it is known as imperfect fungi because only the asexual phases of reproduction are known. All other given classes show both sexual and asexual reproductions.

94. (i) Ascomycetes
(ii) Basidiomycetes

95. *A*–Ascomycetes, *B*–Mycelium aseptate and coenocytic, *C*–*Trichoderma*
D–Basidiomycetes, *E*–*Agaricus*

96. *A*–*Mucor*, *B*–*Saccharomyces*, *C*–*Penicillium*

97. (i) Basidiocarp
(ii) Multinucleate and aseptate
(iii) *Albugo*

98. *Neurospora*

99. Exogenously, basidium

100. Basidiomycetes

101. Deuteromycetes

102. *Penicillium notatum*

103. Conidiospores

104. A–1, B–2, C–3, D–4

105. *Neurospora*

106. *Ascus*

107. Puffball is odd one out as it is Basidiomycetes.

108. (i) Aplanospores,
(ii) Ascospores
(iii) Basidiomycetes

109. Isogamous, anisogamous or oogamous

110. Basidiomycetes

111. 4

112. A-2, B-1, C-4, D-3

113. Insectivorous plant – bladderwort and venus flytrap and parasitic plant- *Cuscuta*

114. Plant phyla

115. Sporophytic, gametophytic

116. Kingdom–Plantae-Chlorophyll, cellulose present and alternation of generation
Kingdom–Animalia - Glycogen present, holozoic nutrition and mobility.

117. (i) True
(ii) False. Higher animals have elaborated sensory and neuromotor mechanisms.
(iii) True

118. Viruses, viroids and prions

119. Non-cellular

120. DNA or RNA (Both *ds* and *ss*)

121. A–3, B–1, C–2

122. *X*-Tobacco masaic virus, *A*-*ss*RNA, *B*-Capsid

123. Venom or poisonous fluid

124. DJ Ivanowsky

125. True

126. Adenovirus

127. True

128. All are caused by virus.

129. (i) Capsid
(ii) Capsomeres

130. 5 (All of the given symptoms are shown by plants infected with virus except leaf spots.) Majority of leaf spots are caused by fungi.

131. (i) Nucleoprotein particle
(ii) *ds*DNA virus
(iii) Bacteriophage

132. *X*–Bacteriophage, *A*–Head, *B*–Collar, *C*–Sheath, *D*–Base plate, *E*-Tail fibre

133. AIDS

134. (i) Plants-*ss*RNA
(ii) Animals-*ss*RNA, *ds*RNA and *ds*DNA

135. Viroid

136. RNA, protein

137. Potato spindle tuber disease, potato spindle tuber viroid

138. Prions

139. A–4, B–3, C–2, D–1

140. Mad cow disease of cattle

141. Creutzfeldt-Jakob Disease

142. Phycobiont, mycobiont

143. (i) Prepare food for fungi,
(ii) Provide shelter, absorb nutrients and water for algae.

144. Less

145. Crustose lichen, foliose lichen, fruticose lichen

146. (i) Virion
(ii) Prions
(iii) Viroid

CHAPTER THREE

PLANT KINGDOM

Topic 1 : Introduction About Plants

1. Blue-green algae are the members of the plant kingdom.

   True ☐ False ☐

2. Find the odd one out with respect to criteria of classification.

Habit, Colour, Shape of leaves, Embryology

   ____

3. Which of the following includes in plant kingdom?

Algae, Fungi, Angiosperm, Gymnosperm, Bryophytes, Pteridophytes, Lichen

   ____

4. Name the scientist who gave artificial system of classification.

   ____

5. Artificial systems gave equal weightage to ____________ and ____________ characters.

6. Name the scientist who gave natural system of classification.

   ____

7. Natural system of classification is based on natural affinities among the organisms and consider both ____________ and ____________features.

8. Phylogenetic classification system is based on ____________________ relationships between various organisms.

9. Organisms belonging to same taxa have a common ancestor.

   True ☐ False ☐

10. Match the following columns.

	Column I (Classification systems)		Column II (Features)
A	Chemotaxonomy	1.	Numbers and codes
B.	Cytotaxonomy	2.	Evolutionary relationship
C.	Numerical taxonomy	3.	Cytological information
D.	Phylogenetic system	4.	Chemical constituents of plants

A. ________________ B. ________________

C. ________________ D.________________

Topic 2 : Algae

11. Chlorophyll bearing, simple, thalloid, autotrophic and largely aquatic organisms are called as ______________________________ .

12. Write a common term for the symbiotic association of algae and fungi.

____

13. Name all the reproductive methods used by algae.

____

14. Write an example for the following.

(i) Unicellular algae ______________________________

(ii) Colonial forms of algae______________________________

(iii) Filamentous forms of algae ______________________________

15. Asexual reproduction in algae occurs by fragmentation.
(*Rewrite the statement by changing the underlined word.*)

____

16. Each fragment after the fragmentation develops into ________________________ .

17. Name the most common spore formed in algae during asexual reproduction.

____

18. Complete the following analogy.

Non-flagellated : Flagellated : : ________________________ : *Chlamydomonas*

19. Write the technical term for the fusion of gametes, which are similar in size.

____

20. ________________________ is the fusion of two gametes dissimilar in size and behaviour.

21. Match the following columns.

	Column I (Types of asexual spores)		Column II (Examples)
A.	Zoospores	1.	*Ulothrix*
B.	Aplanospores	2.	*Microspora*
C.	Autospores	3.	*Polysiphonia*
D.	Tetraspore	4.	*Chlorella*

A. ________________ B. ________________

C. ________________ D.________________

22. Non-motile and similar gametes are found in ______________________ (*Spirogyra*/ *Chlamydomonas*).

23. Fusion between one large, non-motile female gamete and smaller, motile male gamete is termed anisogamous, e.g. *Volvox*.
(*Rewrite the statement by changing the underlined word.*)

____

24. Large amount of CO_2 fixation on the earth is carried out by ________________ through photosynthesis.

25. ________________ are the primary producers of energy-rich compounds which form the basis of the food cycles of all aquatic animals.

26. The examples of marine algae used as food are ________________ and ________________ .

27. Write the scientific term for the water holding substances produced by marine brown and red algae. ______________________________

28. Name the commercial product obtained from algae used to grow microbes.

____

29. Name the two species of algae from which agar is obtained.

____

30. ____________ and ________________ are unicellular algae, rich in proteins and are used as food supplements by space travellers.

31. Name the three main classes of algae.

____

32. Match the following columns.

	Column I (Names of algae)		Column II (Products obtained)
A.	*Chlorella*	1.	Agar-agar
B.	*Chrondrus* spp.	2.	Alginates
C.	*Gelidium*	3.	Food supplement
D.	*Laminaria*	4.	Carrageenin

A. ________________ B. ________________

C. ________________ D. ________________

33. Complete the flowchart given below.

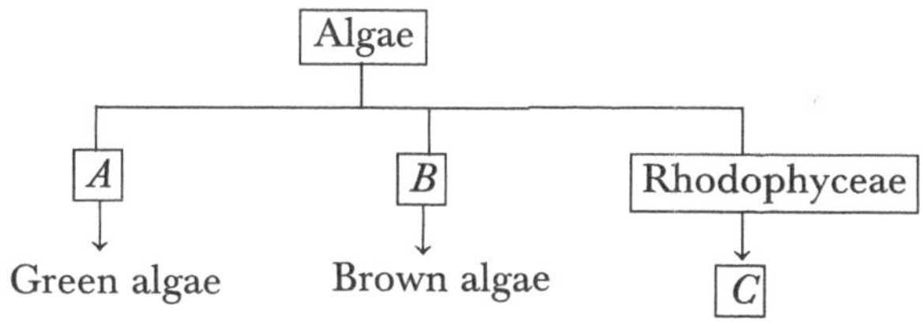

A. ________________ B. ________________ C. ________________

34. Name the chlorophyll pigments in the class–Chlorophyceae.

____

35. Find the odd one out with respect to shape of chloroplast in Chlorophyceae.

Cup-shaped, Plate-like, Discoid, Reticulate, Hexagonal, Spiral

____

36. Name the storage body located in the chloroplast containing starch and protein.

____

37. Green algae usually have a rigid cell wall made of an inner layer of ________________ and an outer layer of ________________.

38. Flagellated zoospores are produced in megasporangia.

True ☐ False ☐

39. Vegetative reproduction in Chlorophyceae takes place by budding.

True ☐ False ☐

40. Algae store food in the form of oil droplets.

True ☐ False ☐

41. Write any two examples of algae belonging to class–Chlorophyceae.

____

42. Study the following figure and answer the questions that follows.

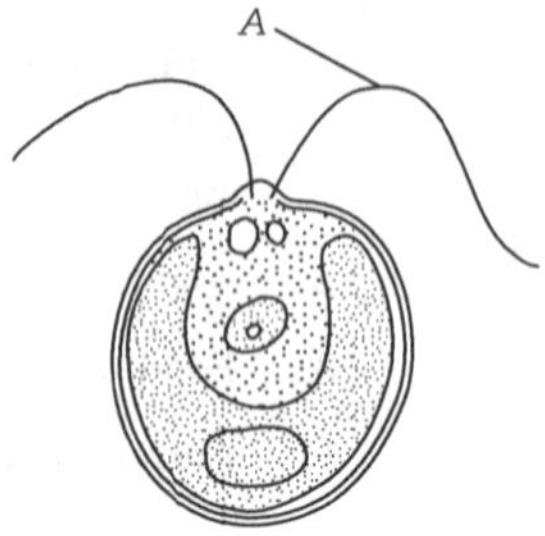

(i) Name the algae.________________________

(ii) Identify *A* ________________________

43. Study the structural details of the figure given below and answer the following questions.

(i) Identify *A* and *B*.

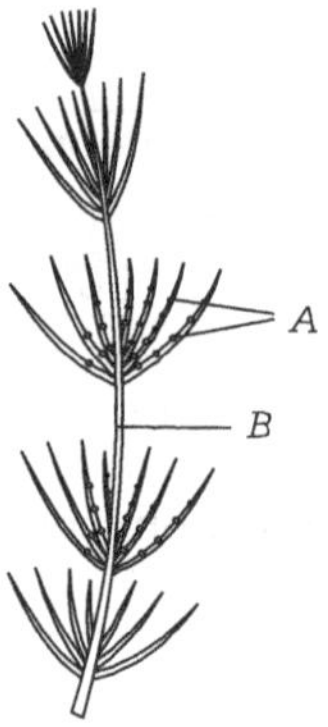

A. ________________________

B. ________________________

(ii) Name this alga.________________________

44. Name the class of algae that is found primarily in marine habitats.

____

45. Write an example for simple branched, filamentous forms of brown algae.

____

46. Name the two pigments other than chlorophyll, present in members of Phaeophyceae.

____

47. Phaeophyceae colour depend upon the amount of pigment named

____

48. Food is stored as proteins in Phaeophyceae.

True ☐ False ☐

49. The carbohydrates can be stored in the form of ________________ and ________________ in brown algae.

50. Name the gelatinous coating covering cellulosic wall in vegetative cells of algae.

____

51. Write the structure attaching the brown algae to the substratum.

____

52. ______________ (Stipe/Holdfast) is the stock attaching the plant body to substratum.

53. In brown algae, a leaf-like photosynthetic organ called stalk.
(*Rewrite the statement by changing the underlined word.*)

____

54. Asexual reproduction in brown algae occurs by monoflagellated zoospores.

True ☐ False ☐

55. The pear-shaped forms of gametes in brown algae gametes are called ______________

56. Write any two examples of brown algae.______________________________

57. Study the figure given below and answer the following questions.

(i) Name the plant. ______________________________

(ii) Identify *A*, *B* and *C*.

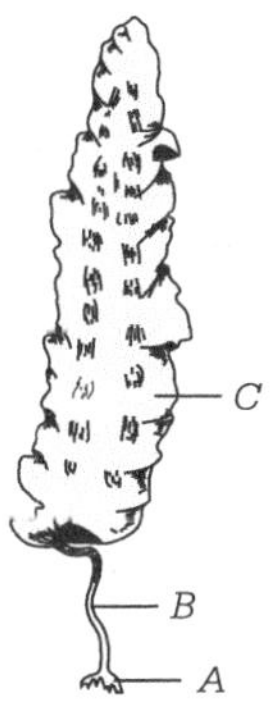

A. ______________

B. ______________

C. ______________

58. Study the figure given below and answer the following questions.

(i) Name the plant. ______________________________

(ii) Identify *A*, *B* and *C*.

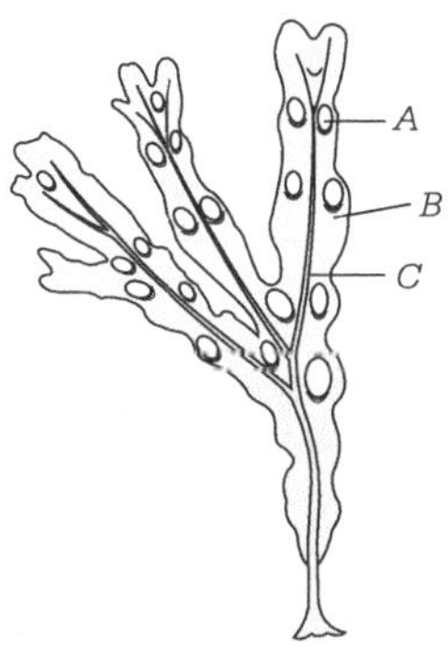

A. ______________

B. ______________

C. ______________

59. Study the figure given below and answer the following questions.

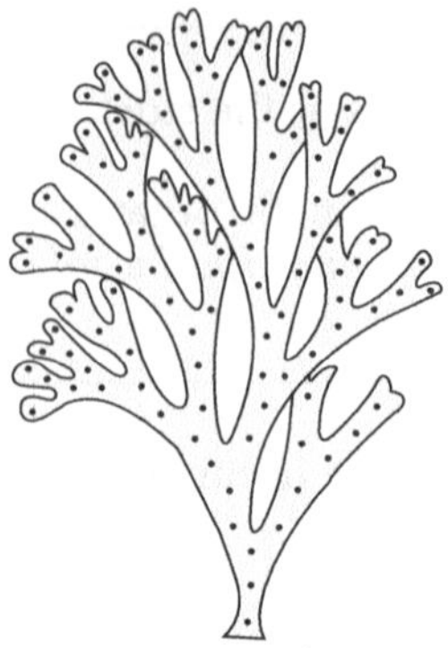

(i) Identify the algal species.______________________________

(ii) Name the pigment found in the cells of this species. ______________________________

60. Name the predominant pigment in red algae.______________________________

61. Red algae do not occur at deep ocean depths.

True ☐ False ☐

62. The food is stored as ______________________________ in red algae which is similar to ______________________________ and ______________________________ in structure.

63. Sexual reproduction is ______________________________ (isogamous/oogamous /anisogamous) in Rhodophyceae.

64. Red algae reproduce sexually by motile gametes.

True ☐ False ☐

65. Write any two examples of Rhodophyceae. ______________________________

66. Study the figure given below and identify the algal species.

____

67. Red algae reproduce asexually by non-motile spores.

True ☐ False ☐

68. The thalli of most red algae are unicellular.

True ☐ False ☐

69. Match the following columns.

	Column I (Thallus organisations)		Column II (Examples)
A	Unicellular (motile)	1.	*Spirogyra*
B.	Colonial	2.	*Chlamydomonas*
C.	Filamentous	3.	*Chlorella*
D.	Unicellular (non-motile)	4.	*Volvox*

A. ________________ B. ________________

C. ________________ D. ________________

70. Match the following columns.

	Column I (Types of sexual reproduction)		Column II (Features)
A.	Isogamy	1.	Two gametes, one larger than the other
B.	Syngamy	2.	Two kinds of morphologically distinct gametes
C.	Anisogamy	3.	Both gametes are morphologically similar
D.	Heterogamy	4.	Union of two gametes

A. ________________ B. ________________

C. ________________ D. ________________

71. Choose the odd one out with respect to cell wall components of algae.

Cellulose, Renin, Algin, Pectin

____

72. Read the table carefully and fill up the blanks.

Classes	Major pigments	Stored food	Cell wall	Flagellar number and position of insertions
Chlorophyceae	(A)	Starch	Cellulose	2-8, equal, apical
Phaeophyceae	Chlorophyll-*a* and *c*, fucoxanthin	Mannitol, laminarin	(C)	2, unequal, lateral
Rhodophyceae	Chlorophyll-*a* and *d*, phycoerythrin	(B)	Cellulose	(D)

A. ________________ B. ________________

C. ________________ D. ________________

Topic 3 : Bryophytes

73. Bryophytes are also called __________ of the plant kingdom.

74. Bryophytes are live in soil, but dependent on water for sexual reproduction.

    True ☐ False ☐

75. Name the structure in bryophytes that attach the plant body to substratum.

    ____

76. The plant body of bryophytes are less differentiated than that of algae.

    True ☐ False ☐

77. Bryophytes lack true root, stem or leaves.

    True ☐ False ☐

78. The main plant body of bryophytes is the gametophyte.

    True ☐ False ☐

79. The gametophyte of bryophyte is __________ (haploid/diploid) structure.

80. Complete the following analogy.

    Female : Male : : __________ : Antheridium

81. Name the biflagellated unicellular structure in antheridium which comes in contact with archegonium. __________

82. In bryophytes, the multicellular body produced by zygote is called __________ .

83. The __________ is not a free-living but attached to the photosynthetic __________ and derives nourishment from it. (sporophyte/gametophyte).

84. Write the collective term for the following.

Elators, Spores, Spore sac, Foot, Seta, Capsule

    ____

85. Sporophyte undergoes __________ to produce haploid spores which germinate to produce __________.

86. Name the moss which had long been used as fuel.__________

87. Mosses and __________ are the first organisms to colonise rocks.

88. Write the two divisions of bryophytes.

    ____

89. The plant body of a liverwort is haploid, e.g. *Marchantia*.

    (*Rewrite the statement by changing the underlined word.*)

    ____

90. Name the specialised structures formed in liverworts for asexual reproduction.

____

91. Gemmae are multicellular, green, asexual buds developed in small receptacles called ______________ located on the thalli.

92. Gemma gets detached from the parent body and germinate to form new individuals.

True ☐ False ☐

93. During sexual reproduction in liverwort, male and female sex organs are produced on same thalli only.

True ☐ False ☐

94. Find the odd one out with respect to sporophyte of liverwort.

Capsule, Seta, Tapetum, Foot

____

95. After meiosis, the spore formation takes place within the ______________

96. Consider the following figure for the structural details of bryophyte. Answer the following questions.

(i) Identify *A*, *B* and *C*.

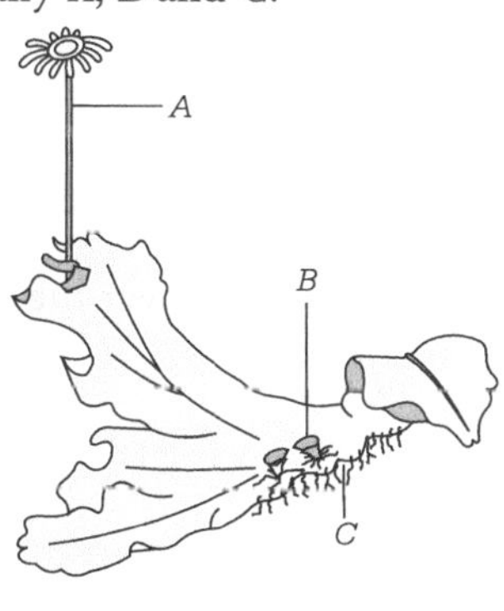

A. ______________________

B. ______________________

C. ______________________

(ii) Identify the ploidy of thallus structure shown. ______________________

97. Name the first stage of life cycle of mosses.______________________

98. ______________ stage in mosses develops from secondary protonema on a lateral bud.

99. Leafy stage of mosses is attached to soil through multicellular and branched buds.
(*Rewrite the statement by changing the underlined word.*)

____

100. Leafy stage does not bear sex organs.

True ☐ False ☐

101. Vegetative reproduction in mosses is by ______________ and ______________ in the ______________.

102. The gametophyte in mosses is more elaborate than that in liverworts.
(*Rewrite the statement by changing the underlined word.*)

____

103. *Funaria* and *Marchantia* are called liverworts.

True ☐ False ☐

104. *Sphagnum* is also called as ____________ (club moss/ peat moss).

105. Consider the following figure for structural details of moss. Answer the following questions.

(i) Identify *A, B, C, D* and *E.*

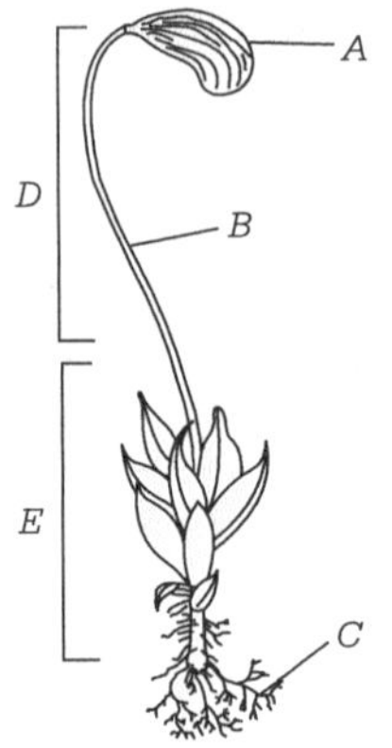

A. ____________

B. ____________

C. ____________

D. ____________

E. ____________

(ii) Write name of the plant. ____________

106. Consider the structural details shown in the following figure. Answer the following questions.

(i) Identify *A* and *B.*

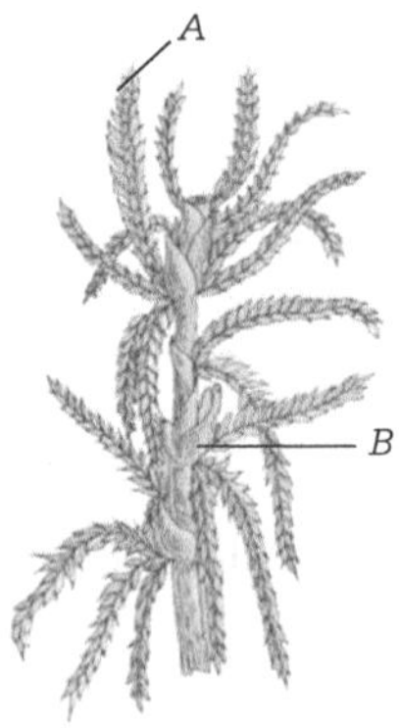

A. ____________

B. ____________

(ii) Name the bryophytes. ____________

Topic 4 : Pteridophytes

107. ______________________ are mainly used for medicinal purposes and as soil binders.

108. Pteridophytes are the first <u>aquatic</u> plants to possess vascular tissues.
(*Rewrite the statement by changing the underlined word.*)

____

109. In pteridophytes, dominant phase in the life cycle is

____

110. The main plant body is a ______________________ which is differentiated into true root, stem and leaves.

111. The leaves in ferns are ______________________ . (macrophyllus/ microphyllus)

112. Write the technical name for the leaf-like appendages found in sporangia.

____

113. Write the compact arrangements of sporophyll found in *Selaginella.*

____

114. Spores are produced in spore mother cells by ____________________ (equational/ reductional) division.

115. Germination of spores gives rise to small, multicellular photosynthetic thalloid gametophytes called ________________ .

116. Write any two reasons which restrict pteridophytic plant growth in limited area.
(i) ____
(ii) ____

117. Complete the following, using the words given in the box below.

Archegonium, Archesporium, Antherozoids, Anthers, Archegonia, Anthrospore.

The gametophytes bear male and female sex organ called ______________________ and ______________________, respectively. Fusion of ______________________ with the egg present in the ______________________ results in the formation of zygote.

118. Arrange the correct sequence of reproductive structures in pteridophytes.

Strobili → Sporophyll → Sporangia → Spores → Spore mother cell

____

119. In bryophytes and pteridophytes, transport of male gametes requires ______________.

120. Mosses and ferns are found in moist and shady places because both <u>require sunlight for fertilisation</u>.
(*Rewrite the statement by changing the underlined word.*)

____

121. Pteridophytes with spores of similar kinds are called ____________

122. Write any two examples of heterosporous pteridophytes. ____________

123. Match the following columns.

	Column I (Names of plants)		Column II (Sexual characteristics)
A.	*Cycas*	1.	Heterosporous
B.	*Marsilea*	2.	Dioecious
C.	*Equisetum*	3.	Monoecious
D.	*Pinus*	4.	Homosporous

A. ____________ B. ____________

C. ____________ D. ____________

124. Heterospory is the production of ____________

125. Megaspore germinates to give rise to female gametophyte.

True ☐ False ☐

126. The female gametophytes are retained in parent ____________ in heterosporous pteridophytes.

127. The development of zygote into young embryo take place within the

____________ .

128. Development of zygote into embryo within the female gametophyte is a precursor to the ____________, which is important for evolution.

129. Seed habit is linked to ____________.

130. Identify *A*, *B* and *C* in the following figure.

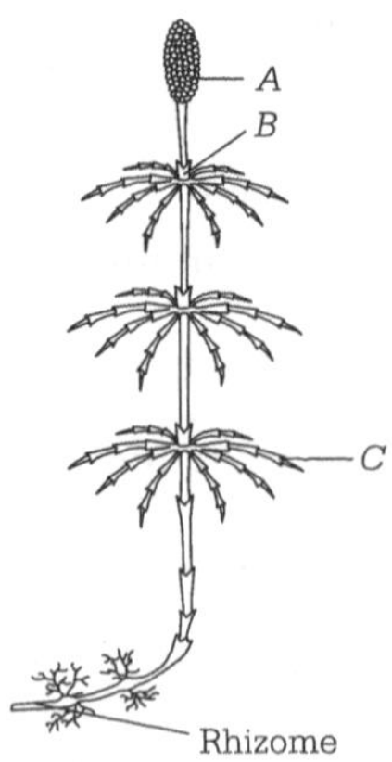

A. ____________

B. ____________

C. ____________

131. Go through the following figures and identify these plants (*A*, *B* and *C*).

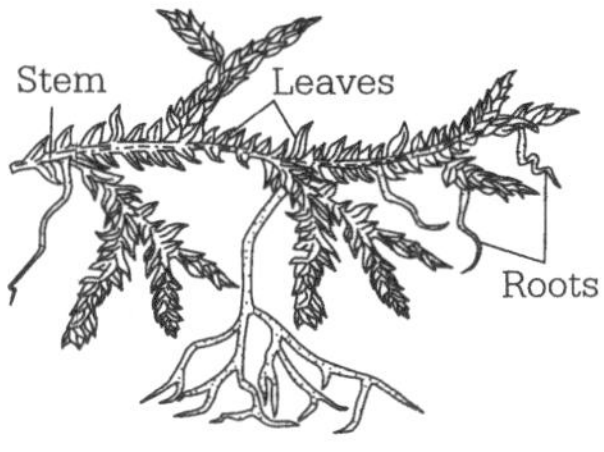

A

B

C

A. ______________________

B. ______________________

C. ______________________

132. Write the four classes of pteridophytes.

____

133. Write any two examples of aquatic pteridophytes. ______________________

134. Match the following columns.

	Column I (Pteridophyte class)		Column II (Examples)
A.	Psilopsida	1.	*Dryopteris, Pteris, Adiantum*
B.	Lycopsida	2.	*Equisetum*
C.	Sphenopsida	3.	*Selaginella*
D.	Pteropsida	4.	*Psilotum*

A. ____________ B. ____________

C. ____________ D. ____________

Topic 5 : Gymnosperms

135. In gymnosperms, ovules are not enclosed by ______________________ .

136. Name the gymnosperm species which is one of the tallest tree species. ____________

137. The roots of gymnosperm are generally ______________ (tap/fibrous) roots.

138. Name the association of roots of gymnosperm and fungi. ______________________

139. Small specialised roots called ____________ roots are associated with N_2-fixing cyanobacteria in *Cycas*.

140. In conifers, the needle-like leaves increases surface area which ultimately reduces water loss.

True ☐ False ☐

141. Write any two adaptations of leaves in gymnosperms that help them reducing water loss.

(i) ______________________ (ii) ______________________

142. The gymnosperms are homosporous.

True ☐ False ☐

143. The __________ microspore and megaspore that are borne on __________ within sporangia are arranged spirally along an axis to form __________.

144. Write other names of cones in gymnosperms. ______________________

145. Name the strobili bearing microsporophylls and microsporangia.

____

146. The reduced male gametophyte present in gymnosperm is called______________________

147. The male or female cones may be borne on the same trees.

True ☐ False ☐

148. The male cones and megasporophylls are borne on same trees in *Cycas.*

True ☐ False ☐

149. Complete the paragraph using the words given below in the box.

Megaspore mother cell, Integument, Nucellus, Ovary, Ovule, Microspore mother cell

The ______________ is differentiated from one of the cells of the ______________. The nucellus is protected by envelopes and the composite structure is called an ______________.

150. The megaspore mother cell divides ______________ (meiotically/ mitotically) to form ______________ (two four eight) megaspores.

151. The megasporangium develops into a multicellular female gametophyte that bears two or more ______________ .

152. The female gametophyte is retained within megasporangium.

True ☐ False ☐

153. The gametophytes have an independent free-living existence in gymnosperms.

True ☐ False ☐

154. Complete the following sequence.

Pollen grain released from (A)	→	(B) carrying male gametes grow towards (C)	→	Following fertilisation, zygote develops into (D) and ovule develops into (E)

A. ________________ B. ________________

C. ________________ D. ________________

E. ________________

155. Consider the following figure and answer the following questions.

(i) Name the plant. ________________

(ii) Identify *A*, *B* and *C*.

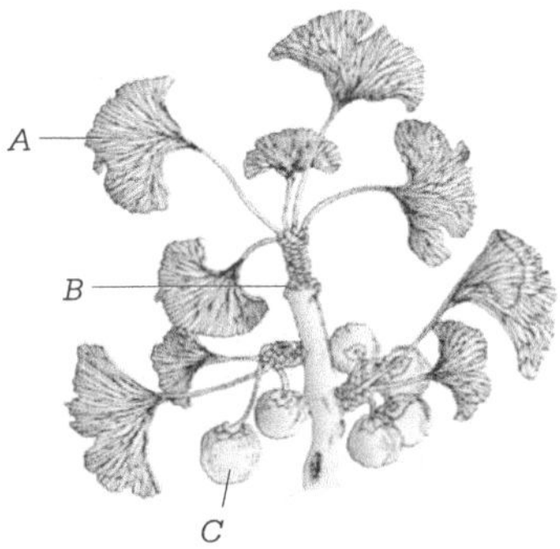

A. ________________

B. ________________

C. ________________

Topic 6 : Angiosperms

156. In angiosperms, pollen grains and ovules are developed in specialised structures called ________________

157. Seeds are enclosed by flowers in angiosperms.

True ☐ False ☐

158. Write the two divisions of angiosperms.

(i) ________________ (ii) ________________

159. Name the class of angiosperm that possess two cotyledons, reticulate venation and tetramerous or pentamerous flowers. ________________

160. Complete the following analogy.

Tallest angiosperm : Smallest angiosperm : : *Eucalyptus* : ________________

161. Name the male sex organ in a flower. ________________

162. Within the anther ________________ undergoes meiosis to produce microspore which mature into pollen grains.

163. The highly reduced female gametophytes in angiosperm are called ________________.

164. The female sex organ in a flower is called pistil or carpel.

True ☐ False ☐

165. Choose the odd one out.

Pistil, Stamen, Anther, Embryo

____

166. Each ovule has a <u>microspore mother cell</u> that undergoes meiosis to form four haploid megaspores.
(*Rewrite the statement by changing the underlined word.*)

____

167. Write the component of three-celled apparatus found in the embryo sac.

____

168. There are two antipodal cells, three polar nuclei in the embryo sac along with one egg cell and two synergids.

True ☐ False ☐

169. The <u>antipodal cells</u> eventually fuse to produce a diploid secondary nucleus.
(*Rewrite the statement by changing the underlined word.*)

____

170. Pollen grains form anthers are carried to the ______________ of a pistil.

171. Name the process of dispersal of pollen grains to the stigma of the pistil. ______________

172. Write the number of male nuclei (gametes) entering embryo sac. ______________

173. Write common term for the union of male gametes and egg cell to form a zygote.

____

174. Among the two male gametes, one fuses with the diploid secondary nucleus to produce the ______________ (triploid / diploid/ tetraploid) structure.

175. Expand PEN. ______________________________

176. Write a common term used to mention the involvement of two types of fusion taking place in angiosperms. ______________________________

177. Complete the following analogy.

Zygote : PEN :: ______________ : Endosperm.

178. Name the structure providing nourishment to the developing embryo.

____

179. ______________ and ______________ degenerate after fertilisation.

180. Complete the following analogy.

Ovules : Ovary : : ____________________ : Fruit

181. The diagram represents the life cycle of angiosperm. Choose the correct combination for labellings (*A-E*).

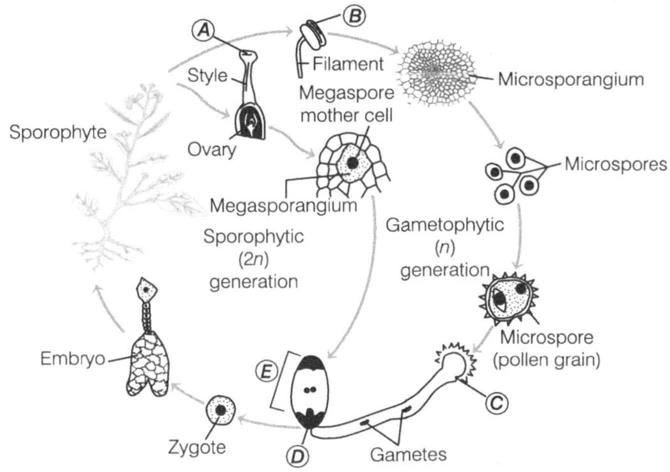

A. ________________ B. ________________ C. ________________

D. ________________ E. ________________

Topic 7 : Alternation of Generation

182. During life cycle of sexually reproducing plants, there is an alternation of generation between gamete producing ____________________ (haploid/ diploid) gametophyte and spore producing ________________ (haploid/diploid) sporophyte.

183. Zygote divides by ________________ (mitosis/meiosis) to produce ________________ (haploid/diploid) sprophytic body, which produce ________________ (diploid/haploid) spores by ______________ (mitosis /meiosis).

184. The haploid spore divides ____________________ (mitotically/meiotically) to form the gametophyte.

185. In haplontic life cycle, sporophytic generation is represented only by the ________________________. There are no free-living ____________________________ .

186. Name the kind of life cycle where dominant photosynthetic phase is a free-living gametophyte.____

187. Write any two examples in which haplontic life cycle occur.

(i) ______________________________ (ii) ______________________________

188. Name the life cycle in which diploid sporophyte is dominant. ______________________

189. Diplontic life cycle occurs in seed bearing plants like *Cycas* and maize.

True ☐ False ☐

190. Name the type of life cycle where diploid sporophyte is dominant, photosynthetic, independent phase of the plant. ____________________

191. The gametophytic, phase is represented by the single to few-celled haploid gametophyte in ____________________.

192. __________ and __________ exhibit haplo-diplontic life cycle.

193. A dominant, independent, photosynthetic thalloid represented by __________ gametophyte in haplo-diplontic cycle.

194. Dominant, independent, photosynthetic vascular plant body of a diploid sporophyte is represented by pteridophyte.

True ☐ False ☐

195. Write a classical example of diplontic algae. ____________________

196. Consider the following figures and identify the type of life cycle patterns.

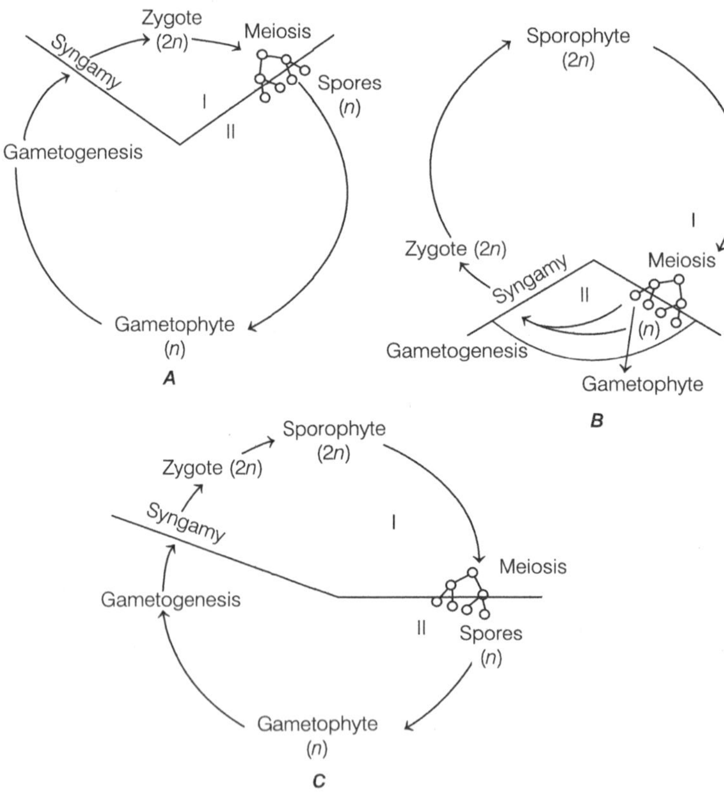

197. Give any two examples of haplo-diplontic algae are ________________________ and ________________________.

198. Match the following columns.

	Column I (Plant life cycle)		Column II (Plant groups)
A.	Haplontic life cycle	1.	Bryophytes and pteridophytes
B.	Diplontic life cycle	2.	Gymnosperms and angiosperms
C.	Haplo-diplontic life cycle	3.	*Volvox*, *Spirogyra* and *Chlamydomonas*

A. ______________ B. ______________ C. ______________

Answers

1. False. Blue-green algae are the members of kingdom–Monera.
2. Embryology is odd one out. Rest of the options were the superficial morphological characters used in earlier system of classification.
3. Kingdom–Plantae includes algae, bryophytes, pteridophytes, angiosperm and gymnosperm.
4. Linnaeus
5. Vegetative, sexual
6. George Bentham and Joseph Dalton Hooker
7. External, internal
8. Evolutionary
9. True
10. A-4, B-3, C-1, D-2
11. Algae
12. Lichen
13. Algae reproduce by vegetative, asexual and sexual methods.
14. (i) *Chlamydomonas*, (ii) *Volvox*, (iii) *Ulothrix*.
15. Vegetative
16. Thallus
17. Zoospores
18. *Spirogyra*
19. Isogamy
20. Anisogamy
21. A-1, B-2, C-4, D-3
22. *Spirogyra*
23. Oogamous
24. Algae
25. Marine algae
26. *Laminaria*, *Sargassum*
27. Brown algae–algin, red algae–carrageen
28. Agar
29. *Chlorella* and *Gracillaria*
30. *Chlorella*, *Spirulina*
31. Chlorophyceae, Phaeophyceae, Rhodophyceae
32. A-3, B-4, C-1, D-2
33. *A*–Chlorophyceae, *B*–Phaeophyceae, *C*–Red algae
34. Chlorophyll-*a* and *b*.
35. Hexagonal is odd one out. Rest of the options are the shapes of chloroplasts in different species of algae and hexagonal shape is not identified.
36. Pyrenoids
37. Cellulose, pectose
38. False. Flagellated zoospores are produced in zoosporangia.
39. False. Vegetative reproduction in Chlorophyceae takes place by fragmentation.
40. True
41. *Volvox* and *Ulothrix*
42. (i) *Chlamydomonas*, (ii) Flagella
43. (i) *A*–Branches, *B*–Axis, (ii) *Chara*
44. Phaeophyceae
45. *Ectocarpus*
46. Xanthophylls, carotenoids
47. Fucoxanthin
48. False. Food is stored as complex carbohydrates in Phaeophyceae.
49. Laminarin, mannitol
50. Algin
51. Holdfast
52. Stipe
53. Frond
54. False. Asexual reproduction in most brown algae occurs by biflagellated zoospores.
55. Pyriform
56. *Fucus*, *Dictyota*
57. (i) *Laminaria*, (ii) *A*–Holdfast, *B*–Stipe, *C*–Frond.
58. (i) *Fucus*, (ii) *A*–Air bladder, *B*–Frond, *C*–Midrib.
59. (i) *Dictyota*, (ii) Fucoxanthin
60. *r*-phycoerythrin
61. False. Red algae occurs at both surfaces of water and also at deep ocean depths.
62. Floridean starch, amylopectin, glycogen
63. Oogamous
64. False. Red algae reproduce sexually by non-motile gametes.
65. *Porphyra* and *Gelidium*
66. *Porphyra*
67. True
68. False. The thalli of most red algae are multicellular.
69. A-2, B-4, C-1, D-3
70. A-3, B-4, C-2, D-1
71. Renin is odd one out. Rest of the options are constituents of cell wall in algae and renin is protease secreted by kidneys.

72. *A*–Chlorophyll-*a* and *b*, *B*–Floridean starch, *C*–Cellulose and algin, *D*–Absent
73. Amphibian
74. True
75. Rhizoids
76. False. Bryophytes are more differentiated than algae.
77. True
78. True
79. Haploid
80. Archegonium
81. Antherozoids
82. Sporophyte
83. Sporophyte, gametophyte
84. Sporophytic capsule
85. Meiosis, gametophyte
86. *Sphagnum*
87. Lichens
88. Liverworts and Mosses
89. Diploid
90. Gemmae
91. Gemma cups
92. True
93. False. During sexual reproduction in liverwort, male and female sex organs are produced on different or same thalli.
94. Tapetum is odd one out. Rest of the options are differentiation of sporophyte whereas tapetum is a nutritive cell of anther in angiosperms.
95. Capsule
96. (i) *A*–Archegoniophore, *B*–Gemma cup, *C*-Rhizoids
    (ii) Haploid
97. Protonema stage
98. Leafy
99. Rhizoids
100. False. Leafy stage bears sex organs.
101. Fragmentation, budding, secondary protonema.
102. Sporophyte
103. False. *Marchantia* and *Porella* are liverworts.
104. Peat moss
105. (i) *A*–Capsule, *B*–Seta, *C*–Rhizoids, *D*–Sporophyte, *E*–Gametophyte.
     (ii) *Funaria*
106. (i) *A*–Antheridial branch, *B*–Archegonial branch.
     (ii) *Sphagnum*
107. Pteridophytes
108. Terrestrial
109. Sporophytic plant body.
110. Sporophyte
111. Macrophyllus
112. Sporophylls
113. Strobili or Cones
114. Reductional
115. Prothallus
116. (i) Need of water for fertilisation.
     (ii) The pteridophytic gametophytes require cool, damp, shady places to grow.
117. Antheridia, archegonia, antherozoids, archegonium.
118. Sporophyll → Strobili → Sporangia → Spore mother cell → Spores
119. Water
120. Require the presence of water for fertilisation.
121. Homosporous
122. *Selaginella* and *Salvinia*
123. A-2, B-1, C-4, D-3
124. Large and small spores
125. True
126. Sporophyte
127. Female gametophyte
128. Seed habit
129. Heterospory
130. *A*–Strobilus, *B*–Node, *C*–Branch
131. *A*–*Selaginella*, *B*–Fern, *C*–*Salvinia*
132. Psilopsida, Lycopsida, Sphenopsida, Pteropsida.
133. *Marsilea* and *Azolla*
134. A-4, B-3, C-2, D-1
135. Ovary wall
136. *Sequoia*
137. Tap
138. Mycorrhiza
139. Coralloid
140. False. Needle-like leaves reduce surface area in conifers, which help to reduce water loss.
141. (i) Thick cuticle, (ii) Sunken stomata
142. False. Gymnosperms are heterosporous.
143. Haploid, sporophylls, cone
144. Lax or compact strobili
145. Male strobili or microsporangiate
146. Pollen grain
147. True
148. False. Male cones and megasporophylls are borne on different trees in *Cycas*.
149. Megaspore mother cell, nucellus, ovule
150. Meiotically, four
151. Archegonia
152. True
153. False. In gymnosperms, the male and female gametophytes do not have an independent free-living existence.
154. *A*–Microsporangium, *B*–Pollen tube, *C*–Archegonia, *D*–Embryo, *E*-Seeds.
155. (i) *Ginkgo*, (ii) *A*–Dwarf shoot, *B*–Long shoot, *C*–Seeds.
156. Flowers
157. False. Seeds are enclosed by fruits in angiosperms.
158. (i) Dicotyledons,
     (ii) Monocotyledons
159. Dicotyledons
160. *Wolffia*
161. Stamen
162. Pollen mother cell
163. Embryo sac
164. True
165. Embryo is odd one out. Rest of the options are the parts of male and female sex organs of angiosperms whereas embryo is the development of zygote after fertilisation.
166. Megaspore mother cell
167. One egg cell and two synergids

168. False. There are three antipodal cells and two polar nuclei in the embryo sac along with one egg cell and two synergids.
169. Polar nuclei
170. Stigma
171. Pollination
172. Two
173. Syngamy
174. Triploid
175. Primary Endosperm Nucleus
176. Double fertilisation
177. Embryo
178. Endosperm
179. Synergids, antipodal cells
180. Seeds
181. *A*–Stigma, *B*–Anther, *C*–Male gametophyte, *D*–Egg, *E*–Embryo sac
182. Haploid, diploid.
183. Mitosis, diploid, haploid, meiosis
184. Mitotically
185. One celled zygote, sporophyte
186. Haplontic
187. *Volvox*, *Spirogyra*
188. Diplontic
189. True
190. Diplontic life cycle
191. Diplontic life cycle
192. Bryophytes and pteridophytes
193. Haploid
194. True
195. *Fucus*
196. *A*–Haplontic, *B*–Diplontic, *C*–Haplo-diplontic
197. *Ectocarpus*, *Polysiphonia*
198. A-3, B-2, C-1

CHAPTER FOUR

ANIMAL KINGDOM

Topic 1 : Basis of Classification

1. All members of kingdom–Animalia are multicellular and exhibit the same pattern of organisation cells.

   True ☐ False ☐

2. Choose the odd one out.

Parazoans, Metazoans, Eumetazoans

   ____

3. Consider the following.

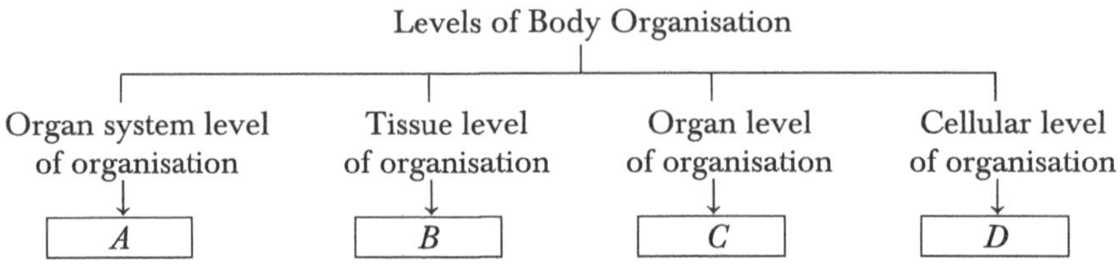

   Name the animal phylums showing the above level of body organisation.

   A.____________________ B. ____________________

   C.____________________ D. ____________________

4. Division of labour is completely absent among the cells in phylum ________________.

5. How many of the examples mentioned in the box given below possess closed (C) and open (O) circulatory system?

Cephalopod molluscs, Non-cephalopod molluscs, Chordates, Annelids, Tunicates, Hemichordates, Arthropods

   ____

6. Names the type of symmetry found in animal world.

   ____

7. In the given diagram, *A* shows the bilateral symmetry and *B* represents radial symmetry. Based on it, identify the phylums at *A, B* and *C*.

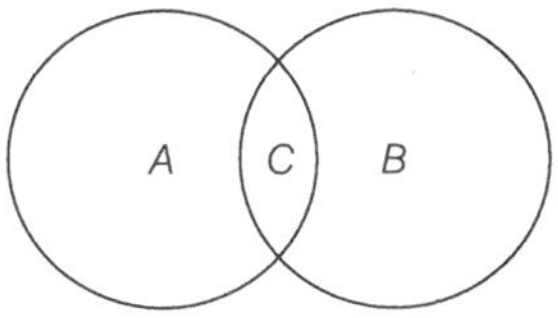

A. ______________________

B. ______________________

C. ______________________

8. Identify the type of symmetry in animals, *A* and *B*.

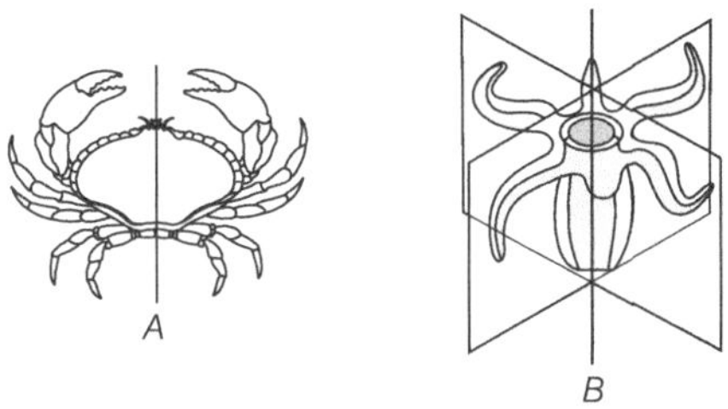

A. ______________________

B. ______________________

9. Identify the germ layers, *A, B, C* and *D* in the diploblastic (I) and triploblastic (II) organisation in animals.

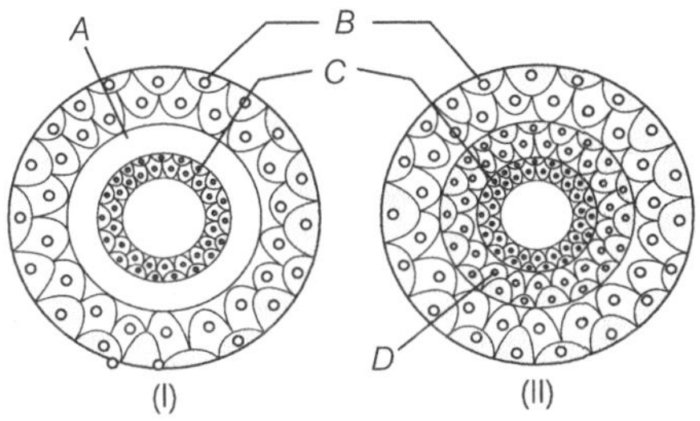

A. ______________________

B. ______________________

C. ______________________

D. ______________________

10. Rewrite the following statements by changing the underlined word.
   (i) Animals in which the cells are arranged into two embryonic layers, i.e. an external ectoderm and an internal endoderm, are called <u>bilateral symmetrical</u>.

   ____

   (ii) These animals in which the developing embryo has a third germinal layer <u>mesoglea</u>, in between ectoderm and endoderm are called triploblastic.

   ____

11. ______________ is the space between the wall of alimentary canal and body wall enclosed by the layer of mesoderm on both sides.

12. The figures given below show the types of coelom. Identify them and write the correct group of organisms which possess them.

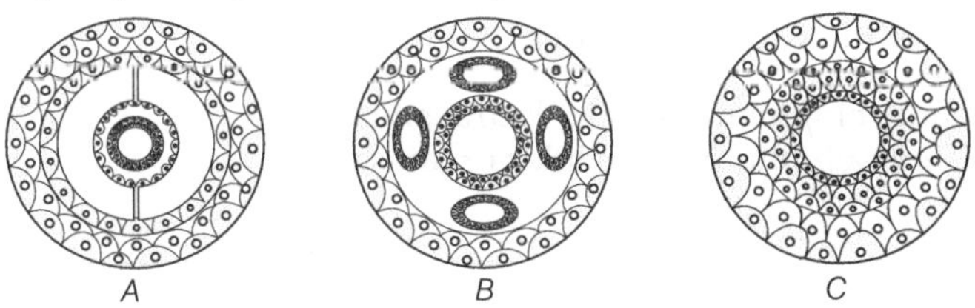

A. ______________________

B. ______________________

C. ______________________

13. Name the type of coelom present in molluscs, arthropods, echinoderms and chordates.

____

14. Give biological term for the following sentences.

(i) Type of coelom in which the mesoderm is present as scattered pouches in between the ectoderm and endoderm. _______________

(ii) The animals in which the body cavity is absent. _______________

15. Division of body with serial repetition of some organs along the length of an animal is known as _______________.

16. Name an animal in which metameric segmentation is seen. _______________

17. Notochord is a mesodermally derived rod-like structure formed on the ventral side during embryonic development in chordates.

True ☐ False ☐

18. Study the flowchart given below and identify the missing parts *A, B, C, D* and *E.*

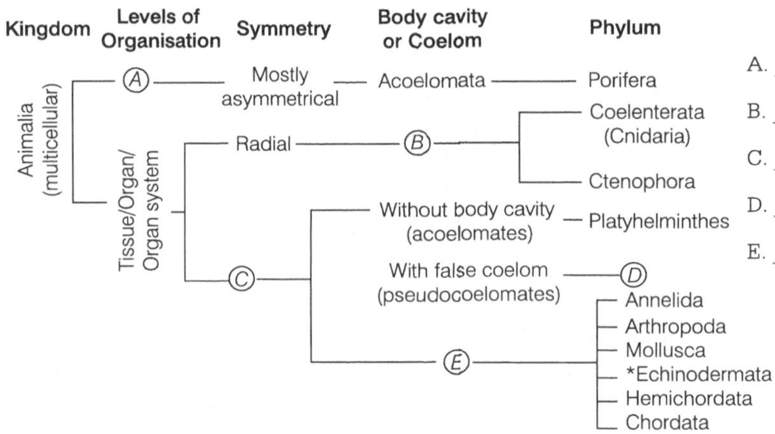

A. _______________

B. _______________

C. _______________

D. _______________

E. _______________

Topic 2 : Classification of Animals–Porifera to Aschelminthes

19. Members of which phylum are commonly known as sponges? _______________

20. The type of symmetry mostly found in sponges is _______________

21. Presence of water transport or canal system is a characteristic of phylum _______________.

22. In sponges, the water enters through minute pores called _______________ into the central cavity, _______________ from where it goes out through the _______________.

23. Out of the given characters, how many are present in sponges?

Internal fertilisation, Indirect development, Sessile, Extracellular digestion, Hermaphrodite, Generally marine, Cellular organisation

____

24. Sponges are diploblastic and contain outer dermal layer of ____________ and inner gastral layer of ____________ .

25. Name three different types of canal system found in sponges.

____

26. Give the scientific name for the collar cells lining the spongocoel and canal in poriferans.

____

27. Undifferentiated totipotent cells of sponges are ____________ (porocytes/archaeocytes).

28. Sponges have a larval stage which is morphologically distinct from the adult.

True ☐ False ☐

29. Complete the following analogy related to sponges.

Sexual reproduction: Gametes :: Asexual reproduction : ____________

30. Name the structures which support the body of sponges as a skeleton.

____

31. Complete the following analogy related to poriferans.

Amphiblastula larva : *Sycon* :: Parenchymula larva : ____________.

32. Write the scientific names for the following.

(i) The venus flower basket ____________

(ii) Glass rope sponge ____________

(iii) Boring sponge ____________

(iv) Deadman's fingers or mermaid's gloves ____________

33. Identify the figures and write their names.

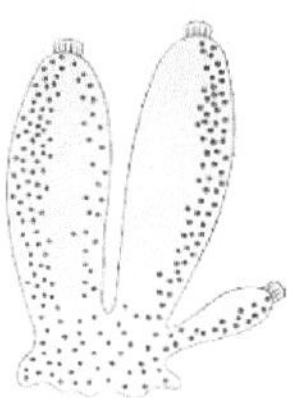

A

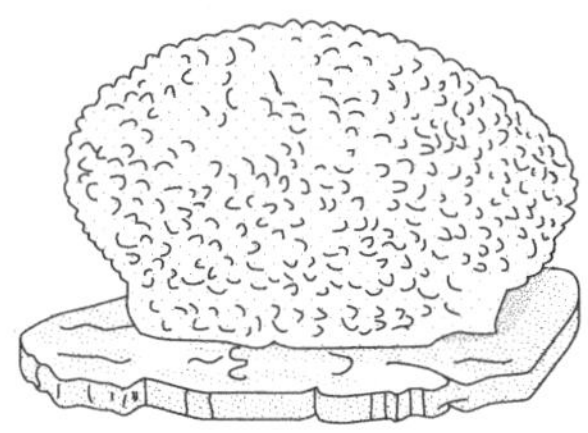

B

C

A. ____________

B. ____________

C. ____________

34. Animals which do not have separate sexes are called as ____________________ e.g. eggs and sperms are produced by the same individual.

35. When development of an organism occurs through larval stage then this type of development is called as ________________________ .

36. Identify the phylum based on given features.

(i) Sessile or free–swimming.

(ii) Tissue level organisation and diploblastic.

(iii) Have central gastro-vascular cavity.

(iv) Digestion is extracellular and intracellular.

____

37. (i) Name the special cell of cnidarians which is shown below and also write its function.

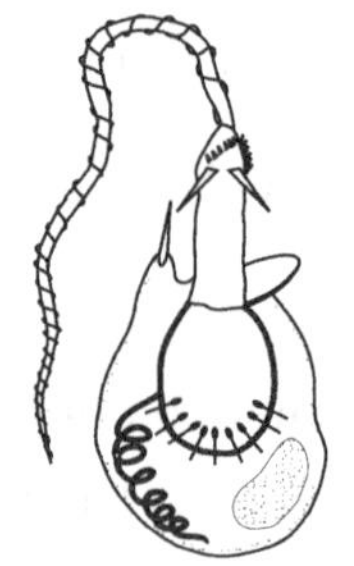

____

(ii) The stinging cell found in the above shown structure is called ________________________ .

38. Answer the following questions.

(i) Name the cnidarian possessing a skeleton made up of calcium carbonate. ____________

(ii) Mention the type of digestion found in coelenterates. ____________________

39. Name the structure marked '*X*' in the given diagram of *Adamsia.*

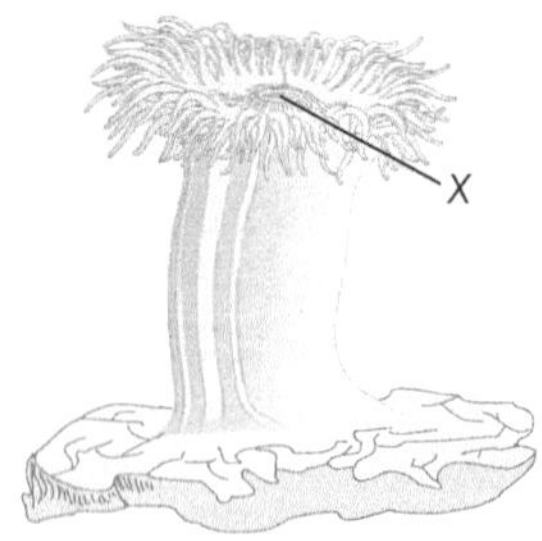

____

40. The body of a cnidarian consists of a body cavity known as ________________ or gastrovascular cavity, containing only one opening, which acts as mouth as well as anus. Thus, the digestive tract is ________________ .

41. Sessile polyps produce medusae ________________ and free-swimming medusae form the polyps ________________ (asexually/sexually).

42. Cnidarians exhibit alternation of generation between their polyp and medusa forms. This process is known as ________________ .

43. Identify the structures *A* and *B* of cnidarians and mention their ploidy correctly.

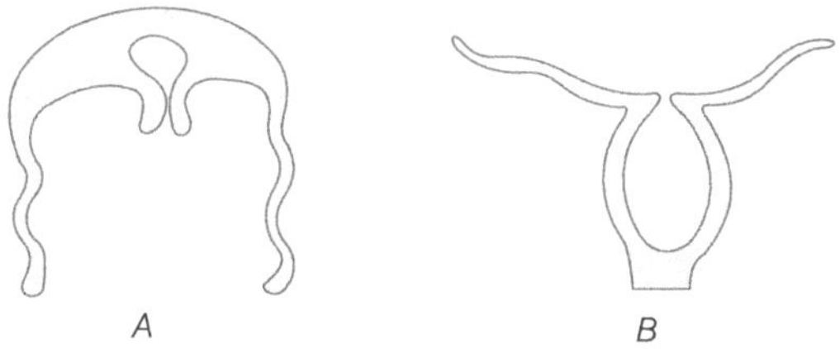

A. ________________

B. ________________

44. Out of the following, how many cnidarians are

(i) Sessile and cylindrical (*Hydra, Obelia, Adamsia, Physalia*)? ________________

(ii) Umbrella-shaped and free-swimming (*Physalia, Obelia, Aurelia*)? ________________

45. Give the scientific name of the following.

(i) Portuguese man of war ________________

(ii) Sea anemone ________________

(iii) Sea pen ________________

(iv) Sea fan ________________

(v) Brain coral ________________

46. Ctenophores are commonly known as ________________ are exclusively marine forms.

47. State True or False.

(i) Ctenophores are exclusively marine, radially symmetrical and diploblastic animals.

True ☐ False ☐

(ii) Ctenophores bear eight external rows of ciliated comb-plates which help in locomotion.

True ☐ False ☐

(iii) Ctenophores exhibit intracellular digestion only.

True ☐ False ☐

48. Choose the correct option for the given animal.

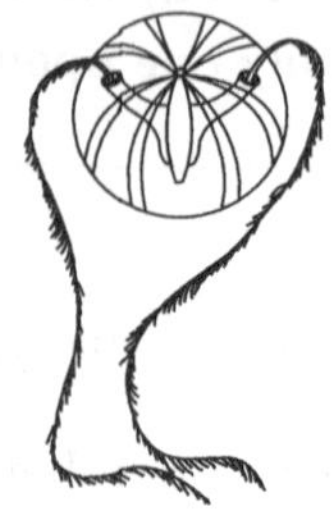

(i) Scientific name is ________________ (*Pleurobrachia/Ctenoplana*).

(ii) Bioluminescence is ________________ (present/absent).

(iii) Reproduction by only ________________ means (sexual/asexual).

(iv) Fertilisation is ________________ (internal/external).

(v) Development is ________________ (direct/indirect).

49. Give the scientific term for the property of emission of light by the body of an organism.

____

50. Complete the following analogy given below.

Cnidarians : Planula larva : : Ctenophores : ________________

51. Name the special adhesive cells present in ctenophores for capturing the prey.

____

52. Choose the odd one out with respect to phylum–Platyhelminthes.

Bilateral symmetry, Coelomate, Triploblastic, Flatworms, Organ level of organisation

____

53. Flatworms are mostly endoparasites, found in animals, having structures like ________________ and ________________ for adhesion .

54. Give the common term for the following.

(i) Onchosphere, Cysticercus, Hexacanth ________________

(ii) Turbellaria, Trematoda, Cestoda ________________

55. Nephridium is present in liver fluke for excretion and osmoregulation.

(*Rewrite the statement by changing the underlined word.*)

____

56. State True or False.

(i) *Planaria* possesses high regeneration capacity and ladder-like nervous system.

True ☐ False ☐

(ii) Common name of *Fasciola* is tapeworm.

True ☐ False ☐

57. Identify the animal phylum based on the given features.

(i) Bilateral symmetry
(ii) Triploblastic
(iii) Internal fertilisation
(iv) Development direct or indirect
(v) Muscular pharynx

____

58. Write three features present in phylum–Aschelminthes but not present in phylum–Platyhelminthes.

____

59. Identify male and female roundworms in the figure given below.

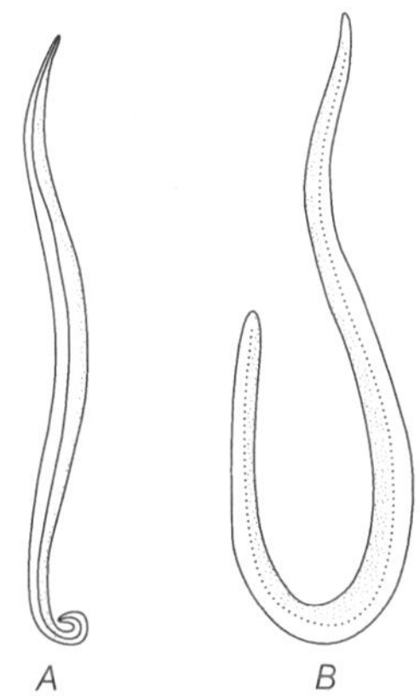

A. ______________________

B. ______________________

60. Name the animal phylum in which organ system level of organisation appeared first.

____

Topic 3 : Classification of Animals–Annelida to Hemichordata

61. Presence of true coelom for the first time was seen in phylum ______________ .

62. The body surface of annelids is distinctly marked out into segments. These segments are called as ______________ .

63. Name the lateral appendages present in *Nereis* which help in swimming.

____

64. Name the structure which help in the osmoregulation and excretion in annelids.

____

65. Choose the odd one out with respect to monoecious annelid.

Earthworm, Leeches, *Nereis*, *Magascolex*

____

66. In annelids, neural system consists of paired ganglia connected by lateral nerves to a double ventral nerve cord.

True ☐ False ☐

67. Identify the organisms, *A*, *B*, *C* and *D* along with their respective phyla.

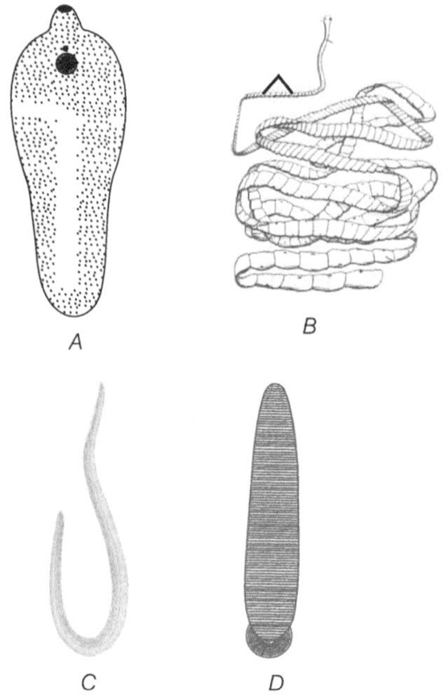

	Organisms	Phyla
A.	__________	__________
B.	__________	__________
C.	__________	__________
D.	__________	__________

68. Match the following columns.

	Column I (Animals)		Column II (Common names)
A.	*Pheretima*	1.	Filarial worm
B.	*Wuchereria*	2.	Hook worm
C.	*Ancylostoma*	3.	Blood sucking leech
D.	*Hirudinaria*	4.	Earthworm

A. __________ B. __________

C. __________ D. __________

69. The largest phylum of animal kingdom is __________

70. All of the following are the exclusive features of members of which phylum?

(i) Chitinous exoskeleton

(ii) Jointed appendages

(iii) Presence of respiratory system

____

71. Body of an arthropod is divided into __________, __________ and __________.

72. Match the following columns.

	Column I (Respiratory structures)		Column II (Animals)
A.	Gills	1.	Horse-shoe crab
B.	Tracheal system	2.	Crab, prawn
C.	Book gills	3.	Butterfly, cockroach
D.	Book lungs	4.	Scorpion, spider

A.________________ B. ________________

C.________________ D. ________________

73. How many features belong to phylum–Arthropoda?

Open circulatory system, Eyes may be simple or compound, Mostly oviparous, Development direct or indirect, Radial symmetry, Triploblastic, Coelomate, Mostly dioecious

____

74. Choose the odd one out with respect to sensory structures of arthropod.

Malpighian tubules, Ocelli, Antenna, Statocysts

____

75. Fill in the blanks.

(i) Balancing organs of aquatic arthropods are ________________ .

(ii) Connecting link between Annelida and Arthropoda is ________________.

76. Complete the following analogy given below.

(i) ____________ : Organism produces silk :: *Laccifer* : Organism produces lac.

(ii) Gregarious pest : *Locusta* : : Vector : ____________

77. Name the excretory organ in arthropods. ________________________

78. Refer to the given pictures and identify the class to which these animals belong.

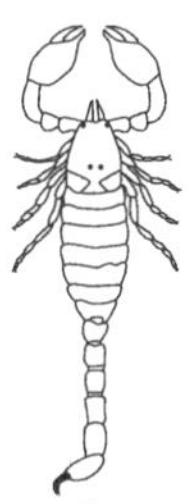

A

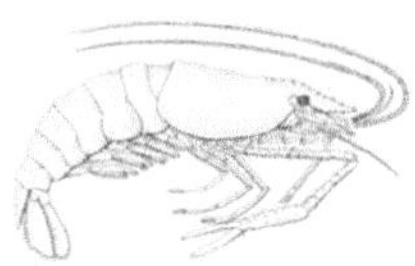

B

A. ________________________

B. ________________________

79. ____________ is the process of development of an egg into a complete individual without fertilisation by a sperm. It is seen in honeybees.

80. The second largest phylum of animal kingdom is ____________ .

81. The soft spongy layer of skin over the visceral hump of molluscs is known as ________ .

82. Members of phylum–Mollusca can be distinguished from other invertebrate by the presence of mantle and unsegmented body.

True ☐ False ☐

83. Body of a mollusc is covered by a ________________ shell with a distinct ____________ , ______________ and visceral hump.

84. Exchange of gases in molluscs occurs through feathery gills present in the visceral hump.
(*Rewrite the statement by changing the underlined word.*)

____

85. In molluscs, the mouth contains a file-like rasping organ for feeding called __________

86. Given below is a hypothetical body plan in molluscs. Label its parts, *A*, *B*, *C* and *D*.

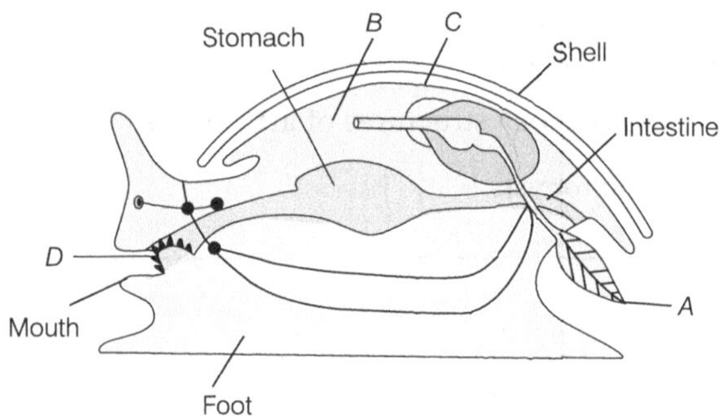

A. ____________________

B. ____________________

C. ____________________

D. ____________________

87. Molluscs have copper containing respiratory pigment in their body known as

____ .

88. (i) Molluscs do not show asexual development, but their sexual phase includes characteristic larvae like ________________ and ________________ .

(ii) An example of mollusc without shell is ________________ .

89. Identify the class to which the animals shown in picture belong.

A

B

A. ____________________

B. ____________________

90. Study the following table carefully and write the correct words for *A, B, C, D* and *E*.

Phylums	Excretory organs	Circulatory systems	Respiratory organs
Ⓐ	Nephridia	Closed	Skin
Arthropoda	Ⓑ	Ⓒ	Lungs/Gills tracheal system
Ⓓ	Metanephridia	Open	Ⓔ

A. ____________________ B. ____________________

C. ____________________ D. ____________________

E. ____________________

91. Match the following columns.

	Column I (Molluscs)		Column II (Scientific names)
A.	Apple snail	1.	*Dentalium*
B.	Pearl oyster	2.	*Pila*
C.	Cuttle fish	3.	*Octopus*
D.	Devil fish	4.	*Sepia*
E.	Tusk shell	5.	*Pinctada*

A. ____________________ B. ____________________

C. ____________________ D. ____________________

E. ____________________

92. Complete the following analogy given below.

__________ : Larvae :: Pentamerous radial : : Adults

93. Choose the odd one out feature of echinoderm

Spiny bodies, All marines, Coelomate, Tissue level organisation.

____

94. Name the most distinctive system in the animals of phylum–Echinodermata.

____

95. Digestive system in echinoderms is complete with mouth on the lower (ventral) side and anus on the upper (dorsal) side.

True ☐ False ☐

96. Write the functions of water vascular system in star fish.

____

97. Match the following columns.

	Column I (Respiratory organs in echinoderms for gaseous exchanges)		Column II (Present in animals)
A.	Dermal branchiae or papulae	1.	Starfish
B.	Peristominal gills	2.	Sea urchin
C.	Genital bursae	3.	Brittle star
D.	Cloacal respiratory trees	4.	*Holothuria*

A.__________ B. __________

C.__________ D. __________

98. Give the scientific names and their respective phyla.

(i) Sea lily __________

(ii) Sea urchin __________

(iii) Sea cucumber __________

(iv) Sea horse __________

(v) Sea mouse __________

(vi) Sea hare __________

99. Choose the correct option for the given animal and fill the blanks given below.

(i) Excretory system is __________ (present/absent)

(ii) Symmetry is __________ (triploblastic/diploblastic)

(iii) Reproduction by __________ (sexual/asexual)

(iv) Sexes are __________ (dioecious/monoecious)

(v) Scientific name is __________ *(Asterias/ Ophiura)*

100. __________ is the phylum which consists of a small group of worm–like marine animals with organ system level of organisation.

101. Write the scientific name of the given organism '*X*' and correctly name its parts, *A*, *B* and *C*.

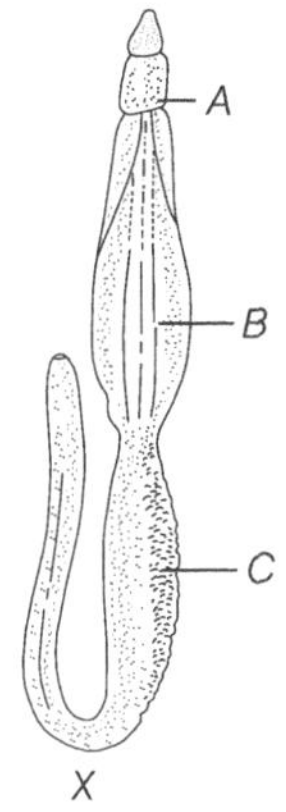

A. ____________________

B. ____________________

C. ____________________

X. ____________________

102. Hemichordates have a rudimentary structure in the collar region called proboscis, that is similar to notochord.
(*Rewrite the statement by changing the underlined word.*)

____

103. Hemichordates are now placed as separate phylum under non-chordates.

True ☐ False ☐

104. Choose the odd one out with respect to feature in *Saccoglossus.*

Triploblastic, Coelomate, Gills, Closed circulation, Bilaterally symmetrical

____

105. Sexes are separate, fertilisation is external and development is indirect in which two phyla of non-chordates? ____________________

Topic 4 : Classification of Animals : Chordata

106. Write the three fundamental chracteristics of phylum–Chordata.

____

107. Recognise and write the names of structures *A*, *B*, *C* and *D* in the diagrammatic representation of chordates.

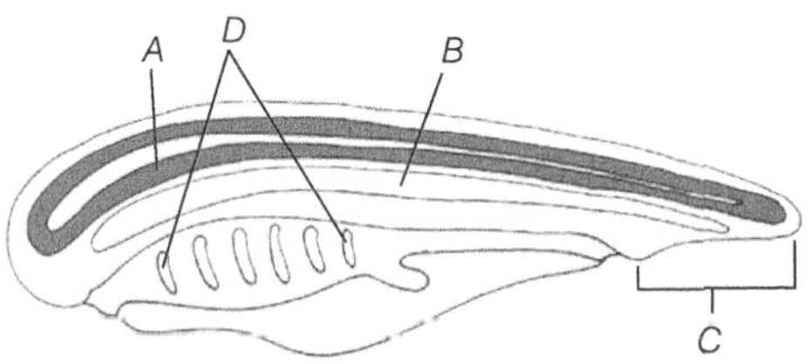

A. ____________________

B. ____________________

C. ____________________

D. ____________________

108. Go through the following flowchart and fill in the box *A-F.*

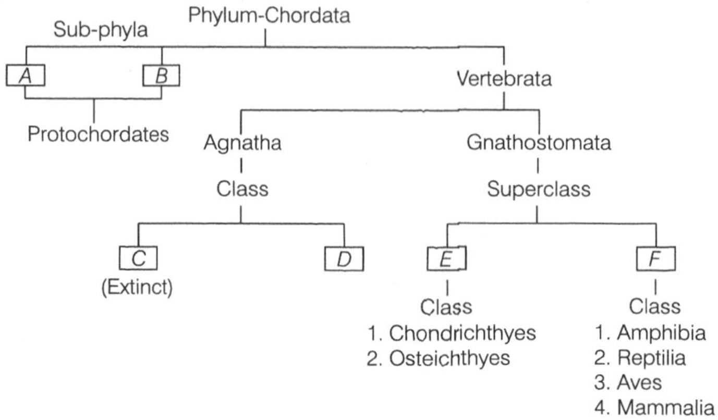

A. ____________________

B. ____________________

C. ____________________

D. ____________________

E. ____________________

F. ____________________

109. __________ possess notochord throughout their life but __________ have it only in larval stage.

110. How many of the following represent urochordates (U) and cephalochordates (C)?

Ascidia, Branchiostoma, Salpa, Doliolum

____

111. The echinoderms, hemichordates and chordates had ____________________ (tornaria/dipleurula) larva as a common ancestral form.

112. State True or False.

(i) All chordates are vertebrates, but all vertebrates are not chordates.

True ☐ False ☐

(ii) Tunicata is an alternative name of urochordata.

True ☐ False ☐

113. Complete the following analogy given below.

Kidneys : Chordates :: ________________ : Hemichordates.

114. A common characteristic of all vertebrates without exception is the presence of ____________________ (exoskeleton/skull).

115. In vertebrates, embryonic notochord is replaced by a ______________ in the adult.

116. Select the correct options for the given animal.

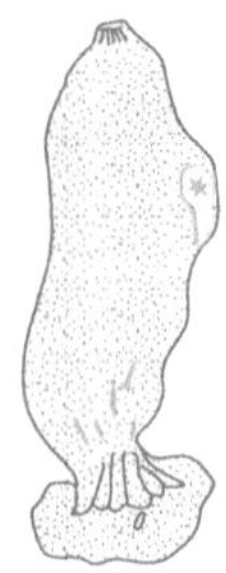

(i) Heart is ______________________ (dorsal heart/ventral heart)

(ii) Colelom is ______________________ (acoelomate/pseudocoelomate/coelomate)

(iii) Position ______________________ (vertebrates/invertebrates)

(iv) Sub-phylum is ______________________ (urochordates/cephalochordates)

(v) Scientific name is ______________________ (*Ascidia/ Doliolum*)

117. Fill in the in the table with correct differences.

Cyclostomes	Chondrichthyes	Osteichthyes
(A)	Ventral mouth	(B)
(C)	(D)	Cycloid/Ctenoid scales
(E)	(F)	Marine and freshwater
(G)	5-7 paired of gills without operculum	(H)

A. ______________________ B. ______________________

C. ______________________ D. ______________________

E. ______________________ F. ______________________

G. ______________________ H. ______________________

118. Identify the diagrams and answer the questions as follows.

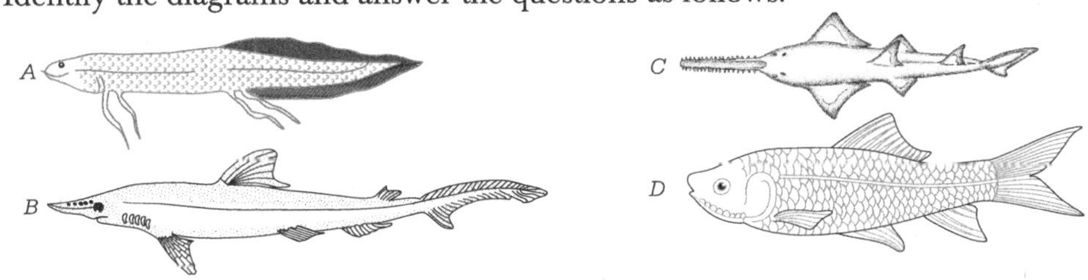

(i) Which fish has air bladder? ______________________

(ii) Which fish is cartilaginous and bony, respectively? ______________________

(iii) Which animal is ectoparasite on true fishes? ______________________

(iv) Which fish has paired fins? ______________________

(v) Which fish lives in freshwater and possesses operculum? ______________________

(vi) Write names of each *A*, *B*, *C* and *D* above? ______________________

119. Match the following columns.

	Fishes		Features
A.	Cartilaginous fishes	1.	Usually external fertilisation
B.	Bony fishes	2.	Internal fertilisation
		3.	Mostly oviparous
		4.	Mostly viviparous

A. ______________________ B. ______________________

120. Name the class of jawless vertebrates which are marine but migrate for spawning to freshwater. ____________

121. Complete the following analogy given below.

Bony fish : : Osteichthyes : Cartilaginous fish : ____________

122. ____________ (*Carcharodon/ Clarias*) are cold-blooded animals in which male pelvic fins bear claspers for copulation.

123. Name the Chondrichthyes with the following characters.

(i) Electric organs ____________

(ii) Poison sting ____________

(iii) Ampulla of Lorenzini ____________

124. Write the scientific names of the following.

(i) Flying fish ____________

(ii) Angel fish ____________

(iii) Dog fish ____________

(iv) Saw fish ____________

(v) Fighting fish ____________

(vi) Hag fish ____________

(vii) Star fish ____________

125. The heart of fishes is _______ chambered with _______ auricle and _______ ventricle.

126. Name the structure that prevents the bony fishes from sinking. ____________

127. Cartilaginous fishes have to swim constantly because they lack ____________.

128. Write true or false about the given animal (vertebrate).

(i) It belongs to class–Reptilia.

True ☐ False ☐

(ii) It has a prehensile tail and exhibits sexual dimorphism.

True ☐ False ☐

(iii) The male bears a brood pouch in which the female lays eggs which remain there till they hatch.

True ☐ False ☐

129. Choose the odd one out.

Neoceratodus, Lepidosiren, Protopterus, Exocoetus

____

130. Name a fish which is a connecting link between fishes and amphibians, also known as living fossil. ______________________________

131. Fill in the blanks with respect to amphibians.

(i) Body of amphibians is divisible into __________ and __________ .

(ii) A __________ represents the ear.

(iii) Alimentary canal, urinary and reproductive tracts open into a common chamber called __________ which opens to outside.

(iv) Heart is __________ chambered with ________ auricles and ________ ventricles.

132. Choose the odd one out with respect to respiration in amphibians.

Gills, Lungs, Skin, Nephridia

____

133. Choose the correct option for the given animal.

(i) Animal is ______________________________ (frog/toad)

(ii) Eyes ______________________________ (with eyelids/without eyelids)

(iii) Skin ______________________________ (with scales/without scales)

(iv) Fertilisation is ______________________________ (internal /external)

(v) Development is ______________________________ (oviparous/viviparous)

(vi) Development is ______________________________ (indirect/direct)

134. Write the names of the amphibian from the box given below based on the characteristics.

Ichthyophis, Ambystoma, Cryptobranchus, Seymouria.

(i) The largest salamander ______________________________

(ii) Limbless amphibian ______________________________

(iii) Amphibian with axolotl larva ______________________________

(iv) A connecting link between amphibian and reptile. ______________________________

135. Choose the odd one out and also write the names of each.

____

136. Name the class of vertebrates possessing dry and cornified skin with epidermal scales or scutes.______________________

137. The mode of locomotion in reptiles is ______________________

138. Out of the below features, which is not present in amphibians but found in reptiles?

Poikilotherms, 3-chambered heart, Oviparous, Tympanum, Two pairs of limbs, Direct development, Internal fertilisation

____

139. Name two reptiles based on the features listed below.

(i) Four-chambered heart ______________________

(ii) Shed their scales as skin cast ______________________

(iii) Limbless reptiles ______________________

140. Give common names for the following.

(i) *Testudo* ______________________

(ii) *Bufo* ______________________

(iii) *Hyla* ______________________

(iv) *Calotes* ______________________

(v) *Hemidactylus* ______________________

(vi) *Bangarus* ______________________

141. Write the names of the class–Tetrapoda induded in

(i) Anamniotes ______________________

(ii) Amniotes ______________________

142. In poisonous snakes, fangs are modified ______________ (mandibular/maxillary) teeth, having poison glands.

143. Complete the following analogy given below.

Herpetology : Study of reptiles : : ______________ : Study of snakes.

144. Find the exception.

(i) Monocondylic skull (Reptiles/Amphibians/Birds) ____________________

(ii) Metanephric kidney (Birds/Reptiles/Amphibians) ____________________

(iii) Lack teeth (Turtles/Lizards/Chameleon) ____________________

145. Name the additional chambers found in the digestive tracts of birds.

____

146. The bones of birds are long and hollow with air cavities. These types of bones are called as

____________________ .

147. Given below are the groups of animals.

Pisces, Amphibians, Reptiles, Aves, Mammals

Write the correct sequence of decreasing order of number of species in them.

____

148. In birds, the hindlimbs are modified into wings and forelimbs are modified for walking, swimming or clasping the tree branches.

True ☐ False ☐

149. The skin of class–Aves is dry without glands except the ________ gland present at the base of tail.

150. Name the structures in birds.

(i) Voice production ____________________

(ii) Respiration ____________________

(iii) Feeding, seed crushing, flesh tearing, etc. ____________________

151. Match the following columns.

	Column I (Aves)		Column II (Common names)
A.	*Corvus*	1.	Pigeon
B.	*Columba*	2.	Crow
C.	*Aptenodytes*	3.	Vulture
D.	*Neophron*	4.	Penguin

A. ____________________ B. ____________________

C. ____________________ D. ____________________

152. Urinary bladder is absent in all birds except ____________________

153. Write the scientific names of birds *A-D* and identify which among the following is a flightless bird?

A. ____________________

B. ____________________

C. ____________________

D. ____________________

____

154. Write three common features present in all mammals.

____

155. How many features are common to both mammals and Aves?

Development direct, Internal fertilisation, 4-chambered heart, Teeth present, Viviparous, Homoiothermous

____

156. Fill in the the table given bleow with correct differences between three groups of mammals.

Prototheria	Metatheria	Eutheria
(A)	(B)	Viviparous
(C)	Pinna present	(D)
No marsupial pouch	(E)	(F)
(G)	Placenta present	(H)

A. ____________________ B. ____________________

C. ____________________ D. ____________________

E. ____________________ F. ____________________

G. ____________________ H. ____________________

157. Write one feature in each given mammals not present in other three.

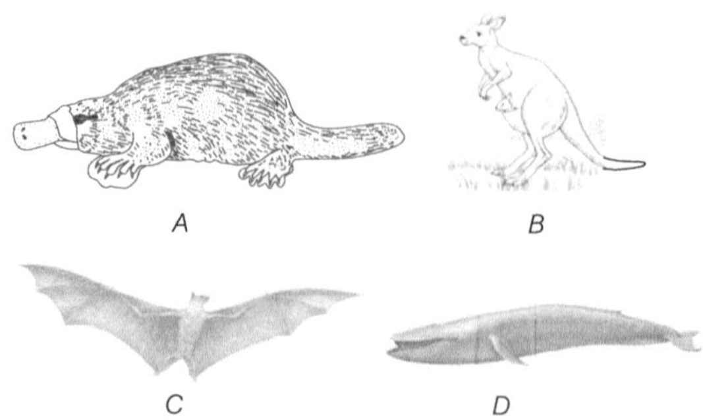

A. ____________________

B. ____________________

C. ____________________

D. ____________________

158. Which character is present in bats but not in birds?

Homoiothermy, Four-chambered, Heart, Syrinx, Diaphragm

____

159. How many organisms in the list given below respire through gills?

Sponges, Coelentrates, Flatworms, Aquatic arthropods, Molluscs, Fishes, Tadpoles, Reptiles, Aves, Mammals

____

160. Complete the missing difference.

	Class–Insecta	Class–Mammalia
Phylums	(A)	(B)
Examples	(C)	(D)

A. ______________ B. ______________
C. ______________ D. ______________

161. Respiration in all mammals occurs *via*. ______________

162. All mammals are viviparous except ______________ which is oviparous.

163. Write the names of any two mammals adapted for the given features habitat.

(i) Flying ______________

(ii) Swimming ______________

(iii) Climbing ______________

(iv) Burrowing ______________

Answers

1. False. All members of Animalia are multicellular, they do not exhibit the same pattern of organisation of cells.
2. Parazoans is odd one out as they lack tissue grade organisation.
3. *A*–Aschelminthes, annelids, arthropods, molluscs, echinoderms, chordates.
   B–Ctenophores, coelenterates
   C–Platyhelminthes
   D–Poriferans
4. Porifera
5. **Closed** - 3(Cephalopod molluscs, chordates, annelids),
   Open - 4(Non-cephalopod molluscs, tunicates, hemichordates, arthropods).
6. 3 types. Asymmetry, radial, bilateral
7. *A*-Platyhelminthes
   B-Coelenteratata
   C-Echinodermata
8. *A*–Bilateral symmetry, *B*–Radial symmetry
9. *A*-Mesoglea, *B*-Ectoderm, *C*-Endoderm, *D*-Mesoderm
10. (i) Diploblastic, (ii) Mesoderm
11. Coelom
12. *A*-Annelids (coelomate), *B*-Aschelminthes (pseudocoelomate), *C*-Platyhelminthes (acoelomate).
13. True coelom
14. (i) Pseudocoelom
    (ii) Acoelomates
15. Segmentation
16. Earthworm
17. False. Notochord is a mesodermally derived rod-like structure formed on the dorsal side during embryonic development in chordates.
18. *A*–Cellular level *B*–Acoelomata, *C*–Bilateral, *D*–Aschelminthes, *E*–Coelomates
19. Porifera
20. Asymmetrical
21. Porifera
22. Ostia, spongocoel, osculum

23. 6(All of them except digestion because it is intracellular)
24. Pinacocyte, choanocyte
25. Asconoid, syconoid and leuconoid
26. Choanocytes
27. Archaeocytes
28. True
29. Fragmentation
30. Spicules or spongin fibres
31. *Leucosolenia*
32. (i) *Euplectella*
 (ii) *Hyalonema*
 (iii) *Cliona*
 (iv) *Chalina.*
33. *A-Sycon* (Scypha), *B-Euspongia* (Bath sponge), *C-Spongilla* (Freshwater sponge)
34. Hermaphrodite
35. Indirect development
36. Phylum–Coelenterata
37. (i) Cell – Cnidoblast
 Function – Help in defence and capture of prey
 (ii) Nematocyst
38. (i) Coral
 (ii) Both (intracellular and extracellular)
39. Hypostome
40. Coelenteron, incomplete
41. Asexually, sexually
42. Metagenesis
43. *A*-Medusa ($2n$), *B*-Polyp ($2n$)
44. (i) 3 (*Hydra, Obelia, Adamsia*)
 (ii) 2 (*Physalia, Aurelia*)
45. (i) *Physalia*, (ii) *Adamsia*, (iii) *Pennatula*, (iv) *Gorgonia* (v) *Meandrina*
46. Sea walnuts or comb jellies.
47. (i) True
 (ii) True
 (iii) False. Ctenophores exhibit both extracellular and intracellular digestion.
48. (i) *Pleurobrachia*
 (ii) Present
 (iii) Sexual
 (iv) External
 (v) Indirect
49. Bioluminescence
50. Cydippid larva
51. Colloblast or lasso cells
52. Coelomate is odd one out. Platyhelminthes are acoelomates. Others are characteristics of phylum–Platyhelminthes.
53. Hooks, suckers
54. (i) Larval forms in tapeworm (Platyhelminthes)
 (ii) Classes of Platyhelminthes
55. Flame cells or solenocytes
56. (i) True
 (ii) False. Common name of *Fasciola* is liver fluke. Tapeworm is the name of *Taenia.*
57. Phylum-Aschelminthes
58. Dioecious, pseudocoelomates, complete alimentary canal
59. *A* is male roundworm, *B* is female roundworm.
60. Phylum–Aschelminthes
61. Annelida
62. Metamers
63. Parapodia
64. Nephridia
65. *Nereis* is odd one out as, it is dioecious, rest are monoecious.
66. True
67. *A* - Liver fluke, Platyhelminthes, *B* - Tapeworm, Platyhelminthes, *C* - Roundworm, Aschelminthes, *D* - *Hirudinaria*, Annelida
68. A-4, B-1, C-2, D-3
69. Arthropoda
70. Phylum–Arthropoda
71. Head, thorax and abdomen
72. A-2, B-3, C-1, D-4
73. 7(All of them except radial symmetry). Arthropods exhibit bilateral symmetry.
74. Malpighian tubules is odd one out as they are not a sensory structure in arthropods. These are excretory structure.
75. (i) Statocysts
 (ii) *Peripatus*
76. (i) *Bombyx*
 (ii) *Culex*
77. Malpighian tubules
78. *A*-Arachnida
 B-Crustacea
79. Parthenogenesis
80. Mollusca
81. Mantle
82. True
83. Calcareous, head, muscular foot
84. Mantle cavity
85. Radula
86. *A*-Gill, *B*-Visceral hump, *C*-Mantle, *D*-Radula
87. Haemocyanin
88. (i) Veliger, trochophore/ glochidium
 (ii) *Octopus, Doris* (Any one)
89. *A*-Gastropoda, *B*-Cephalopoda
90. *A*-Annelida, *B* - Malpighian tubules, *C* - Open, *D* - Mollusca, *E* - Feather-like gills
91. A–2, B–5, C–4, D–3, E–1
92. Bilateral
93. Tissue level organisation is odd one out. Echinoderms are with organ system level of body organisation.
94. Presence of water vascular system (ambulacral system).
95. True
96. Locomotion, capture and transport of food and respiration.
97. A-1, B-2, C-3, D-4
98. (i) *Antedon* (Echinodermata)
 (ii) *Echinus* (Echinodermata)
 (iii) *Cucumaria* (Echinodermata)
 (iv) *Hippocampus* (Chordata)
 (v) *Aphrodite* (Annelida)
 (vi) *Aplysia* (Mollusca)
99. (i) Absent
 (ii) Triploblastic
 (iii) Sexual reproduction
 (iv) Dioecious
 (v) *Asterias*
100. Hemichordata
101. *X-Balanoglossus*, *A*-Proboscis, *B*-Collar, *C*-Trunk
102. Stomochord
103. True
104. Closed circulation is odd one out as in it, circulatory system is of open type.
105. Phya–Echinodermata, Hemichordata
106. A notochord, a dorsal hollow nerve cord and paired pharyngeal gill slits.

107. *A* - Nerve cord, *B* - Notochord, *C* - Post-anal tail, *D* - Gill slits

108. *A* - Urochordata
B - Cephalochordata
C - Ostracodermi
D - Cyclostomata
E - Pisces
F - Tetrapoda

109. Cephalochordates, Urochordates

110. U – 3 (*Ascidia, Salpa, Doliolum*), C – 1 (*Branchiostoma*)

111. Dipleurula

112. (i) False. All vertebrates are chordates, but all chordates are not vertebrates.
(ii) True

113. Proboscis

114. Well-developed skull

115. Cartilaginous or bony vertebral column.

116. (i) Ventral heart
(ii) Coelomate
(iii) Vertebrates
(iv) Urochordates
(v) *Ascidia*

117. *A*-Sucking mouth, *B*-Terminal mouth, *C*-Scales absent, *D*-Placoid scales, *E*-Marine, *F*-Marine, *G*-6-15 paired of gills, *H*-4 paired of gills with operculum

118. (i) *D*
(ii) *B* and *C* are cartilaginous fishes, *D* is bony fish
(iii) *A*
(iv) *B, C, D*
(v) *D*
(vi) *A-Petromyzon, B- Scoliodon, C-Pristis, D-Catla*

119. A-2, 4, B-1, 3,

120. Cyclostomes

121. Chondrichthyes

122. *Carcharodon*

123. (i) *Torpedo,*
(ii) *Trygon,*
(iii) *Scoliodon*

124. (i) *Exocoetus*
(ii) *Pterophyllum*
(iii) *Scoliodon*
(iv) *Pristis*
(v) *Betta*
(vi) *Myxine*
(vii) *Asterias*

125. 2, 1 and 1

126. Air bladder

127. Air bladder

128. (i) False. It belongs to class-Osteichthyes (Pisces)
(ii) True
(iii) True

129. *Exocoetus* is odd one out as it is not a lung fish.

130. *Latimeria*

131. (i) Head, trunk
(ii) Tympanum
(iii) Cloaca
(iv) 3, 2, 1

132. Nephridia is odd one out as it helps in excretion in annelids.

133. (i) Frog
(ii) With eyelids
(iii) Without scales
(iv) External
(v) Oviparous
(vi) Indirect

134. (i) *Cryptobranchus*
(ii) *Ichthyophis*
(iii) *Ambystoma*
(iv) *Seymouria*

135. *A*–Chameleons *B*–Crocodile, *C*–*Salamandra, D*–Chelone, C–is odd one out as it is amphibian and rest are reptiles.

136. Class–Reptilia

137. Creeping or crawling

138. Internal fertilisation and direct development are features specific to reptiles.

139. (i) Crocodile, alligators
(ii) Snakes, lizards
(iii) Viper, cobra

140. (i) Tortoise, (ii) Toad, (iii) Tree frog, (iv) Garden lizard, (v) Wall lizard, (vi) Krait

141. (i) Amphibians, (ii) Reptiles, Aves mammals

142. Maxillary

143. Ophiology

144. (i) Amphibians have dicondylic skull.
(ii) Mesonephric kidney is present in amphibians.
(iii) Turtles

145. Crop and gizzard

146. Pneumatic bones

147. Pisces > Aves > Reptiles > Mammals > Amphibians

148. False. In birds, the forelimbs are modified into wings. The hindlimbs are modified for walking, swimming or clasping.

149. Oil

150. (i) Syrinx
(ii) Lungs
(iii) Beak

151. A-2, B-1, C-4, D-3

152. *Rhea americana*

153. *A*-Neophron, *B-Struthio, C-Psittacula, D-Pavo*
Out of these, *B* is a flightless bird.

154. Hairs on their body, warm-blooded, presence of mammary glands.

155. 4(All of them exupt viviparous and teeth). Teeth are present in mammals only and birds are oviparous.

156. *A*-Oviparous, *B*-Viviparous, *C*-No pinna, *D*-Pinna usually present, *E*-Marsupial pouch usually present, *F*-Marsupial pouch absent, *G*-Placenta absent, *H*-Placenta present (well-developed)

157. *A*-Platypus (it is oviparous), *B*-Kangaroo (has marsupial pouch), *C*-Bat (has wings for flying), *D*-Whale (has flippers and no hindlimbs)

158. Diaphragm is present in bats, but not present in birds.

159. 4 (Molluscs, fishes, aquatic arthropods, tadpoles)

160. *A*-Arthropoda, *B*- Chordata, *C*-Butterfly, *D*-Shrews

161. Lungs

162. *Ornithorhnynchus*

163. (i) *Pteropus, Scotophilus,*
(ii) *Balaenoptera, Delphinus*
(iii) *Macaca*, Lemurs
(iv) *Rattus*, Rabbit

CHAPTER FIVE

MORPHOLOGY OF FLOWERING PLANTS

Topic 1 : The Root

1. The angiosperms show a large diversity in ______________. They all characterised by the presence of ______________, ______________, ______________, ______________ and ______________.

2. The underground part of the flowering plant is the ________ system, while the portion above the ground is the ________ system.

3. The direct elongation of the radicle leads to the formation of primary root in majority of the monocotyledonous plants.

   True ☐ False ☐

4. The primary root bears lateral roots of several orders that are referred to as ______________, ______________ roots.

5. Refer to the given figure and answer the following questions.

   (i) Identify *A*, *B* and *C* parts.

A. ______________

B. ______________

C. ______________

   (ii) Root *A* and *B* constitute which type of root system? ______________

   (iii) In ______________ (monocot/dicot) root, *B* is short-lived.

6. Match the following columns.

	Column I (Types of root system)		Column II (Features)		Column III (Examples)
(i)	Tap root	a.	Root arises from parts of the plant other than the radicle	1	*Monstera*
(ii)	Adventitious	b.	Replaced by large number of roots	2.	Mustard
(iii)	Fibrous	c.	Consists of primary and secondary root	3.	Wheat

(i) ______________ (ii) ______________ (iii) ______________

7. How many of the following functions given are related to the root system?

Absorption of water, Food production, Food storage, Plant growth regulator production

____

8. Root is covered at the apex by a thimble-like structure called ______________ .

9. Match the following columns with the labelled parts of the regions of the root tip.

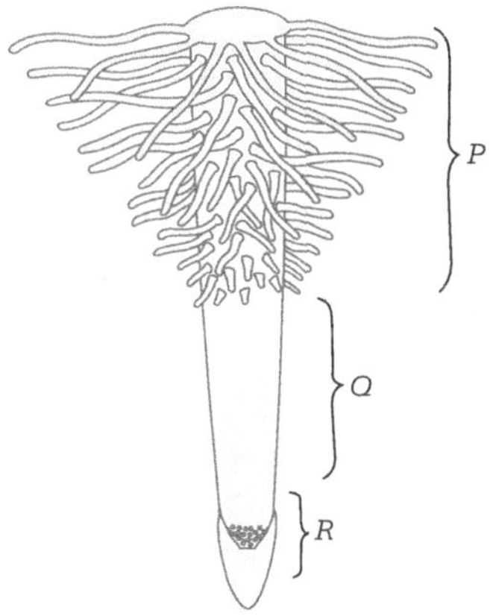

	Column I (Regions of the root)		Column II (Regions of the root)		Column III (Features)
(i)	Region of meristematic activity	a.	*Q*	1.	Form very fine and delicate, thread like structure called root hairs
(ii)	Region of elongation	b.	*P*	2.	Cells are small, thin-walled with dense cytoplasm
(iii)	Region of maturation	c.	*R*	3.	Cells undergo rapid elongation and enlargement

(i) ______________ (ii) ______________ (iii) ______________

10. ______________ (Tap/Adventitious) roots of carrot, turnip and ______________ (tap/adventitious) roots of sweet potato get swollen and store food.

11. Select the odd one out from the following terms with reference to types of roots.

Nodulated, Moniliform, Annulated, Palmate

____

12. Match the following columns.

	Column I (Types of roots)		Column II (Characteristics)
A.	Fusiform	1.	These are thick and fleshy roots, with irregular shape
B.	Napiform	2.	Roots are spindle-shaped
C.	Tuberous	3.	Roots swell in the middle and narrow towards the apical region

A. ______________ B. ______________ C.______________

13. How many of the following roots are adventitious root?

Nodulated, Buttress, Stilt, Prop, Respiratory root, Nodulose

____

14. The hanging structure of banyan tree is called ______________________ .

15. The stems of maize and sugarcane have supporting roots coming out of the lower nodes of the stem is called clinging roots.
(*Rewrite the statement by changing the underlined word.*)

____

16. Prop roots come out of the ground and grow vertically upwards.

True ☐ False ☐

17. Consider the following diagram of modification of root and answer the following questions.

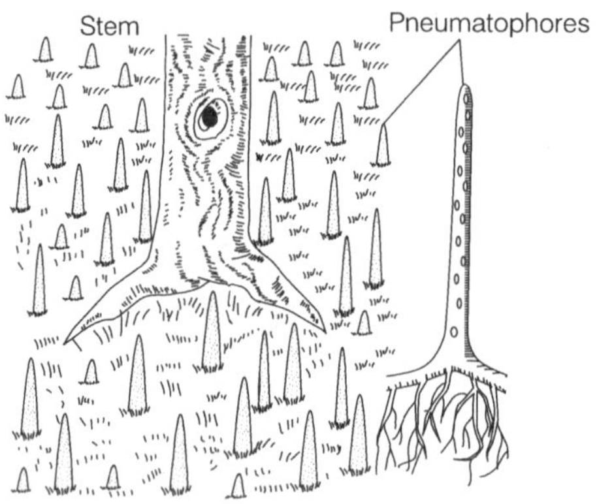

(i) These roots are found in mangroove species, e.g. ______________________ growing in swampy areas.

(ii) What is the function of these roots? ______________________

18. Identify the type of root arising from the plant shown in the following figure.

____

19. Identify the mismatched pair.

(i) Prop roots – Provide support

(ii) Stilt roots – Originate from stem's nodal part

(iii) Pneumatophores – Help in storing food

(iv) Edible roots – Sweet potato

____

20. A type of modified ______________ root that arises from the internode of stem, grows and becomes fleshy and tuberous, serving as storage for reserve food material is observed in ______________ .

21. Consider the following figure of *Asparagus* root and answer the given questions.

(i) These type of root modification are known as ______________ .

(ii) They occur in cluster at the base of the stem.

True ☐ False ☐

Topic 2 : The Stem

22. The stem is the ______________ part of the axis bearing branches, leaves, flowers and fruits.

23. Where does the stem develop from?

____

24. The region of the stem where leaves are borne is called ______________ while ______________ are the portion between two nodes.

25. The stem bears buds, which may be terminal or axillary.

True ☐ False ☐

26. Give any two main functions of stem.

(i) ______________________________

(ii) ______________________________

27. Give a common collective term for the following plant species.

Potato, Ginger, Turmeric, Jimikand, *Colocasia*

____

28. Match the following columns.

Column I (Types of stem modifications)	Column II (Characteristics)
A. Rhizome	1. It possesses a number of spirally arranged buds
B. Tuber	2. It grows horizontally, bears nodes and internodes
C. Bulb	3. It is short, thick and highly condensed
D. Corm	4. Its internodal part becomes highly reduced

A. ______________ B. ______________

C. ______________ D. ______________

29. Stem <u>sukers</u> develop from axillary buds are slender, spirally coiled and help plants to climb.

(*Rewrite the statement by changing the underlined word.*)

____

30. Give a common collective term for the structure possessed by the following plant species.

Cucumber, Pumpkins, Watermelon, Grapevine

____

31. Axillary buds of stems are modified into woody, straight and pointed ______________ (bulbils/thorns). These protect plants from browsing animals.

32. Which type of stem modification is found in *Citrus* and *Bougainvillea*?

____

33. Some plants of arid regions modify their stems into ______________________________ or ______________________________ structures, containing chlorophyll.

34. Identify the odd one out from the following species.

Asparagus, Opuntia, Coccoloba, Euphorbia, Casuarina

____

35. Name the stem modification in which underground stems of some plants spread to new niches and when older parts die new plants are formed.

____

36. Which type of stem modification is shown in the following figure of *Oxalis* sp.?

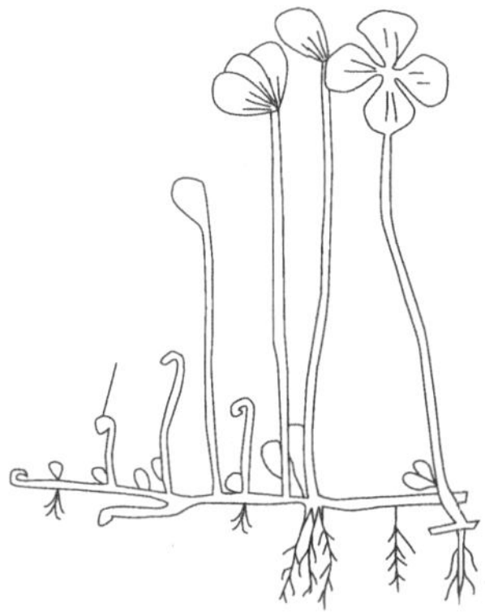

____

37. In suckers, a slender lateral branch arises from the base of the main axis and after growing aerially it downwards to touch the ground.
(*Rewrite the statement by changing the underlined word.*)

____

38. ____________________ (Runners/Offsets) are lateral branch with short internodes and each node bearing a rosette of leaves and a taft of roots.

39. Stolons are the lateral branches originate from the basal and underground portion of the stem, grow horizontally and give rise to leafy shoots.

True ☐ False ☐

40. Match the following columns.

	Column I (Types of stem modification)		**Column II** (Examples)
A.	Runners	1.	Mint, jasmine
B.	Sucker	2.	*Pistia, Eichhornia*
C.	Stolon	3.	Banana, pineapple, *Chrysanthemum*
D.	Offset	4.	Grass, strawberry

A. ________________ B. ________________ C. ________________

D. ________________

41. Complete the following analogy with respect to stem modification.

Cladodes : Bulbil :: ____________ : *Dioscorea*

42. ____________ is a modified stem developed either from a vegetative bud or a flower bud.

43. How many of the following are correctely matched group?

(i) Stem tendril - *Passiflora*

(ii) Leaf tendril - *Lathyrus*

(iii) Floral bud tendril - *Smilax*

____

Topic 3 : The Leaf

44. Name the structure of plant which is lateral, flattened borne on the stem.

____

45. Leaf develops at the node and bears a bud in its axil. The ________________ later develops into a branch.

46. ____________ originate from ______________________ and are arranged in an ________________ order.

47. Consider the diagram of leaf and answer the following questions.

(i) Identity the labelled parts *A*, *B*, *C*, *D* and *E*.

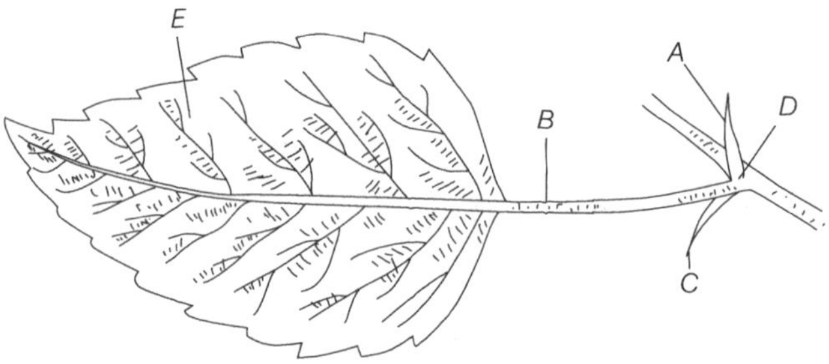

A. ____________________

B. ____________________

C. ____________________

D. ____________________

E. ____________________

(ii) Name the structure which attach leaf to the stem. ____________________

(iii) In ________________, structure *D* expands into a sheath covering the stem partially or wholly.

48. ______________________ (Stipules/Petioles) are the small leaf-like structure.

49. In which plants, swollen leaf base pulvinus is found?

____

50. Name the structure of leaf which allow leaf blades to flutter in wind.

____

51. The ________________ or the ________________ is the green expanded part of the leaf with veins and veinlets and middle prominent vein is called ________________.

52. Pulvinus provides rigidity to the leaf blade and acts as channels of transport for water, minerals and food materials.
(*Rewrite the statement by changing the underlined word.*)

____

53. The arrangement of veins and the veinlets in the lamina of leaf is termed as ______________________________.

54. In ______________ venation, the veinlet forms a network and in ______________ venation, the vein runs parallel to each other (parallel/reticulate).

55. Palmate reticulate venation consists of a single midrib in its centre.
(*Rewrite the statement by changing the underlined word.*)

____

56. The ______________ reticulate venation consists of a number of midribs arising from the top of the petiole.

57. Give a common characteristic feature of a following species.

Smilax, Dioscorea, Alocasia, Colocasia

____

58. Leaf with the entire lamina or when incised, the incisions do not touch the midrib is known as ______________ leaf.

59. In ______________ leaf, the incisions of the lamina reach up to the midrib breaking it into a number of leaflets.

60. A bud is present in the axil of petiole in both simple and compound leaves, but not in the axil of leaflets of simple leaf.

True ☐ False ☐

61. The number of leaflets are present on a rachis in ______________ compound leaf.

62. In ______________ compound leaves, the leaflets are attached at the tip of petiole.

63. Identify the type of leaves and species in which these are found from the given figures.

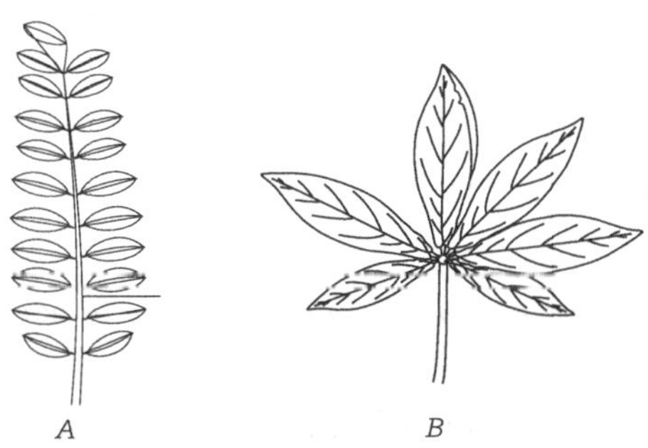

A. ______________________________

B. ______________________________

64. Venation is the pattern of arrangement of leaves on the stem or branch. (*Rewrite the statement by changing the underlined word.*)

____

65. In alternate type of phyllotaxy, a single leaf arises at each node in __________ manner.

66. In opposite type of phyllotaxy, a pair of leaves arise at each __________ and lies __________ to each other.

67. In opposite phyllotaxy, more than two leaves arise at a node and form a whorl.

True ☐ False ☐

68. Match the following columns.

	Column I (Types of phyllotaxy)		Column II (Figures of phyllotaxy)		Column II (Examples)
(i)	Opposite	a.		1.	*Alstonia*
(ii)	Alternate	b.		2.	Guava
(iii)	Whorled	c.		3.	China rose

(i) __________ (ii) __________ (iii) __________

69. The opposite leaves of two successive nodes lie at right angle to each other, this type of phyllotaxy is known as __________.

70. Name the type of phyllotaxy in which opposite leaves are superimposed on each other.

____

71. Name the specialised modification of leaf present at the base of flower axis.

____

72. The tendrils of cucumber and spines of citrus are the modification of leaves.

True ☐ False ☐

73. The ____________ of Australian *Acacia* expand, become green and synthesise food, because their ____________ are small and short-lived.

74. ____________ and ____________ modified their leaves for trapping insects.

75. Identify the type of leaf modification in the given diagram (*A* to *C*).

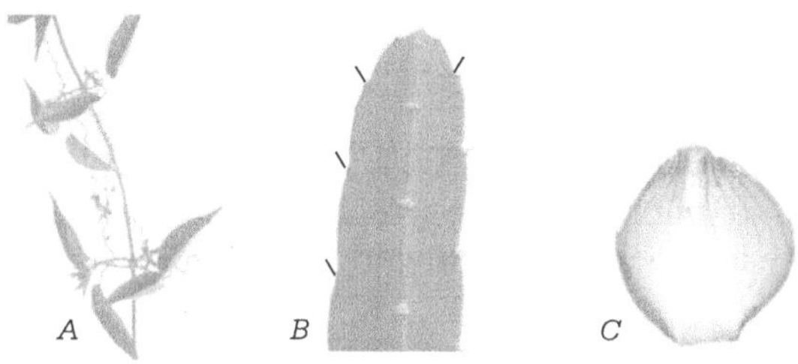

A. ____________

B. ____________

C. ____________

Topic 4 : Inflorescence and the Flower

76. Name the modified shoot wherein the shoot apical meristem changes to floral meristem.

____

77. In flower, ____________ do not elongate and axis get condensed.

78. The apex of flower produces different kinds of floral appendages vertically at successive nodes instead of leaves.
(*Rewrite the statement by changing the underlined words.*)

____

79. When a shoot tip transforms into a flower it is called as solitary flower.

True ☐ False ☐

80. The arrangement of flower on the floral axis is known as ____________ .

81. The main axis continues to grow in ____________ (acropetal/basipetal) succession in racemose inflorescence.

82. The main axis terminates into a flower and flowers are borne in a ____________ (acropetal/basipetal) succession in cymose inflorescence.

83. Identify the type of inflorescence in the given figures.

A

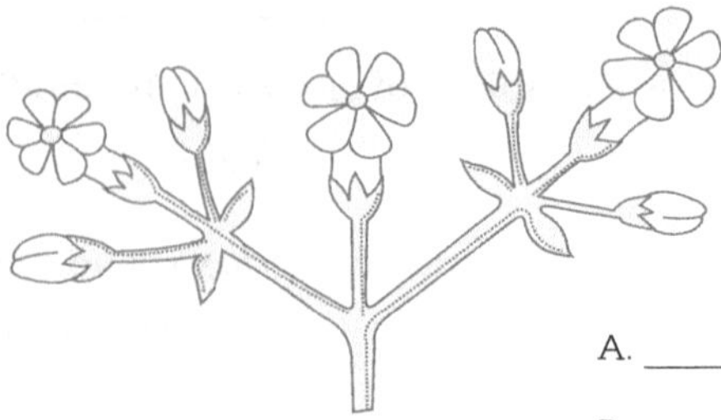

B

A. ____________________

B. ____________________

84. How many of the following inflorescences are the types of simple racemose?

Compound spike, Scapigerous head, Scorpioid cyme, Raceme, Spike, Umbel, Corymb, Capitulum

____

85. How many of the given species have inflorescence in which peduncle is drastically reduced, bears numerous bracteate and pedicellate flowers.

Capradla, Colocasia, Centella, Populus, Dill

____

86. Name the inflorescence in which peduncle is modified into a disc-like receptacle, bears sessile, bracteolate centripetally arranged florets.

____

87. The peduncle is weak and the whole inflorescence drops down in <u>spike</u> inflorescence. (*Rewrite the statement by changing the underlined word.*)

____

88. In ____________________, peduncle is fleshy and contains female, male and neutral flowers, terminal part of the peduncle is called ____________________ .

89. A typical flower has three different kinds of whorls arranged successively on the swollen end of the pedicel called thalamus.

True ☐ False ☐

90. Name the accessory and reproductive organs of a flower.

____

91. A ____________________ flower has both androecium and gynoecium, while ____________________ flowers have either only stamens or carpels (unisexual/bisexual).

92. Complete the following analogy with respect to floral symmetry.

Actinomorphic : Zygomorphic :: ____________________ : Bilateral symmetry

93. How many of the following flowers can be divided into two equal radial halves in any radial plane passing through the centre?

Mustard, bean, *Datura*, Chilli

____

94. Actinomorphic flower can be divided into two similar halves only in one particular vertical plane.

True ☐ False ☐

95. Pea, gulmohur, bean and *Cassia* have _______________ flower.

96. The flower of _______________ is _______________ which cannot be divided into two similar halves by any and vertical plane passing through the centre.

97. A flower may be trimerous, tetramerous or pentamerous when the floral appendages are in multiple of __________ (3/4), __________ (4/5), __________ (5/6), respectively.

98. Flowers with bracts-reduced leaf found at the base of the pedicel are called ebracteate and without bracts are called bracteate.
(*Rewrite the statement by changing the underlined words.*)

____

99. Match the following columns.

	Column I (Types of flower)		Column II (Positions of ovary)		Column III (Structural details)
(i)	Hypogynous	a.	Half inferior	1.	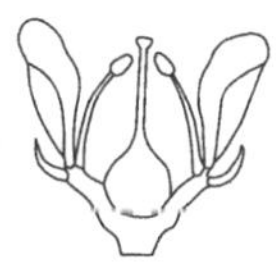
(ii)	Epigynous	b.	Inferior	2.	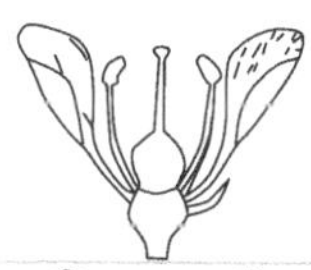
(iii)	Perigynous	c.	Superior	3.	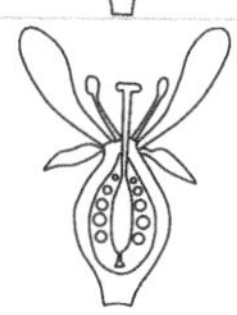

(i) _______________ (ii) _______________ (iii) _______________

100. In the _______________ (perigynous/hypogynous) flower, the gynoecium occupies the highest position, while the other parts are situated below it.

101. In _______________ flower, the gynoecium is situated in the centre and other parts of the flower are located on the rim of the thalamus.

102. In perigynous flowers, the margin of thalamus grows upward enclosing the ovary completely and getting fused with it.

True ☐ False ☐

103. Some groups of species are given below, differentiate the following on the basis of position of floral parts on thalamus.

Guava, China rose, Cucumber, Peach, Plum, Brinjal, Mustard, Ray floret of sunflower

____

104. The ____________________ is the outermost whorl of the flower and the members are called ____________________ .

105. When sepals are united calyx is known as polysepalous and when it is free termed as gamosepalous.

True ☐ False ☐

106. Identify the flower parts *A* to *E* in the given diagram.

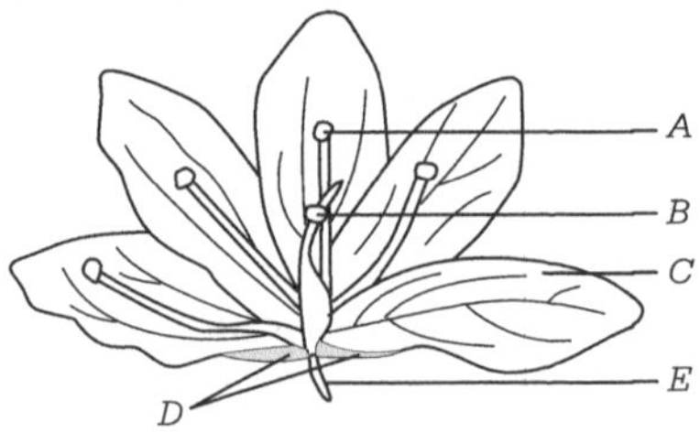

A. ____________________

B. ____________________

C. ____________________

D. ____________________

E. ____________________

107. Match the following columns.

	Column I (Parts of a flower)		Column II (Structures)
A.	Calyx	1.	
B.	Androecium	2.	
C.	Corolla	3.	
D.	Gynoecium	4.	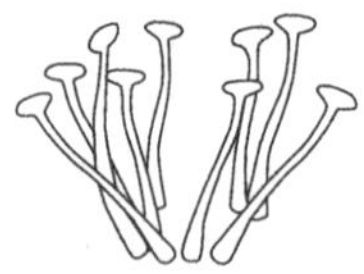

A. ______________ B. ______________ C.______________

D.______________

108. Petals are usually brightly coloured to attract ______________________________ for ____________________.

109. Match the following columns.

	Column I (Shapes of corolla)		Column II (Examples)
A.	Tubular	1.	*Ocimum*
B.	Infundibulum	2.	Brinjal
C.	Wheel-shaped	3.	*Physalis*
D.	Bilabiate	4.	Disc florets of sunflower
E.	Ligulate	5.	Ray floretes of sunflower
F.	Bell-shaped	6.	*Ixora*
G.	Star-shaped	7.	*Datura*

A. ________________ B. ________________ C.________________

D.________________ E. ________________ F.________________

G.________________

110. Placentation is the mode of arrangement of sepals or petals in a floral bud.

(*Rewrite the statement by changing the underlined word.*)

____

111. In ______________ (imbricate/valvate) aestivation, sepals or petals in a whorl just touch one another at the margin without overlapping.

112. Name the aestivation in which one margin of the appendage overlaps that of the next one.

____

113. How many of the following species possess aestivation in which the margins of sepals/petals overlaps one another but not in any particular direction.

Artabotrys, Cassia, Callistemon, Brassica, Gulmohar

____

114. Name the aestivation in which the largest petal (standard) overlaps the lateral petal (wings) which in turn overlap the two smallest anterior petals (keel).

____

115. In ______________ aestivation, two petals are completely external, two are completely internal and other one is partially external and internal.

116. Identify the types of aestivation *A-E* in the given diagram.

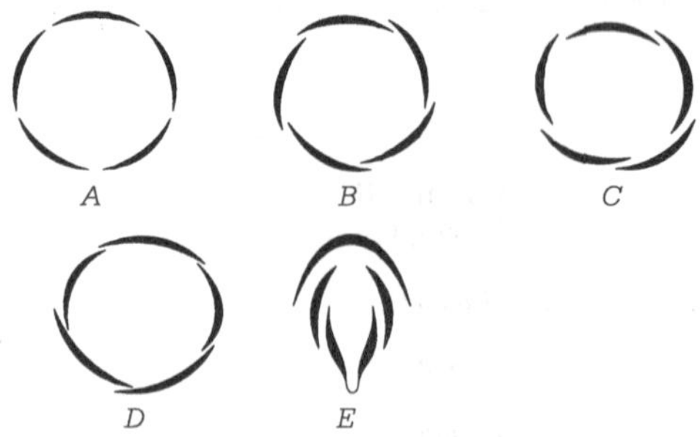

A. ______

B. ______

C. ______

D. ______

E. ______

117. In angiosperms each anther is usually trilobed and each lobe has two chambers.

True ☐ False ☐

118. Where is pollen grains are produced?

______

119. The sterile stamen is called ______ in which pollen grains are absent.

120. Complete the following analogy with reference to anther.

Dithecous : Monothecous :: Pea : ______

121. ______ is the condition of stamens in which it is attached to the petals.

122. In ______, stamens attached to the perianth.

123. Complete the following analogy with respect to stamen position.

Epipetalous : Epiphyllous :: Brinjal : ______

124. In which situation, stamens are free?

______

125. The stamens may be united into one bunch, i.e. ______ or two bundles, i.e. ______ or into more than two bundles, i.e. ______ .

126. Complete the following analogy.

Monoadelphous : Diadelphous : Polyadelphous :: ______ : ______ : *Citrus*

127. In ______ (*Cucurbita/Citrus*), stamens are united throughout their whole length to form a complete single structure is known as ______ (syngenesious/synandrous).

128. The three parts of a carpel are ______ , ______ and ______.

129. Stigma is the enlarged basal part, ovary is the elongated tube which connects the ovary to the style.
(*Rewrite the statement by changing the underlined words.*)

______

130. The ovary is usually at the tip of the style and is the receptive surface for pollen grains.

True ☐ False ☐

131. The ovary bears one or more ovules attached to a flattened cushion-like structure called ______________ (ovary wall/placenta).

132. Complete the following analogy with respect to number of carpels.

Monocarpellary : Bicarpellary : Tricarpellary :: ___________ : *Helianthus* : ___________ .

133. How many carpels are found in *Melia* and *Papaver*, respectively?

____

134. In ______________ (syncarpous/apocarpous) situation, more than one carpel is present and they are free.

135. In ______________ situation, more than one carpel are fuse to each other.

136. Identify the types of style from the following statements and give one example of each.

(i) It occurs upright and arised on the tip of ovary. ____________________

(ii) It arises from one lateral side of an ovary. ____________________

(iii) It arises directly from the centre of ovary. ____________________

137. The arrangement of ovules within the ovary is known as ______________ .

138. In parietal placentation, the placenta forms a ridge along the ventral suture of the ovary and the ovules are borne on this ridge forming two rows.

True ☐ False ☐

139. The placenta is axial and the ovules are attached to it in a multilocular ovary, the placentation is said to be ______________ (marginal/axile).

140. Name the placentation in which the ovules develop on the inner wall of the ovary or on peripheral part.

____

141. Ovary is one chambered but it becomes two chambered due to the formation of the ______________ in ___________ placentation.

142. In <u>basal</u> placentation, the ovules are borne on central axis and septa are absent.
(*Rewrite the statement by changing the underlined word.*)

____

143. Name the placentation in which the placenta develops at the base of ovary and contains single large ovule.

____

144. Match the following columns.

	Column I (Diagrammatic representation of placentation)	Column II (Types of placentation and examples)
A.	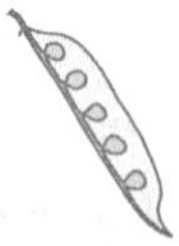	1. Free central (*Dianthus*, primrose)
B.		2. Parietal (mustard, *Argemone*)
C.	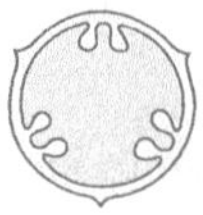	3. Marginal (*Pisum*, *Lathyrus*)
D.	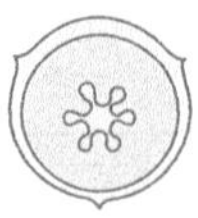	4. Basal (sunflower, marigold)
E.		5. Axile (China rose, tomato)

A. ________________ B. ________________ C.________________

D.________________ E.________________

Topic 5 : The Fruit

145. Name the structure which is a mature ovary developed after fertilisation.

____

146. The ______________ fruits are formed without fertilisation of the ovary.

147. Consider the diagram given below and answer the following questions.

(i) Identify the labelled parts *A*, *B*, *C* and *D*.

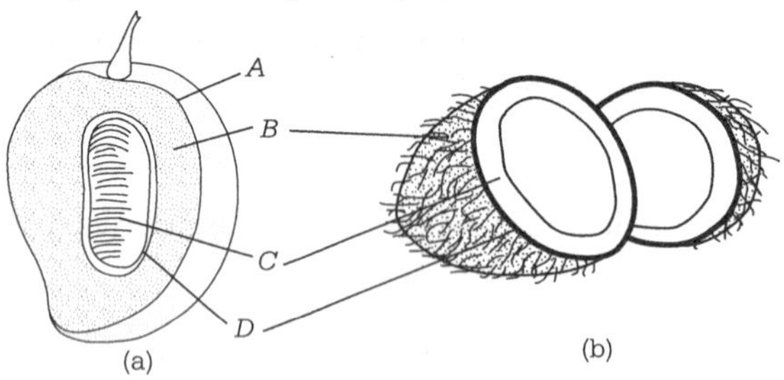

A. ______________________________

B. ______________________________

C. ______________________________

D. ______________________________

(ii) In which of the given fruit pericarp is fleshy? ______________________________

(iii) Identify the type of fruit represented by (a) and (b). ________________

148. The ______________ fruits develop from monocarpellary superior ovaries and are one seeded.

149. In mango, the pericarp is well-differentiated into an outer thin epicarp, a middle hard mesocarp and an inner soft endocarp.

True ☐ False ☐

150. Name the species in which mesocarp is fibrous in drupe fruit. ____________

151. ____________ (Follicle/Capsule) is a dry fruit develops from multicarpellary and superior ovary with axile placentation.

152. Answer the following questions, considering the given figure.

(i) Identify the labelled parts *A*, *B* and *C*.

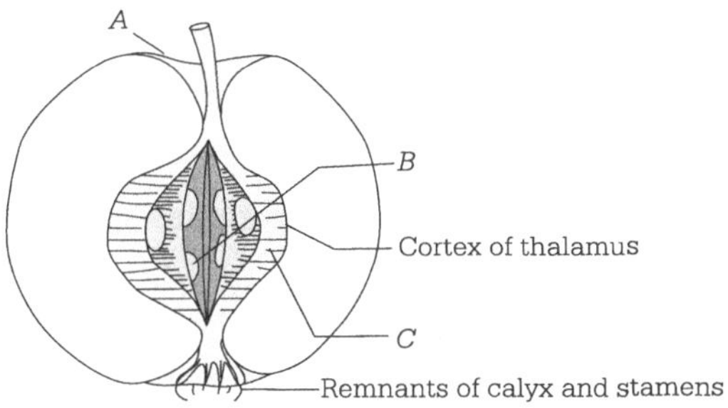

A. ____________

B. ____________

C. ____________

(ii) Their pericarp is not differentiated into epicarp, mesocarp and endocarp.

True ☐ False ☐

(iii) Identify the type of fruit. ____________

153. Name the fruit in which outer skin is thick and leathery represents the epicarp which contains oil glands. ____________

154. Aggregate fruit is a group of ____________ develop from the ____________, ____________ ovaries of a single flower.

155. Complete the following analogy.

Sorosis : Syconus :: ____________ : *Ficus religiosa*

156. Complete the following table.

	Column I (Edible parts)		Column II (Types of fruit)		Column III (Fruits)
(i)	Endosperm	a.	(*A*)	1.	Maize
(ii)	Fleshy aril	b.	(*B*)	2.	(*C*)
(iii)	Mesocarp and endocarp	c.	Pepo	3.	(*D*)
(iv)	(*E*)	d.	Hesperidium	4.	(*F*)
(v)	(*G*)	e.	Sorosis	5.	Mulberry
(vi)	Fleshy receptacle	f.	(*H*)	6.	Peepal

A. ____________ B. ____________ C. ____________

D. ____________ E. ____________ F. ____________

G. ____________ H. ____________

Topic 6 : The Seed

157. A seed is the ____________ ovule made up of a ____________ and an ____________ .

158. The embryo is made up of a ______________ , an ________________ and one or two ________________ .

159. The outermost covering of a seed is the testa it has two layers, the outer hilum and the inner tegmen.
(*Rewrite the statement by changing the underlined words.*)

____

160. Give the biological term for the scar on the seed coat through which the developing seeds were attached to the fruit ________________________.

161. A small pore above the hilum is called ______________ .

162. Name the region found between cotyledonary node and plumule. ______________

163. Answer the following questions, considering the given figures.
(i) Identify the labelled parts of dicot seed (*A-F*).

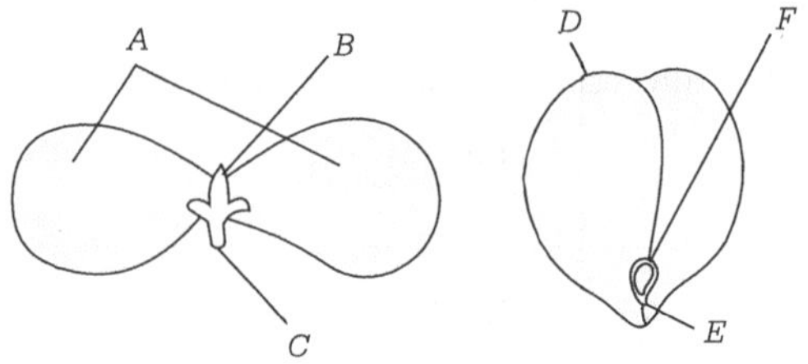

A. ______________
B. ______________
C. ______________
D. ______________
E. ______________
F. ______________

(ii) Which of the following part is fleshy and full of reserve food? ______________

164. Complete the following analogy with respect to seed structure.
Endospermic : Non-endospermic : : Castor : ______________

165. Give a common collective term for the following species.

Pisum sativum, Mango, Orchids, Gram

____

166. In ________________ (dicotyledonous/ monocotyledonous) seed, seed coat is membranous and generally fused with the fruit wall.

167. The outer covering of endosperm separates the embryo by a proteinaceous layer called ________________ .

168. Dicotyledonous seeds consist of one large and shield-shaped cotyledon known as scutellum.

True ☐ False ☐

169. The plumule and radicle are enclosed in sheaths known as ________________ and ________________ , respectively.

170. In the figure given below, some parts are wrongly labelled. Rewrite the wrongly labelled parts in front of their correct labels.

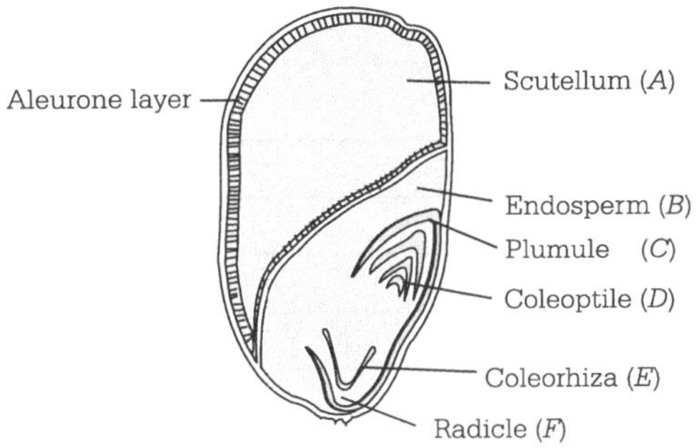

A. ____________

B. ____________

C. ____________

D. ____________

E. ____________

F. ____________

____

171. Name any two species in which labelled part '*A*' of the given figure is present in mature seed.

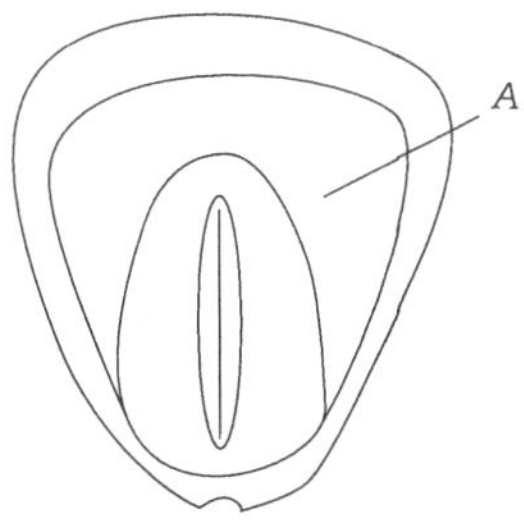

____

172. Name the seed dispersal mechanism in which, the fruit matures, some cells surrounding the seeds produce mucilage to create turgor pressure, hence seeds are thrown away ____________.

173. In ____________ (bursting/jaculator) mechanism, the values of pod twist spirally and coil with a force.

174. Name the seed dispersal mechanism in which the dry and capsular fruits burst suddenly when they come in contact with water.

____

175. In ____________ mechanism, seeds come out through the pores as the fruits are shaken by air.

176. The internal inhibition of germination of a normal or viable seed even when it is placed under most favourable conditions is called ____________ .

177. In epigeal germination, the cotyledons are first pushed above the ground by the rapid elongation of ____________ .

178. Name the process in which germination of seed takes place inside the fruit, that is still attached to the parent tree.

____

Topic 7 : Semi-Technical Description of a Typical Flowering Plant

179. Write the symbol of the following character of a flower.

(i) Bracteate ______

(ii) Calyx ______

(iii) Corolla ______

(iv) Perianth ______

(v) Androecium ______

(vi) Gynoecium ______

(vii) Superior ovary ______

(viii) Inferior ovary ______

(ix) Actinomorphic ______

(x) Zygomorphic ______

180. A ______ provides information about the ______ of the parts of a flower, their ______ and the relation they have with each another.

181. The position of the mother axis with respect to the flower is represented by a dot on the bottom of the floral diagram.

True ☐ False ☐

182. The floral formula shows ______ and ______ within parts of whorls and between whorls.

183. Write the floral formula of the given floral diagram.

______

Topic 8 : Description of Some Important Families

184. Answer the following questions considering the given figure.

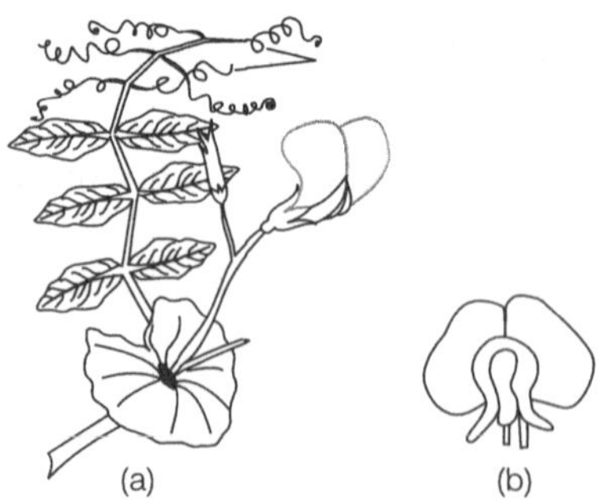

(a) (b)

(i) Write the feature of their leaves.

____

(ii) Plant (a) shows _______________ inflorescence.

(iii) Flowers are unisexual and actinomorphic.

True ☐ False ☐

(iv) Identify plant twig (a).

____

185. How many of the following features of corolla are not related to corolla of pea?

5 petals, Gamopetalous, Papilionaceous, Twisted aestivation, Vexillary aestivation

____

186. The calyx of pea flower is five in number with polysepalous or valvate/imbricate condition.

True ☐ False ☐

187. Consider the given floral diagram and answer the following questions.

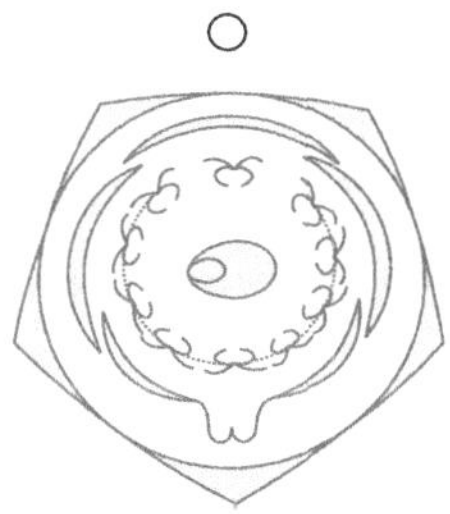

(i) Write all the features of androecium of the given floral diagram.

____

(ii) Write all the features of gynoecium of the given floral diagram.

____

(iii) Write their floral formula.

____

188. Give example of each from Fabaceae family as (i) dye, (ii) fibres, (iii) fodder, (iv) medicine, (v) ornamentals.

____

189. Give a common collective feature for the following species.

Gram, Arhar, Sem, Moong, Soyabean, Groundnut

____

190. Solanaceae is also known as _______________.

191. Select the odd one out from the following features with respect to Solanaceae family

Herbaceous, Aerial, Erect, Non-endospermic, Cylindrical branched, Solid or Hollow

____

192. What are the features of the leaves of given flowering twig of *Solanum nigrum*?

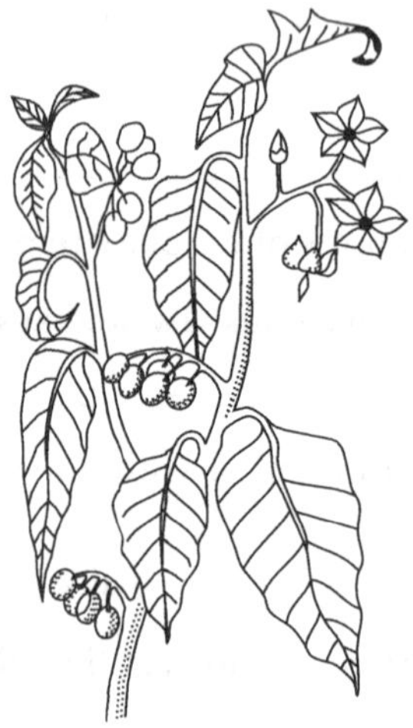

____

193. The inflorescence of *Solanum nigrum* is solitary or ____________ (racemose/cymose).

194. The calyx and corolla of *Solanum nigrum* are five in number, united and with imbricate aestivation.

True ☐ False ☐

195. Answer the following questions after considering the floral parts and floral diagram given below.

(i) The floral diagram belongs to ____________ plant.

(ii) The gynoecium is ____________ (monocarpellary/ bicarpellary), ____________ (apocaropus/ syncarpous), ovary superior with axile placentation.

(iii) There are ____________ stamens with ____________ condition.

(iv) Write their floral formula.____________

196. How many of the following species belong to Solanaceae family?

Belladonna, Ashwagandha, Tomato, Brinjal, Lily, Chilli, Tobacco

____

197. Complete the following analogy with respect to economic importance of Solanaceae family.

(i) Medicine : Fumigatory :: ________________ : Tobacco

(ii) Ornamental : Spice :: ________________ : Chilli

198. The basal, alternate, linear, exstipulate with parallel venation leaves are found in ________________________________ plant.

199. Identify the type of inflorescence shown is the figure and give the species name in which these are found.

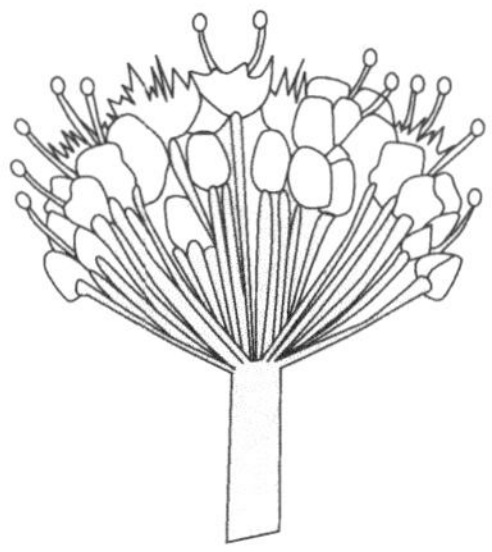

____

200. Answer the following questions considering the given figure of flower.

(i) The flower is ____________________ and ____________________.

(ii) This type of flower is found in <u>Solanaceae</u> family.

(*Rewrite the statement by changing the underlined word.*)

____

(iii) Write the feature of perianth. ________________________________

201. How many of the given species are belongs to the given floral diagram.

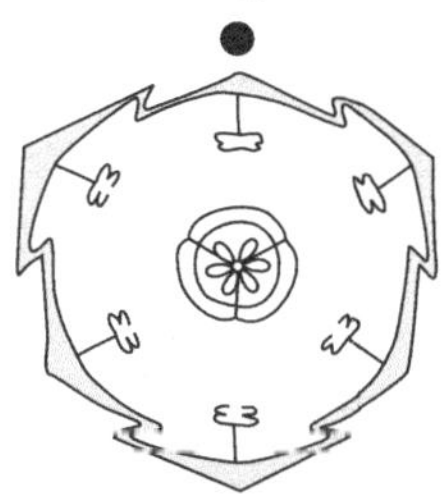

Aloe, Tulip, *Asparagus, Althaea, Colchicum, Nicotiana*

____

202. Write the floral formula of species *Allium cepa.*

____

203. Write the features of gynoecium of *Allium cepa.*

____

204. In which species, stamens are found in 3 + 3, epitepalous situation?

____

205. Match the following columns.

	Column I (Diagnostic features)		**Column II** (Floral diagrams)
A.	Stamens 6 (3 + 3) epiphyllous	1.	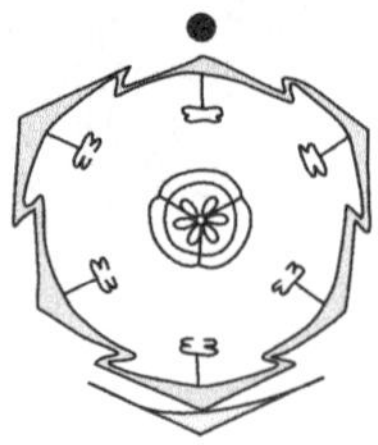
B.	Epipetalous, twisted aestivation in petal	2.	
C.	Polypetalous, cruciform	3.	

A. ____________ B. ____________ C.____________

Answers

1. Morphology, roots, stems, leaves, flowers, fruits.
2. Root, shoot
3. False. The direct elongation of the radicle leads to the formation of primary root in majority of the dicotyledonous plant.
4. Secondary, tertiary
5. (i) *A*–Primary root, *B*–Secondary root, *C*–Internode, (ii) Tap root system, (iii) Monocot
6. i-c-2, ii-a-1, iii-b-3
7. 3 (Absorption of water, food storage, plant growth regulator production).
8. Root cap
9. i-b-3, ii-c-1, iii-a-2
10. Tap, adventitious
11. All of them are adventitious roots except nodulated which is tap root.
12. A-2, B-3, C-1
13. 4 (Buttress, stilt, prop, nodulose).
14. Prop roots
15. Stilt
16. False. Pneumatophores come out of the ground and grow vertically upwards.
17. (i) *Rhizophora*, (ii) These roots help to get oxygen for respiration.
18. Prop root
19. (iii) is mismatched pair. Pneumatophores help to get oxygen for respiration.
20. Adventitious, sweet potato
21. (i) Fasciculated, (ii) True
22. Ascending
23. Stem develops from the plumule of the embryo of a germinating seed.
24. Nodes, internodes
25. True
26. (i) It conducts water, minerals and photosynthates.
    (ii) It stores food.
27. All of them are underground stem and are modified to store food.
28. A-2, B-1, C-4, D-3
29. Tendrils
30. All of them possess stem tendril.
31. Thorns
32. Thorns are found in *Citrus* and *Bougainvillea.*
33. Flattened, fleshy cylindrical
34. All of them possess phylloclades type of stem modification except *Asparagus* which possesses cladodes.
35. Runner
36. This is the species of *Oxalis* which shows runner type of modification.
37. Stolons
38. Offsets
39. False. Suckers are the lateral branches originate from the basal and underground portion of the stem, grow horizontally and give rise to leafy shoots.
40. A-4, B-3, C-1, D-2
41. *Asparagus*
42. Bulbil
43. 2 (i, ii). (iii) is incorrect and can be corrected as Floral bud tendril is found in *Antigonon.*
44. Leaf
45. Axillary bud
46. Shoot, apical meristem, acropetal
47. (i) *A*–Stipule, *B*–Petiole, *C*–Axillary bud, *D*–Leaf base, *E*–Lamina
    (ii) Leaf base
    (iii) Monocotyledons
48. Stipules
49. Leguminous plants
50. Petiole
51. Lamina, leaf blade, midrib
52. Vein
53. Venation
54. Reticulate, parallel
55. Pinnate
56. Palmate
57. All of them are monocotyledons with reticulate venation.
58. Simple
59. Compound
60. False. A bud is present in the axil of petiole in both simple and compound leaves, but not in the axil of leaflets of compound leaf.
61. Pinnately
62. Palmately
63. *A*–It is pinnately compound leaf found in neem. *B*–It is palmately compound leaf found in silk cotton.
64. Phyllotaxy
65. Alternate
66. Nodes, opposite
67. False. In whorled phyllotaxy, more than two leaves arise at a node and form a whorl.
68. i-b-2, ii-a-3, iii-c-1
69. Opposite decussate
70. Opposite superimposed
71. Bracts
72. False. The tendrils of peas and spines of cacti are the modification of leaves.
73. Petiole, leaves
74. Venus flytrap, pitcher plant
75. *A*–Support (tendril), *B*–Protection (spine), *C*–Storage (fleshy leaves)
76. Flower
77. Internodes
78. Floral meristem laterally
79. True
80. Inflorescence
81. Acropetal
82. Basipetal
83. *A*–Racemose inflorescence, *B*–Cymose inflorescence
84. 5 (Raceme, spike, umbel, corymb, capitulum).
85. Inflorescence is umbel type, found in *Centella* and dill
86. Capitulum
87. Down in catkin
88. Spadix, appendix
89. False. A typical flower has four different kinds of whorls arranged successively on the swollen end of the pedicel called thalamus.

90. Calyx and corolla are accessory organs while androecium and gynoecium are the reproductive organs of a flower.
91. Bisexual, unisexual
92. Radial symmetry
93. 3 (Mustard, *Datura*, chilli), all of them are actinomorphic flower.
94. False. Zygomorphic flower can be divided into two similar halves only in one particular vertical plane.
95. Zygomorphic
96. *Canna*, asymmetric
97. 3, 4, 5
98. Bracteate, ebracteate.
99. i-c-2, ii-b-3, iii-a-1
100. Hypogynous
101. Perigynous
102. False. In epigynous flowers, the margin of thalamus grows upward enclosing the ovary completely and getting fused with it.
103. Hypogynous flower – Mustard, China rose and brinjal.

 Perigynous flower – Plum and peach.

 Epigynous flower – Guava, cucumber and floret of sunflower.
104. Calyx, sepals
105. False. When sepals are united calyx is known as gamosepalous when it is free termed as polysepalous.
106. *A*–Androecium, *B*–Gynoecium, *C*–Corolla, *D*–Calyx, *E*–Pedicel
107. A-3, B-4, C-2, D-1
108. Insects, pollination
109. A-4, B-7, C-2, D-1, E-5, F-3, G-6
110. Aestivation is the mode of arrangement of sepals or petals in a floral bud.
111. Valvate
112. Twisted aestivation,
113. 3 (*Cassia*, *Callistemon*, Gulmohur), these shows imbricate aestivation.
114. Vexillary
115. Quincuncial
116. *A* – Valvate, *B* – Twisted, *C* – Imbricate, *D* – Quincuncial, *E* – Vexillary
117. False. In angiosperms, each anther is usually bilobed and each lobe has two chambers.
118. Pollen sacs
119. Staminode
120. China rose
121. Epipetalous
122. Epiphyllous
123. Lily
124. Polyandrous situation,
125. Monoadelphous, diadelphous, polyadelphous
126. China rose, pea
127. *Cucurbita,* Syngenesious
128. Stigma, style, ovary
129. Ovary, stigma
130. False. The stigma is usually at the tip of the style and is the receptive surface for pollen grain.
131. Placenta
132. Pea, *Allium*
133. In *Melia*, gynoecium has four carpels and *Papaver* has many carpels.
134. Apocarpous
135. Syncarpous
136. (i) Terminal style, e.g. China rose, (ii) Lateral style, e.g. Mango, (iii) Gynobasic style, e.g. *Ocimum.*
137. Placentation
138. False. In marginal placentation, the placenta forms a ridge along the ventral suture of the ovary and the ovules are borne on this ridge forming two rows.
139. Axile
140. Parietal placentation
141. Septum, parietal
142. Free central
143. Basal placentation
144. A-3 B-5 C-2 D-1 E-4
145. Fruit
146. Parthenocarpic
147. (i) *A*–Epicarp, *B*–Mesocarp, *C*–Seed, *D*–Endocarp, (ii) Fruit (a) i.e. mango, (iii) Drupe.
148. Drupe
149. False. In mango, the pericarp is well-differentiated into an outer thin epicarp, a middle fleshy edible mesocarp and an inner stony hard endocarp.
150. Coconut
151. Capsule
152. (i) *A*–Pericarp, *B*–Seed, *C*–Thalamus, (ii) True, (iii) Pome fruit
153. Hesperidium
154. Fruitlets, multicarpellary, apocarpous
155. Pineapple
156. *A*–Caryopsis, *B*–Nut, *C*–Litchi, *D*–Cucumber, *E*–Endocarpic juicy hair, *F*–Orange, *G*–Perianth, *H*–Syconus
157. Fertilised, seed coat, embryo
158. Radicle, embryonal axis, cotyledons
159. Seed coat, testa
160. Hilum
161. Micropyle
162. Epicotyl
163. (i) *A*–Cotyledons, *B*–Plumule, *C*–Radicle, *D*–Seed coat, *E*–Micropyle, *F*–Hilum

 (ii) Cotyledons
164. Bean
165. Examples of non-endospermic seeds. Their endosperm is completely consumed in embryo nourishment.
166. Monocotyledonous
167. Aleurone layer
168. False. Monocotyledonous seeds consist of one large and shield-shaped cotyledon known as scutellum.
169. Coleoptile, coleorhiza
170.

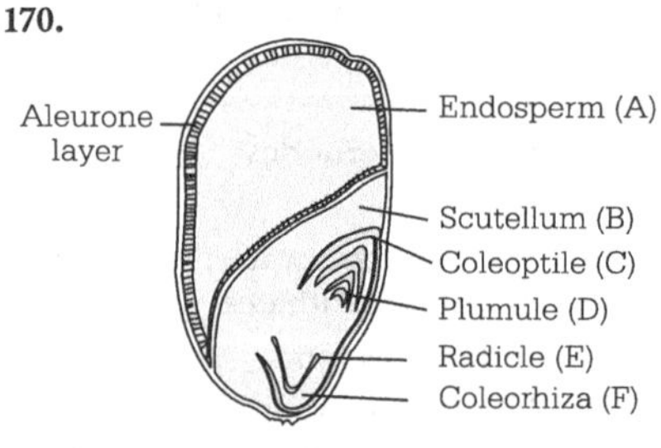

171. The labelled part *A* is endosperm. It is present in *Ricinus* and maize seeds.

172. Piston mechanism

173. Bursting

174. Jaculator mechanism

175. Censor

176. Seed dormancy

177. Hypocotyl

178. Vivipary

179. i. Br, ii. K, iii. C, $\underline{\text{iv.}}$ P, v. A, vi. G, vii. $\underline{G}$, viii. $\overline{G}$, ix. ⊕, x. %

180. Floral diagram, number, arrangement

181. False. The position of the mother axis with respect to the flower is represented by a dot on the top of the floral diagram.

182. Cohesion, adhesion

183. $\oplus$ ⚥ $K_{2+2}\ C_4\ A_{2+4}\underline{G}_{(2)}$

184. (i) Leaves are alternate, pinnately compound, pulvinate leaf base, stipulate, reticulate venation.
(ii) racemose, (iii) False. Flowers are bisexual and zygomorphic.
(iv) This is the plant twig of *Pisum sativum.*

185. 2 (Gamopetalous, twisted aestivation).

186. False. The calyx of pea flower is five in number with gamosepalous and valvate/imbricate condition.

187. (i) Androecium – Anther is ten in number, diadelphous, anther dithecous.
(ii) Gynoecium – Ovary superior, monocarpellary, unilocular, style single.
(iii) % ⊕ ⚥ $K_5 C_{1+2+(2)}\ A_{(9)+1}\ \underline{G}_1$

188. (i) Dye - *Indigofera,* (ii) Fibres - Sunhemp, (iii) Fodder - *Sesbania, Trifolium,*
(iv)Medicine - Muliathi,
(v) Ornamentals - Lupin, sweet pea

189. Members of Fabaceae family.

190. Potato family

191. All of them are the features of stem of Solanaceae family except non-endospermic character.

192. Alternate, simple, rarely pinnately compound, exstipulate and reticulate venation.

193. Cymose

194. False. The calyx and corolla of *Solanum nigrum* are five in number, united and with valvate aestivation.

195. (i) *Solanum nigrum*
(ii) Bicarpellary, syncarpous
(iii) Five, epipetalous
(iv) ⊕ ⚥ $K_{(5)}\overset{\frown}{C_{(5)}\ A_5}\ \underline{G}_{(2)}$

196. 6 (Belladonna, ashwagandha, tomato, brinjal, chilli, tobacco)

197. (i) Belladonna, (ii) *Petunia*

198. *Allium cepa*

199. The given inflorescence is of umbellate type and found in *Allium cepa.*

200. (i) Bisexual, actinomorphic, (ii) Liliaceae, (iii) Tepal (3 + 3), united into tube, valvate aestivation.

201. 4 (*Aloe,* Tulip, *Asparagus, Colchicum*)

202. ⊕ ⚥ $\overset{\frown}{P_{3+3}\ A_{3+3}}\ \underline{G}_{(3)}$

203. Tricarpellary, syncarpous, ovary superior, trilocular, axile placentation.

204. *Allium cepa*

205. A-1, B-3, C-2

CHAPTER SIX

ANATOMY OF FLOWERING PLANTS

Topic 1 : Introduction and Meristematic Tissues

1. Study of internal structure of plant is called ____________________

2. ____________________ deals with the ontogeny, structure and distribution of various types of tissues in roots, stems, leaves, flowers and fruits.

3. On the basis of anatomy monocots and dicots are different.

   True ☐ False ☐

4. Can you explain plant adaptations to diverse environment by studying the plant anatomy? (Yes/No). ____________________

5. Name the two major groups of plant tissue. ____________________

6. In plants, some regions show active cell division, these regions are called ____________ .

7. Mark the given features as present or absent considering the features of meristematic tissue.

   (i) Undifferentiated cells ____________

   (ii) Intercellular spaces ____________

   (iii) Thick cell wall ____________

8. Name the meristem which is found on the tips of roots and shoots.

   ____

9. In the figure of shoot apex meristem, name the structures *A*, *B*, *C* and *D*.

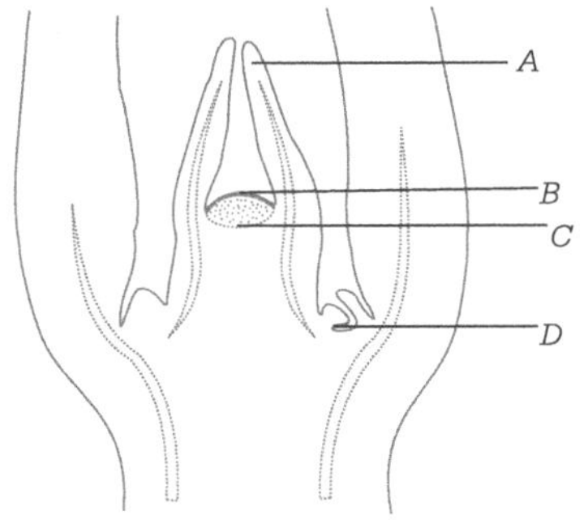

A. ____________________

B. ____________________

C. ____________________

D. ____________________

10. The growth of axillary bud is usually under the influence of the shoot apex.

True ☐ False ☐

11. Consider the following figure of root apical meristem, identify its parts *A* – *E*.

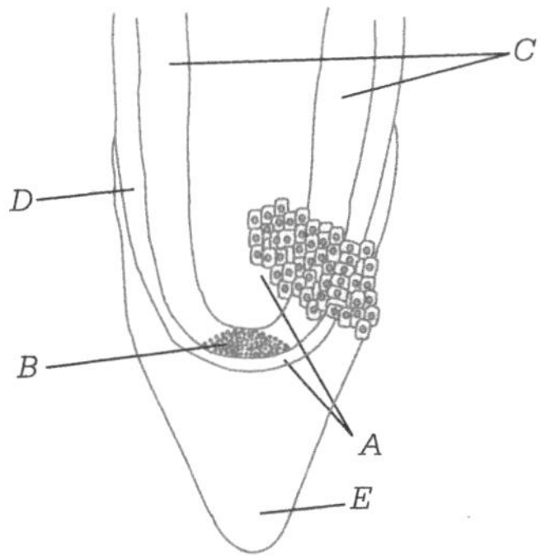

A. ____________________

B. ____________________

C. ____________________

D. ____________________

E. ____________________

12. Bud that develops in the axil of a leaf of a plant is known as ____________________. These buds develop from nodes give rise to ____________________.

13. Name the meristems which are present between mature tissue, that help in longitudinal growth of plants. ____________________

14. ____________________ and ____________________ furnish the development of primary plant body.

15. Intercalary meristem occurs at the base of the ____________________ and ____________________.

16. Bamboo can grow very fast up to approximately 60 cm/day. (*Give reason*)

____

17. If we pluck the upper part of grasses, which meristem helps the grasses to rejuvenate their parts. ____________________

18. Name the meristem which is known as cylindrical meristem, increase thickness of plant and their cells divide periclinally.

____

19. Name the meristem which is also known as secondary meristem. ______________

20. You are given a slide of woody axis in which parts do you see lateral meristems.

____

21. Length of petiole increases due to the division of lateral meristem.

True ☐ False ☐

22. Complete the following flowchart with suitable words.

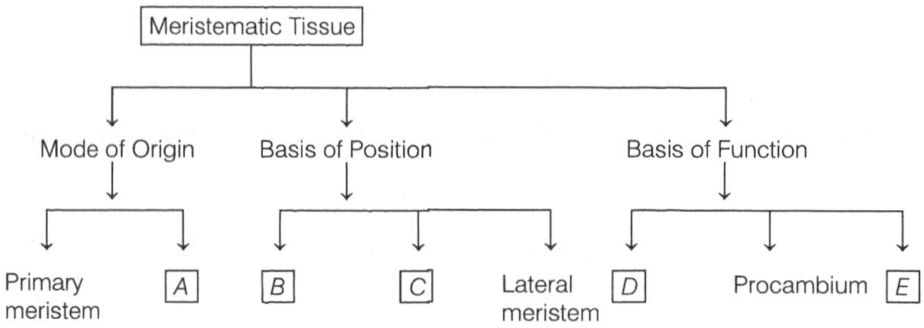

A. ______________ B. ______________ C. ______________

D. ______________ E. ______________

23. The secondary medullary rays are formed from the ______________ (ray/fusiform) initia of vascular cambium and provides ______________ (radial/stellate) conduction of water and food.

24. Following divisions of cells in both primary and secondary meristems, the newly formed cells become structurally and functionally specialised and lose the ability to divide. Such cells are termed as ______________ .

25. During the formation of the primary plant body, specified region of the apical meristem produces ______________, ______________ and ______________ tissues.

Topic 2 : Permanent Tissues

26. The cells of permanent tissue have definite shape, size and function and are actively divide.

True ☐ False ☐

27. Name the type of permanent tissue on the basis of sentences given below.

(i) Permanent tissues having all cells similar in structure and function. ______________

(ii) Permanent tissues having many different types of cells. ______________

28. From the given box, identify how many of these are related to simple tissues?

Aerenchyma, Parenchyma, Xylem, Phloem, Angular collenchyma

____

29. You are given a simple tissue slide in which cells are isodiametric, thin cell wall and have small intercellular spaces. Name the tissue on the basis of above illustration.

____

30. Parenchyma played various roles in plants, e.g. ______________, storage, ______________, etc.

31. Parenchymatous cells are ______________ (living/dead).

32. In the given figure identify '*X*'.

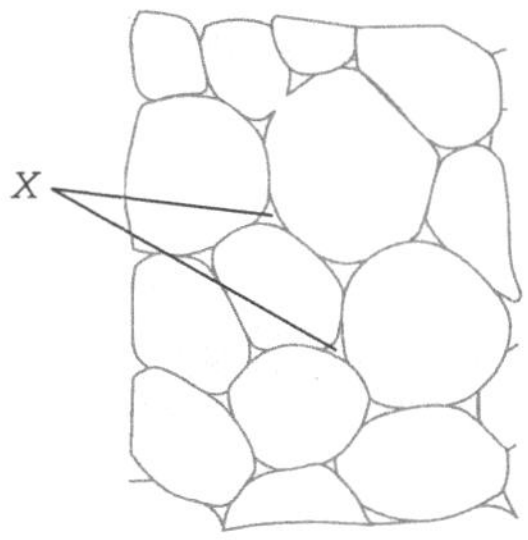

X. ______________________________

33. Parenchymatous cell walls are __________ and made up of __________.

34. Parenchymatic cells undergo dedifferentiation process to give rise of cork cambium.

True ☐ False ☐

35. In dicotyledonous plants, ______________ (sclerenchyma/collenchyma) is present below the epidermis.

36. In the given figure of collenchyma cell identify the material which is deposited on the corner of the cells.

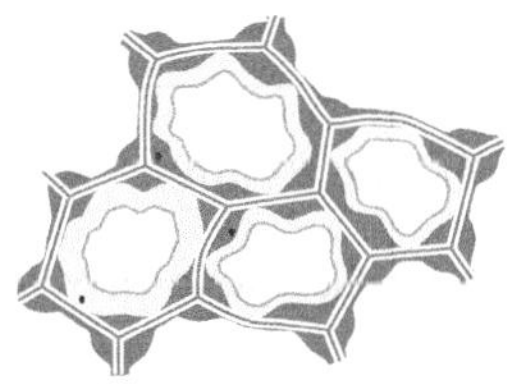

____

37. Name the tissue which provides mechanical support to the plant and are oval, spherical and polygonal in shape.

____

38. Some biological terms are given below, write a single collective term for each group.

(i) Chlorenchyma, Aerenchyma, Prosenchyma______________

(ii) Fibres, Sclereids ______________

39. How many of these characters are not related to sclerenchyma?

Lignified cell walls, Oval cells, Dead cells, Having protoplasts, Containing chloroplast

____

40. The sclerenchyma tissue provides mechanical support. (*Give reason*)

____

41. Name the type of sclerenchyma which are thick walled, elongated and pointed cells, generally occurring in groups in various parts of plants. ______________

42. The ______________ (fibres/sclereids) have reduced lumen and thick cell wall and cells are oval, spherical and cylindrical in shape.

43. Sclereids are found in the mesocarp of coconut.
(*Rewrite the statement by changing the underlined word.*)

____

44. The pulp of guava has gritty texture. (*Give reason*)

____

45. Consider the given figure of sclereid, answer the questions that follows.
(i) Identify *A* and *B*.

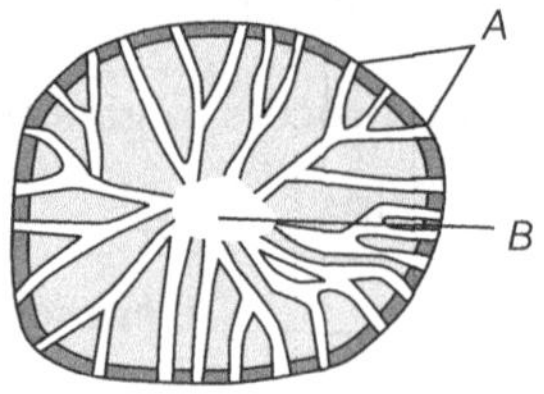

A. ______________

B. ______________

(ii) These are found in leaves of ______________ and seed coats of ______________.

46. When more than one type of cells work together and formed a unit is known as

______________________________.

47. A leaf is not able to receive water due to injury, which organ is injured? ______________

48. Xylem is a conducting tissue in plants, it also provides ______________

49. Name the xylem element which is not present in gymnosperms. ______________

50. Choose odd one out from the following terms.

Tracheids, Vessels, Xylem fibres, Xylem parenchyma, Sieve tube

____

51. ______________ (Companion cell/Tracheid) cells are dead with lignified cell wall and are elongated with tapering ends.

52. Tracheids are the most advance type of conductive element in xylem.

True ☐ False ☐

53. Which of the following figure represents tracheid and vessel?

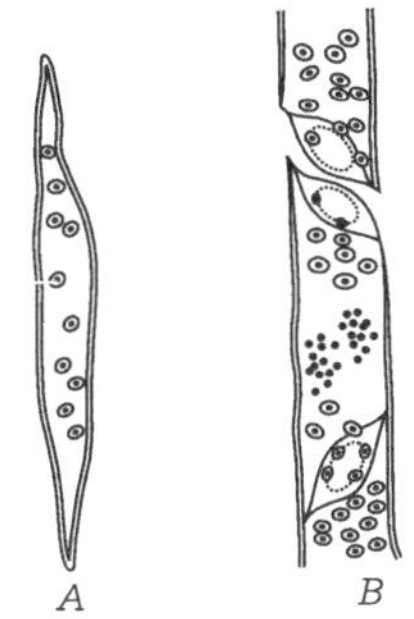

A. ______________________

B. ______________________

54. Vessels are long cylindrical xylem elements, their walls are ______________ and ______________ of protoplasm.

55. The side wall of the vessel is ______________

56. Name the angiosperm in which vessels are absent. ______________

57. Name the xylem element in which central lumen is void. ______________

58. The only living element of xylem is ______________

59. Vessels of xylem are mainly involved in the storage of starch, fat and water conduction. (*Rewrite the statement by changing the underlined word.*)

____

60. Which of the xylem element is responsible for radial conduction of water?

____

61. Identify the type of xylem on the basis of arrangement of primary xylem.

(i) Primary xylem formed first. ______________

(ii) Primary xylem formed later. ______________

62. In endarch condition of primary xylem, metaxylem is found at ______________ and protoxylem at______________.

63. Name all the four components of xylem.

____

64. Complete the following analogy.

______________ : Root : : Endarch : Exarch

65. In exarch condition, protoxylem lies towards the centre whereas metaxylem lies towards the periphery.

True ☐ False ☐

66. Write four components of phloem in angiosperm.

____

67. ________________ cells in gymnosperms are analogous to the companion cells of angiosperm.

68. Consider the following figure, identify parts *A*, *B* and *C*.

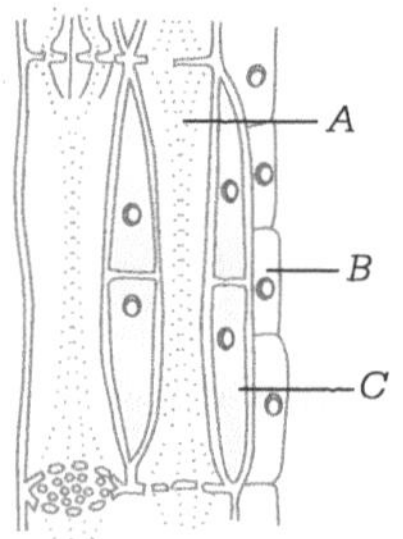

A. ______________________

B. ______________________

C. ______________________

69. Gymnospermic cells lack phloem fibres.
(*Rewrite the statement by changing the underlined word.*)

____

70. The nucleus of companion cells controls the functions of ________________

71. The mature sieve element is unique in having ________________ cytoplasm, a large vacuole, but lacks a ________________.

72. Match the following columns.

	Column I (Types of phloem element)		Column II (Functions)
A.	Companion cells	1.	Provide strength
B.	Phloem parenchyma	2.	Maintain pressure gradient
C.	Phloem fibres	3.	Reserve food material

A. ____________ B. ____________ C. ____________

73. Name the phloem element which is conjoint to sieve tube elements. ____________

74. A mature sieve element is ________________ (living/dead).

75. How many of these are parenchymatous in nature?

Companion cells, Phloem parenchyma, Phloem fibres

____

76. Name the element which is present in secondary phloem, but absent in primary phloem. ____________

77. The hemp's phloem fibres are ________________ (lignified/silicate), whereas fibres of flax are made up of ________________ (calcium/cellulose).

78. Name the phloem element which is absent in most of the monocotyledons.

____

79. Protophloem is first formed primary phloem, it has narrow ____________________

80. A primary phloem in which bigger sieve tubes are found. Name the phloem.

____

81. The cell wall of phloem fibres is ______________, these fibres lose their ______________ and become __________________.

82. Name the tissue which is concerned with secretion of gums, resins, volatile oils, nectar, latex, etc. ______________________________

83. Write a suitable biological term for the following groups.

(i) Latex cell, Latex vessel ______________________________

(ii) Hydathodes, Glandular hair, Oil gland______________________________

Topic 3 : Tissue System

84. Name the all types of tissue system on the basis of their structure and location.

____

85. The epidermal tissue system forms the outermost covering of the whole plant body and comprises __________________, ___________________ and _________________________.

86. The outermost layer of the primary plant body is <u>mesophyll</u>.
(*Rewrite the statement by changing the underlined word.*)

____

87. Identify the odd onc out from the following terms.

Epidermis, Mesophyll, Stomata, Trichomes

____

88. Epidermis is made up of round, compactly arranged cells, it forms a continuous layer.

True ☐ False ☐

89. Epidermal cells are _______________________ with a small amount of cytoplasm lining the cell wall and a large _________________________.

90. The outside of the epidermis is covered with a waxy thick layer that prevents the loss of water is known as ______________________________

91. Cuticle is absent in roots. (*Give reason*)

____

92. Stomata regulate the process of ____________________ and ________________________.

93. Bean-shaped guard cells are found in monocot and dumb-bell-shaped guard cells are found in dicot.

True ☐ False ☐

94. In dicot leaves, stomata are scattered whereas in monocot, these are arranged in rows.

True ☐ False ☐

95. The outer walls of guard cells are ________________ and the inner walls are highly ________________.

96. The ________________ (subsidiary cells/guard cells) possess chloroplast and regulate the opening and closing of stomata.

97. A few epidermal cells in the vicinity of the guard cells become specialised in their shape and size are known as guard cells.
(*Rewrite the statement by changing the underlined word.*)

____

98. Which stomata given in figure belongs to monocot leaves?

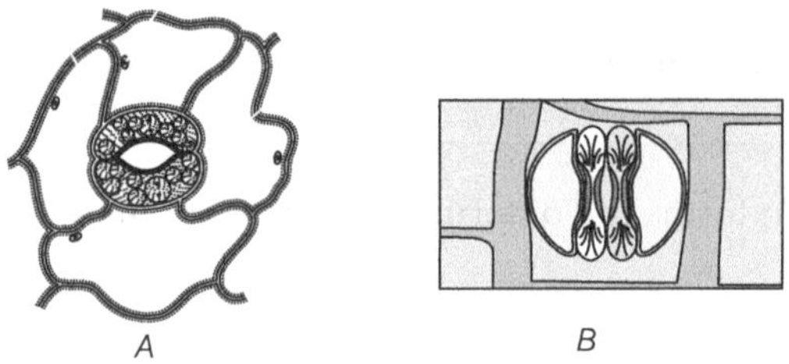

____

99. Write a collective term for the following words.

Stomatal aperture, Guard cell, Subsidiary cell

____

100. The unicellular elongations of the epidermal cells that absorb water and minerals from the soil is known as ________________

101. Name the epidermal hair which is present on stem. ________________

102. The trichomes in the shoot system are usually unicellular.

True ☐ False ☐

103. In some plants, trichomes may be secretory in nature.

True ☐ False ☐

104. The trichomes help in preventing water loss due to ________________.

105. In insectivorous plants, trichomes act as secretory gland.
(*Rewrite the statement by changing the underlined word.*)

____

106. Ground tissue constitutes epidermis and vascular bundle.

True ☐ False ☐

107. Ground tissue consists of simple tissue such as ________________ , ________________ and ________________.

108. Name the cell of ground tissue which contains chloroplast. ____________________

109. The vascular bundles are present in a central cylinder is known as extrastelar.
(*Rewrite the statement by changing the underlined word.*)

____

110. Give a common collective term for the following words.

Pericycle, Medullary rays, Pith

____

111. The thin-walled cells in endodermis are known as ________________ (transfusion cells/tyloses).

112. Pericycle is absent in roots of parasitic plants and hydrophytes.

True ☐ False ☐

113. The vascular system consists of complex tissues, the ____________________ and the ________________.

114. Identify types of cells found in cambium.

(i) Elongated cells with tapering ends ________________________ .

(ii) Cells are more or less isodiametric ________________________ .

115. Cambium is absent in dicot plants.

True ☐ False ☐

116. Cambium is present between xylem and phloem is known as closed vascular bundle.
(*Rewrite the statement by changing the underlined word.*)

____

117. Secondary growth is absent in monocots. (*Give reason*)

____

118. Study the figures *A* and *B* given below and identify the characteristics which correctly describe them.

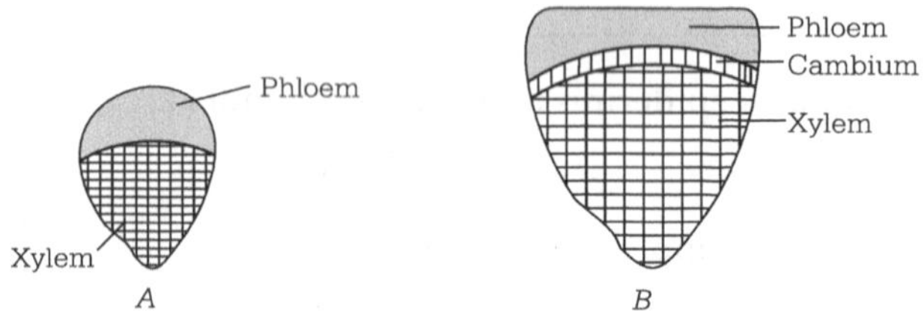

(i) Open vascular bundle________________

(ii) Closed vascular bundle ________________

(iii) Present in monocots ________________

(iv) Forms secondary tissues ________________

(v) Do not form secondary tissues ________________

(vi) Xylem and phloem are situated on the same radius ________________

119. In radial arrangement, xylem and phloem arranged in ________________ manner along the different ________________.

120. Name the vascular bundle in which xylem and phloem are jointly situated along the same radii. ________________

121. Complete the following table.

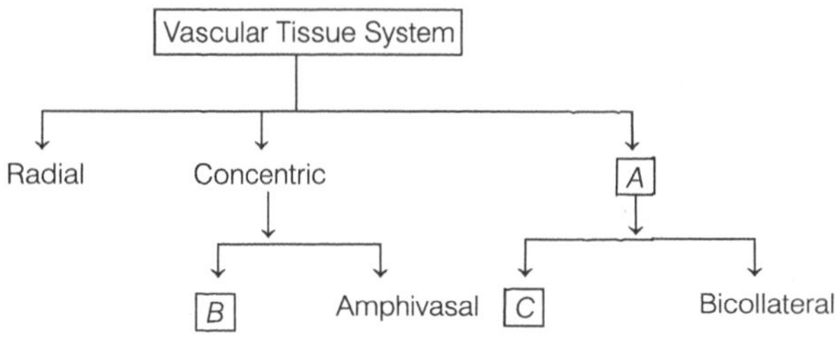

A. ________________

B. ________________

C. ________________

122. Identify the types of vascular bundle on the basis of given statements.

(i) Vascular bundle common in stems and leaves. ________________

(ii) Vascular bundle common in root. ________________

(iii) Phloem located only on the outer side of xylem. ________________

Topic 4 : Anatomy of Dicotyledonous and Monocotyledonous Root

123. Rearrange the sequence of internal tissue of dicot root from outer side to inner side.

Epidermis, Root hair, Endodermis, Cortex, Phloem, Pith, Pericycle

____

124. The outermost layer of dicotyledonous root is <u>cortex</u>.
(*Rewrite the statement by changing the underlined word.*)

____

125. Consider the following figure of TS of dicot root, which of the following parts represent cortex, endodermis and xylem?

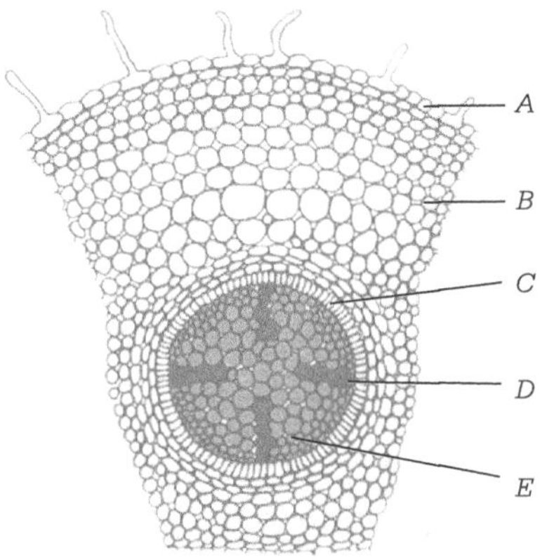

____

126. Unicellular root hairs protrude from ______________________.

127. Name the region of dicotyledonous root where several layers of thin-walled parenchyma cell with intercellular spaces are present. ______________________

128. Complete the following analogy.

Cortex : ______________ : : Several layers : Single layer

129. Which layer of dicot roots possess barrel-shaped cell without any intercellular spaces?

____

130. The tangential as well as radial wall of the endodermal cells possess deposition of water impermeable material, i.e. suberin is known as <u>starch sheath</u>.
(*Rewrite the statement by changing the underlined word.*)

____

131. Lateral roots and vascular cambium originate from ______________________. These are thick-walled ______________________ cells.

132. In dicot root, pith is large.

True ☐ False ☐

133. Name the tissue which is present between xylem and phloem.

____

134. Your are given a slide of root in which, vascular bundles are radial, xylem is exarch, vascular bundle vary from diarch to hexarch. Identify the root.________________

135. Identify the odd one out from the given terms.

Cortex, Endoderm, Pericycle, Vascular bundle, Pith

____

136. Vascular bundles in monocot root are ________________ and xylem is in ________________ condition.

137. Consider the following figure of TS of monocot root. In which of the following part, Casparian strip is present?

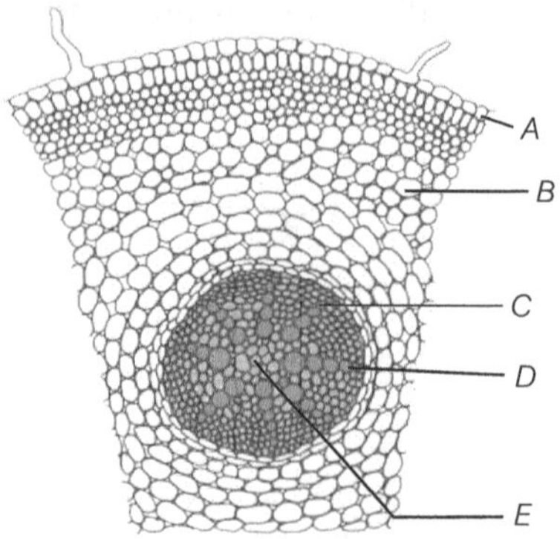

____

138. Read the following characteristics and mark them as dicot or monocot root.

(i) There are usually more than six xylem bundles________________.

(ii) Pith is large and well-developed________________.

(iii) Secondary growth is present________________.

Topic 5 : Anatomy of Dicotyledonous and Monocotyledonous Stem

139. Epidermis of dicot stem bears trichomes and few stomata.

True ☐ False ☐

140. In dicot stem, cortex is categorised into three zones, i.e. ________________, ________________ and ________________.

141. Match the following columns.

	Column I (Zones in dicot roots)		Column II (Characteristics)
A.	Hypodermis	1.	Rounded thin-walled parenchymatous
B.	Cortical layer	2.	Provides mechanical strength
C.	Endodermis	3.	Semilunar patches of sclerenchyma
D.	Pericycle	4.	Innermost layer of cortex

A. ________________ B. ________________ C. ________________ D. ________________

142. Consider the following sequence, of tissues of stem identify *A* and *B*.

Epidermis → ______________ → Cortical layer → ____________________ → Pericycle

143. The endodermis is rich in ________________________ and the layer is also known as ________________________________.

144. The hypodermis of a dicot stem is made up of __________________________ cells.

145. In a dicot stem, the endodermis is present above phloem in the form of semilunar patches of sclerenchyma.
(*Rewrite the statement by changing the underlined word.*)

____

146. Consider the following figure of dicot stem, identify parts *A, B, C* and *D*.

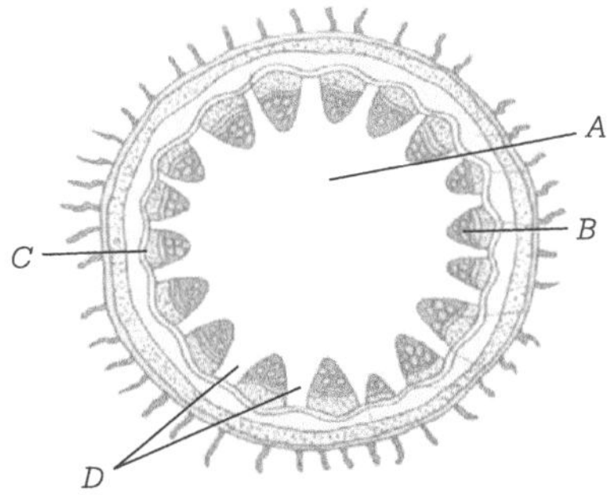

A. ________________________________

B. ________________________________

C. ________________________________

D. ________________________________

147. Some parenchymatous cells are radially placed between the vascular bundle, these are known as xylem rays.
(*Rewrite the statement by changing the underlined word.*)

____

148. The vascular bundles in dicot stem are scattered.

True ☐ False ☐

149. Consider the following figure of TS of stem and identify the parts *A–F*.

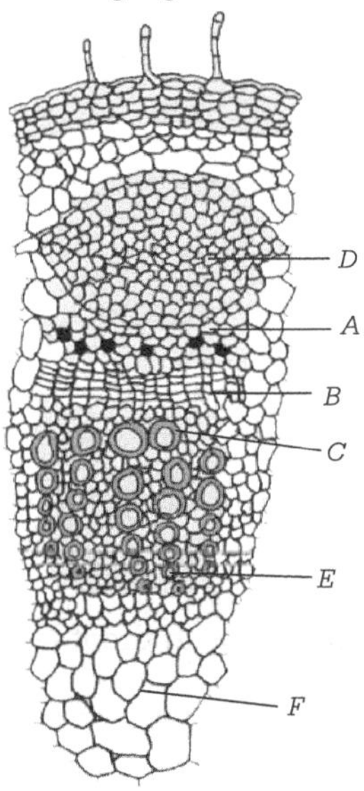

A. ________________________________

B. ________________________________

C. ________________________________

D. ________________________________

E. ________________________________

F. ________________________________

150. The vascular bundle in dicot stem is conjoint, closed and exarch.

True ☐ False ☐

151. ________________ (Pericycle/Pith) is a large number of rounded, parenchymatous cells with large intercellular spaces that occupy central portion of the dicot stem.

152. Complete the following analogy.

Monocot stem hypodermis : Dicot stem hypodermis : : ________________ : Collenchymatous.

153. Vascular bundles in monocot stem are conjoint and ________________.

154. Consider the following figure of TS of stem. Answer the questions that follows.

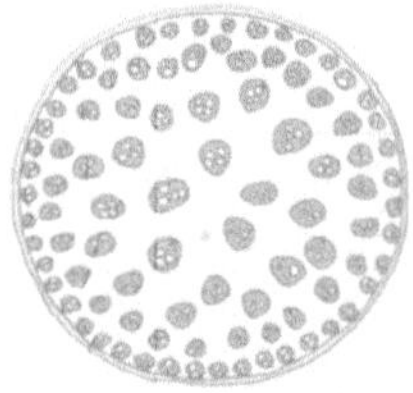

(i) This figure represents ________________ (monocot/dicot) stem.

(ii) Protoxylem and metaxylem are arranged in which shape in the given stem? __________.

Topic 6 : Dorsiventral (Dicotyledonous) and Isobilateral (Monocotyledonous) Leaf

155. Dicotyledonous leaf is also known as isobilateral leaf.

True ☐ False ☐

156. Complete the following analogy of leaf surface.

Upper surface : Lower surface : : ____________ : Abaxial.

157. Consider the following diagram of TS of a dicot leaf identify its parts *A*, *B* and *C*.

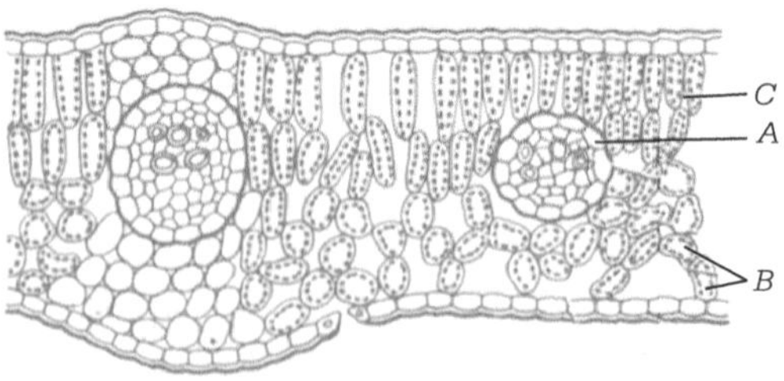

A. ________________

B. ________________

C. ________________

158. The abaxial surface of leaf bears less stomata than the adaxial in dicot.

True ☐ False ☐

159. The tissue between the upper and the lower epidermis in dicot leaves is called as

____

160. Identify the type of parenchyma in the following sentences.

(i) Parenchyma placed on adaxial surface. ______________________

(ii) Oval or round and loosely arranged parenchyma. ______________________

161. Vascular bundles in leaves found at ______________ and ______________.

162. The vascular bundles are surrounded by a layer of thick-walled ______________________ cells.

163. Monocot leaf is also known as bulliform leaf.
(*Rewrite the statement by changing the underlined word.*)

____

164. The epidermis on both side contains ______________ and the mesophyll is usually not differentiated into ______________ and ______________ parenchyma in isobilateral leaf.

165. Name the cells that are large, bubble-shaped, colourless and occur in groups on the upper surface in grasses. ______________________

166. The ______________ in the leaves absorbed water and become turgid, the leaf surface is ______________.

167. The monocot leaves curl inward during water stress condition due to secondary xylem.
(*Rewrite the statement by changing the underlined word.*)

____

168. Similar sizes of vascular bundles are seen in monocot leaves due to the

____.

169. Consider the following figure of TS of monocot leaf, identify its parts *A*, *B* and *C*.

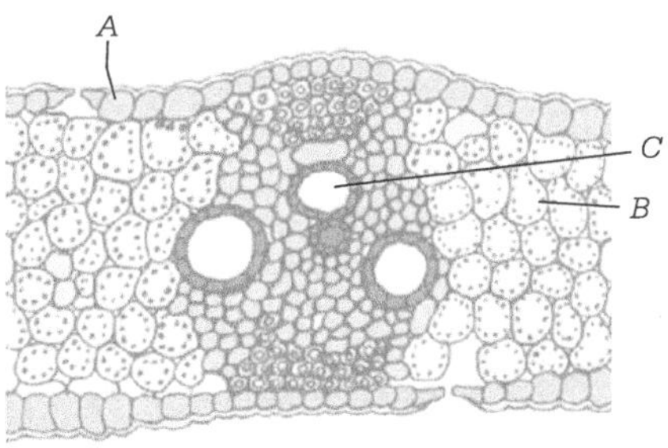

A. ______________________

B. ______________________

C. ______________________

Topic 7 : Secondary Growth

170. Most dicotyledonous plants increase in girth due to primary growth.
(*Rewrite the statement by changing the underlined word.*)

____

171. The tissues involved in secondary growth are ______________ and

____.

172. In young stem, vascular cambium is present in ________________, later it forms a ________________________________.

173. Name the cambium which is present between primary xylem and primary phloem.

____

174. Interfascicular cambium is formed by the cells of xylem and phloem.

True ☐ False ☐

175. The amount of secondary xylem produced is more than secondary phloem. (*Give reason*)

____

176. Which part of the given figure represents interfascicular cambium?

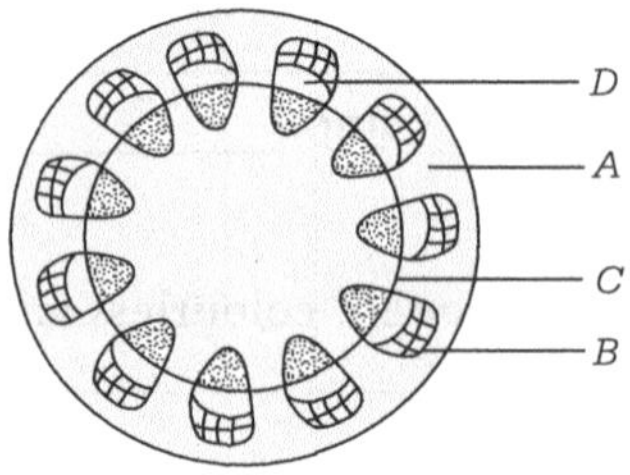

____

177. The cambium forms a narrow band of parenchyma in radial direction which passes through secondary xylem and secondary phloem is known as secondary cambium.

True ☐ False ☐

178. Identify parts *A*, *B*, *C* and *D* from the following figure.

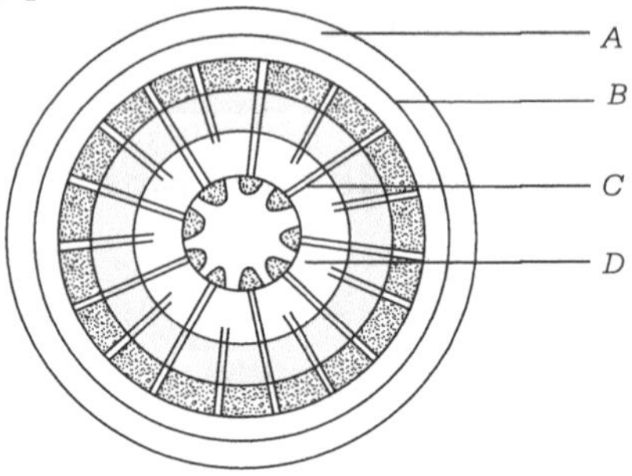

A. ________________

B. ________________

C. ________________

D. ________________

179. Match the following columns.

	Column I		Column II
A.	Secondary xylem	1.	Narrow vessels
B.	Secondary phloem	2.	Highly lignified walls
C.	Spring wood	3.	Lighter in colour
D.	Autumn wood	4.	Cells cut off towards pith
E.	Heart wood	5.	Wider vessels
F.	Sap wood	6.	Cells cut off towards periphery

A. ________________ B. ________________ C. ________________

D. ________________ E. ________________ F. ________________

180. Vascular cambium originates from the portion of pith in dicot root.

True ☐ False ☐

181. During the spring season, the activity of cambium increases hence it produces large number of xylary elements. The wood formed during this season is called ____________ or ____________.

182. Due to less activity of cambium in winter, it produces fewer xylary elements having broad vessels.

True ☐ False ☐

183. Complete the following analogy.

Early wood : ____________ : : Lower density : Higher density

184. The softer, more porous portion of an annual ring are late wood.

True ☐ False ☐

185. As the stem grows older, its heart wood ________ (increases/ decreases) in thickness.

186. Heart wood is involved in the conduction of water and minerals.
(*Rewrite the statement by changing the underlined word.*)

____

187. The age of a tree can be estimated by the spring ring.
(*Rewrite the statement by changing the underlined word.*)

____

188. Name the wood in which the greater part of secondary xylem is dark brown due to the deposition of tannins, resins, etc., and it is hard and durable.

____

189. Complete the following analogy.

Heartwood : Sapwood : : Duramen : ____________ : : Porous : ____________

190. Complete the following analogy.

Cork cambium : Cork : Secondary cortex : : Phellogen : ____________ : ____________

191. Correct the following sequence from inner side to outer side.

Epidermis → Phellogen → Cortex → Phellem → Phelloderm

____

192. ____________ is made up of narrow, thin-walled and ____________ cells.

193. Phellogen cuts off cells on both sides, i.e. outer cell is known as phelloderm and inner cell is known as phellem.

True ☐ False ☐

194. Give a collective term for the following words.

Phellogen, Phellem, Phelloderm

____

195. ____________ is a non-technical term that refers to all tissues exterior to the vascular cambium.

196. Find the odd one out from the following.

Phellogen, Phellem, Phelloderm, Pith, Secondary phloem

____

197. Soft bark is formed at the end of the season.
(*Rewrite the statement by changing the underlined word.*)

____

198. Name the structure in which parenchymatous cells rupture the epidermis to form the lens-shaped opening. ______________________

199. Consider the following figure of lenticel, identify parts *A* and *B*.

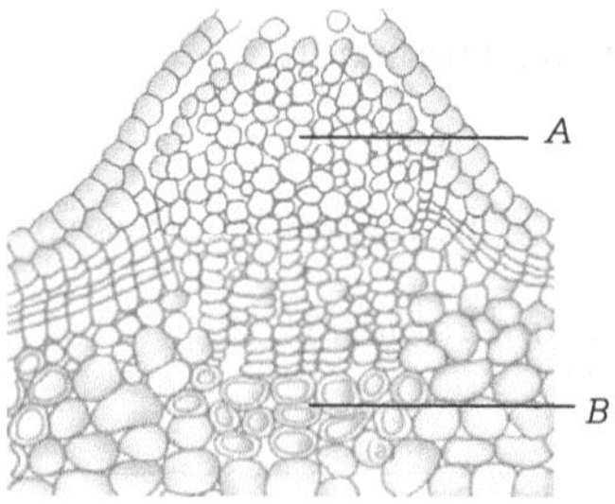

A. ______________________

B. ______________________

200. Hydathodes permit the exchange of gases between the outer atmosphere and internal tissue system of the stem.
(*Rewrite the statement by changing the underlined word.*)

____

201. Secondary growth in stelar region occurs by ______________ and in extrastelar region occurs by ______________ in dicot roots.

202. Consider the following figure of a typical dicot root. Identify parts *A*, *B*, *C* and *D*.

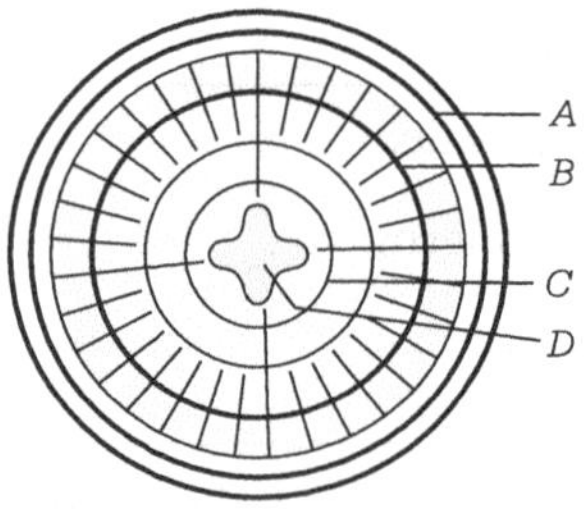

A. ______________________

B. ______________________

C. ______________________

D. ______________________

203. The roots and stems of gymnosperms also show secondary growth.

True ☐ False ☐

204. Monocotyledons show secondary growth.

True ☐ False ☐

Answers

1. Anatomy
2. Plant anatomy
3. True
4. Yes
5. Meristematic and permanent tissues
6. Meristems
7. (i) Present
   (ii) Present
   (iii) Absent
8. Apical meristem
9. *A*–Leaf primordium, *B*–Shoot apical, *C*–Meristematic zone, *D*–Axillary bud
10. True
11. *A*–Initials of central cylinder and cortex, *B*–Root apical meristem, *C*–Cortex, *D*–Protoderm, *E*–Root cap.
12. Axillary bud, branch or a flower
13. Intercalary meristem.
14. Intercalary meristem, apical meristem
15. Internodes, nodes.
16. Bamboo can grow about 60 cm/day because of the presence of intercalary meristem.
17. Intercalary meristem
18. Lateral meristem
19. Lateral meristem
20. Fascicular vascular cambium, interfascicular cambium and cork cambium.
21. True
22. *A*–Secondary meristem, *B*–Apical meristem, *C*–Intercalary meristem, *D*–Protoderm, *E*–Ground meristem.
23. Ray, Radial
24. Permanent or mature cells.
25. Dermal tissues, ground tissues, vascular
26. False. The cells of permanent tissue have definite shape, size and function and they loose the power of division.
27. (i) Simple tissues
    (ii) Complex tissues
28. 3 (Aerenchyma, parenchyma, angular collenchyma).
29. Parenchyma
30. Photosynthesis, secretion
31. Living
32. *X*–Intercellular space
33. Thin, cellulose
34. True
35. Collenchyma
36. Cellulose, hemicellulose and pectin.
37. Collenchyma
38. (i) Parenchyma tissues,
    (ii) Sclerenchyma tissues
39. 3 (Having protoplasts, oval cells, contain chloroplast)
40. Sclerenchymatous cells provide mechanical support because these are lignified, dead and without protoplast.
41. Fibres
42. Sclereids
43. Fibres
44. The gritty texture of guava is due to presence of sclereids.
45. (i) *A*–Pits
       B–Lumen
    (ii) Tea, legumes
46. Complex tissue
47. Xylem
48. Mechanical strength
49. Vessel
50. Seive tube is odd one out, as it is a part of phloem. Other four are the parts of xylem.
51. Tracheid
52. False. Vessels are the most advance type of conductive element in xylem.
53. *A*–Tracheid, *B*–Vessel
54. Lignified, devoid
55. Perforated
56. *Drimys*
57. Xylem fibres
58. Xylem parenchyma
59. Parenchyma cells
60. Ray parenchymatous cell
61. (i) Protoxylem
    (ii) Metaxylem
62. Periphery, centre
63. Tracheids, vessels, xylem fibres and xylem parenchyma.
64. Stem
65. False. In exarch condition, protoxylem lies towards the periphery whereas metaxylem lies towards the centre.
66. Sieve tube elements, companion cells, phloem parenchyma and phloem fibres.
67. Albuminous
68. *A*–Sieve tube element, *B*–Phloem parenchyma, *C*–Companion cell
69. Companion cells and sieve tubes
70. Sieve tubes
71. Peripheral, nucleus
72. A-2, B-3, C-1
73. Companion cells
74. Living
75. 2 (Companion cells, phloem parenchyma)
76. Phloem fibres
77. Lignified, cellulose
78. Phloem parenchyma
79. Sieve tube
80. Metaphloem
81. Thick, protoplasm, dead
82. Secretory tissue
83. (i) Laticiferous tissue,
    (ii) Glandular tissue
84. Epidermal tissue system, ground or fundamental tissue system and vascular tissue system.
85. Epidermal cells, stomata, epidermal appendages.
86. Epidermis
87. Mesophyll is odd one out. All of them are epidermal tissue system except mesophyll.

88. False. Epidermis is made up of elongated, compactly arranged cells, it forms a continuous layer.

89. Parenchymatous, vacuole

90. Cuticle

91. Roots have to absorb water and nutrient freely from the soil and the presence of cuticle would prevent it.

92. Transpiration, gaseous exchange

93. False. Bean-shaped guard cells are found in dicot and dumb-bell-shaped guard cells are found in monocot.

94. True

95. Thin, thickened

96. Guard cells

97. Subsidiary cells.

98. B–Dumb-bell-shaped stomata

99. Stomatal aperture

100. Root hairs

101. Trichomes

102. False. Trichomes in the shoot system are usually multicellular.

103. True

104. Transpiration

105. Digestive gland

106. False. Ground tissue constitutes all tissues except epidermis and vascular bundles.

107. Parenchyma, collenchyma, sclerenchyma

108. Mesophyll

109. Stele

110. Intrastellar ground tissue

111. Transfusion cells

112. True

113. Phloem, xylem

114. (i) Fusiform initial, (ii) Ray initial

115. False. Cambium is absent in monocot plants.

116. Open

117. Cambium is absent in monocotyledons, hence they do not show secondary growth.

118. (i) *B*, (ii) *A*, (iii) *A*, (iv) *B*, (v) *A*, (vi) Both *A* and *B*.

119. Alternate, radii

120. Conjoint

121. *A*–Conjoint, *B*–Amphicribal, *C*–Collateral

122. (i) Conjoint
(ii) Radial
(iii) Conjoint

123. Root hair → Epidermis → Cortex → Endodermis → Pericycle → Phloem → Pith

124. Epiblema

125. *B*–Cortex, *C*–Endodermis, *D*–Xylem

126. Epiblema

127. Cortex

128. Endodermis

129. Endodermal layer

130. Casparian strips

131. Pericycle, parenchymatous

132. False. In dicot root, pith is usually absent.

133. Conjunctive tissue

134. Dicot root

135. Cortex is odd one out. All of them constitute stele except cortex.

136. Radial, polyarch

137. Casparian strip is present in endodermal cell C.

138. (i) Monocot root
(ii) Monocot root
(iii) Dicot root

139. True

140. Hypodermis, cortical layer endodermis

141. A-2, B-1, C-4, D-3

142. Hypodermis, endodermis

143. Starch grain, Starch sheath

144. Collenchymatous

145. Pericycle

146. *A*–Pith, *B*–Vascular bundle, *C*–Pericycle, *D*–Medullary ray

147. Medullary rays

148. False. The vascular bundles in dicot stem are arranged in a ring.

149. *A*–Phloem, *B*–Cambium *C*–Metaxylem, *D*–Pericycle, *E*–Protoxylem, *F*–Pith.

150. False. The vascular bundles in dicot stem are conjoint collateral, endarch and open.

151. Pith

152. Sclerenchymatous

153. Closed

154. (i) Monocot, (ii) Y-shaped

155. False. Dicotyledonous leaf is also known as dorsiventral leaf.

156. Adaxial

157. *A*–Bundle sheath cells, *B*–Spongy mesophyll, *C*–Palisade mesophyll

158. False. The abaxial surface of leaf bears more stomata than adaxial in dicot.

159. Mesophyll

160. (i) Palisade parenchyma,
(ii) Spongy parenchyma

161. Veins, midrib

162. Bundle sheath

163. Isobilateral leaf

164. Stomata, palisade, spongy

165. Bulliform cell

166. Bulliform cell, exposed

167. Bulliform cell

168. Parallel venation

169. *A*–Adaxial epidermis, *B*–Mesophyll cell, *C*–Xylem

170. Secondary

171. Vascular cambium, cork cambium

172. Patches, complete ring

173. Intrafasccular cambium

174. False. Interfascicular cambium is formed by the medullary rays.

175. The cambium is generally more active on the inner side than on the outer.

176. Part *C* represents interfascicular cambium.

177. False. The cambium forms a narrow band of parenchyma in radial direction which passes through secondary xylem and secondary phloem is known as secondary medullary rays.

178. *A*–Phellem, *B*–Phellogen, *C*–Medullary rays, *D*–Secondary xylem

179. A-4, B-6, C-5, D-1, E-2, F-3

180. False. Vascular cambium originates from the tissue located just below the phloem bundles, a portion of pericycle tissue.

181. Spring wood, early wood

182. False. Due to less activity of cambium in winter, it produces fewer xylary elements having narrow vessels.

183. Late wood

184. False The softer, more porous portion of an annual ring are spring wood.

185. Increases

186. Sapwood

187. Annual

188. Heartwood

189. Alburnum, non-porous

190. Phellem, phelloderm

191. Cortex → Phelloderm → Phellogen → Phellem → Epidermis

192. Phellogen, rectangular

193. False. Phellogen cuts of cells on both sides, i.e. outer cell is known as phellem and inner cell is known as phelloderm.

194. Periderm

195. Bark

196. Pith is odd one. All of them constitute bark except pith.

197. Hard

198. Lenticels

199. *A*–Complimentary cells, *B*–Secondary cortex

200. Lenticels

201. Vascular cambium, cork cambium

202. *A*–Cortex, *B*–Vascular cambium, *C*–Secondary xylem, *D*–Primary xylem

203. True

204. False. Monocotyledons do not show secondary growth.

CHAPTER SEVEN

STRUCTURAL ORGANISATION IN ANIMALS

Topic 1 : Introduction and Epithelial Tissues

1. Give the scientific term for this statement.

   A group of similar cells in multicellular animals that perform a specific function.

   ____

2. In animals, tissues are broadly classified into ____________ types while in plants, tissues are of ____________ types.

3. Animal tissues are broadly classified in different types on the basis of

   ____

4. Consider the following.

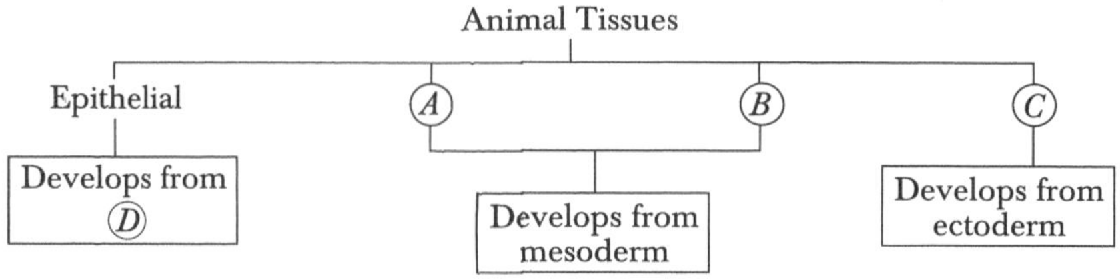

   Complete the flowchart based on the information given above.

   A.________________________ B. ________________________

   C.________________________ D. ________________________

5. Identify the type of animal tissue based on the statements given below.

(i) Cells are compactly packed with little intercellular space.

(ii) It forms covering or lining of external and internal surfaces.

(iii) It helps in protection, secretion, absorption, sensation, etc.

____

6. Fill in the blanks for the types of epithelium.

(i) The ________________ consists of two or more cell layers and has protective function as it does in our skin.

(ii) The ________________ is composed of a single layer of cells and functions as a lining for body cavities, ducts and tubes.

7. On the basis of structural modification of the cells, simple epithelium is further divided into three types *viz*. ________________________

8. Recognise each cell type shown below.

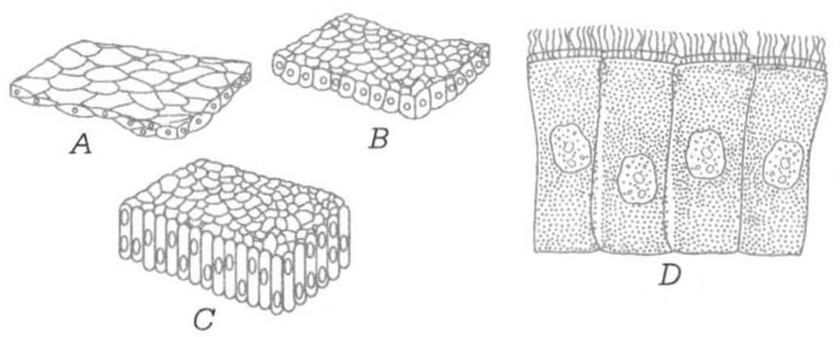

A. ________________ B. ________________

C. ________________ D. ________________

9. Match the following columns.

	Column I (Structures)		Column II (Names)		Column III (Locations)
A.	Tall and slender cells	1.	Cuboidal epithelium	K.	Bronchioles
B.	Flattened cells with irregular boundaries	2.	Columnar epithelium	L.	Tubular parts of nephrons
C.	Cilia on their free surface	3.	Squamous epithelium	M.	Lining of intestine
D.	Cube-like cells	4.	Ciliated epithelium	N.	Walls of blood vessels

A. ____________ B. ____________

C. ____________ D. ____________

10. Based on the function given below, write the names of epithelial tissue.

(i) Function in forming a diffusion boundary. ________________

(ii) Helps in secretion and absorption. ________________

(iii) Function to move particles or mucus in a specific direction over the epithelium.

____

11. Based on the location given below, write the names of epithelial tissue.

(i) Ducts of glands ______

(ii) Fallopian tubes ______

(iii) Air sacs of lungs______

(iv) Lining of stomach ______

12. Some of the columnar or cuboidal cells get specialised for secretion and are called ______ epithelium. It is of two types, i.e. ______ .

13. From the list of characters (i), (ii) and (iii), choose the appropriate characteristics for glands *A* and *B*.

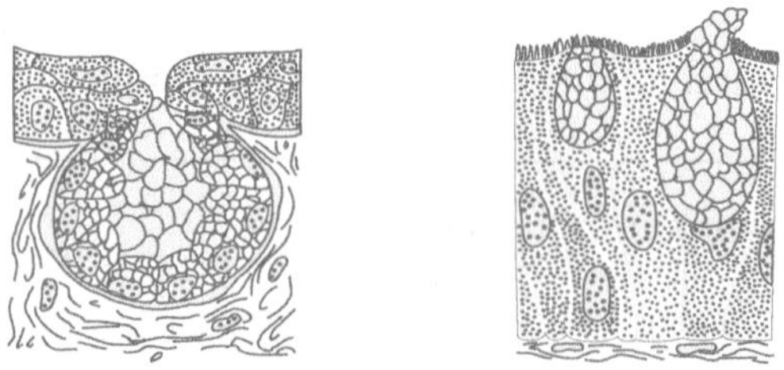

A *B*

(i) **Cellular level** Unicellular/Multicellular

(ii) **Organisation** Isolated cells /Cluster of cells

(iii) **Type and location** Goblet cells of alimentary canal / Cells of salivary gland

A.______

B.______

14. Vagina, oesophagus and urethra contain which type of epithelium? ______

15. Ductless glands in human beings produce ______ (mucus/hormones).

16. Give examples of saccular glands.

______

17. Read the following statements and find out how many of these are related to the figure given below?

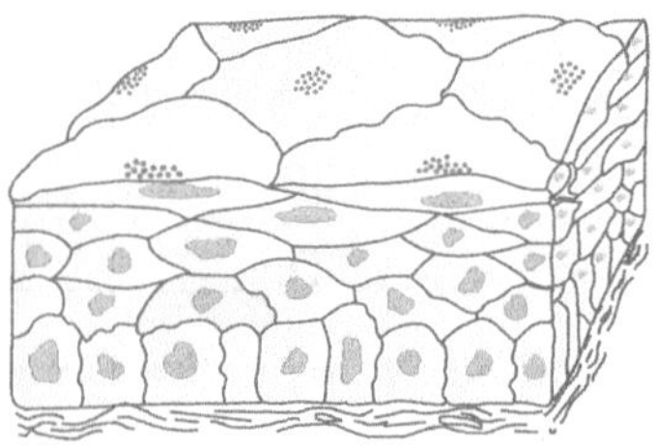

(i) Multilayered epithelium.

(ii) Play major role role in secretion and absorption.

(iii) Main function to provide protection against chemical and mechanical stresses.

(iv) They cover the dry surface of skin, moist surface of buccal cavity and pharynx.

______

18. On the basis of the mode of pouring of their secretions, glands are divided into two categories, namely ______________________

19. Complete the following analogy given below.

Exocrine glands : Glands with ducts : : ______________ : Ductless glands.

20. How many of the following are the secretions of exocrine glands?

Earwax, Mucus, Saliva, Oil, Milk, Digestive enzymes, Hormones

____

21. State, whether the following statements are True or False.

(i) All cells in epithelium are held together with little intercellular material.

True ☐ False ☐

(ii) In nearly all animal tissues, specialised junctions provide both structural and functional links between individual cells.

True ☐ False ☐

22. Give common term for the following terms related to tissues.

Tight, Adhering, Gap

____

23. Match the following columns.

	Column I (Cell junctions)		Column II (Functions)
A.	Adhering junction	1.	Helps to stop substances from leaking across a tissue
B.	Gap junction	2.	Performs cementing to keep neighbouring cells together
C.	Tight junction	3.	Facilitates the cells to communicate with each other

A. ______________ B. ______________

C. ______________

Topic 2 : Connective Tissues

24. In all connective tissues except ______________, the cells secrete fibres of structural proteins called collagen or elastin.

25. Choose the right option/s for statements in the given bracket.

(i) It provides strength, elasticity and flexibility to the tissue. (Fibres, Blood, Bones)

____

(ii) These cells secrete modified polysaccharides, which accmulate between cells and fibres and act as a matrix or ground substances. (Fibres, Bone, Adipose, Cartilage).

____

26. Which tissue has cells and fibres loosely arranged in a semi-fluid ground substance?

____

27. Which connective tissue often serves as a support framework for epithelium, contains fibroblast (cells that produce and secrete fibres), macrophages and mast cells?

____

28. The excess nutrients which are not used immediately are converted into fats and stored in

____

29. Read the statements carefully and names the respective tissue.

(i) Fibres and fibroblasts are compactly packed. _______________

(ii) The collagen fibres are present in rows between many parallel bundles of fibres.

____

(iii) It has fibroblasts and many fibres (mostly collagen) that are oriented differently.

____

30. Match the following columns.

Column I	Column II
A. Tendon	1. Loose connective tissue
B. Cartilage	2. Dense regular connective tissue
C. Areolar tissue	3. Dense irregular connective tissue
D. Perichondrium of bones	4. Specialised connective tissue

A. _______________ B. _______________

C. _______________ D. _______________

31. Identify the tissue '*X*' and write its structures labelled as *A* to *D*.

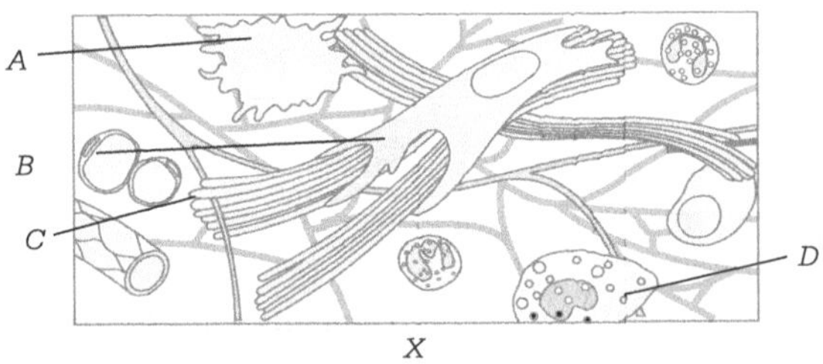

X. _______________

A. _______________

B. _______________

C. _______________

D. _______________

32. Identify the tissue '*X*' and write its structures labelled as *A* to C.

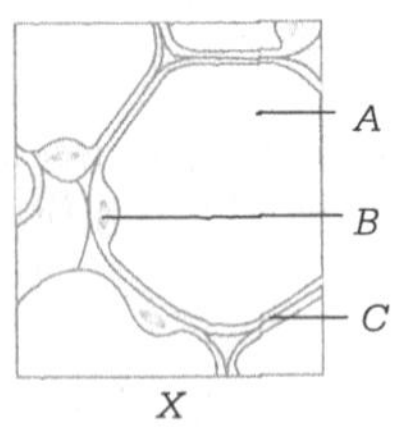

X. _______________

A. _______________

B. _______________

C. _______________

33. Identify the figure *A* and *B*, respectively.

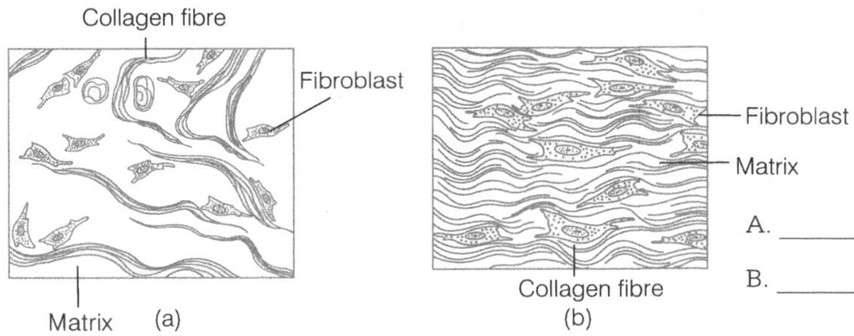

A. ______________________

B. ______________________

34. The intercellular material of bone is solid and pliable and resists compression. Cells of this tissue are enclosed in small cavities within the matrix.
(*Rewrite the statement by changing the underlined word.*)

____

35. Give three locations where cartilage is present in the body.

____

36. Give the scientific term for the following statement.

A hard and non-pliable ground substance rich in calcium salts and collagen fibres which provides structural framework to the body. ______________________

37. Complete the following analogy given below.

Bone forming cells : Osteoblast : : Bone destroying cells : ______________ .

38. The bone cells, i.e. ______________ are present in the spaces called ______________ .

39. The site of production of Red Blood Cells (RBCs) in adult humans is

____

40. In the given diagram of TS of cartilage, identify the parts labelled as *A* and *B*.

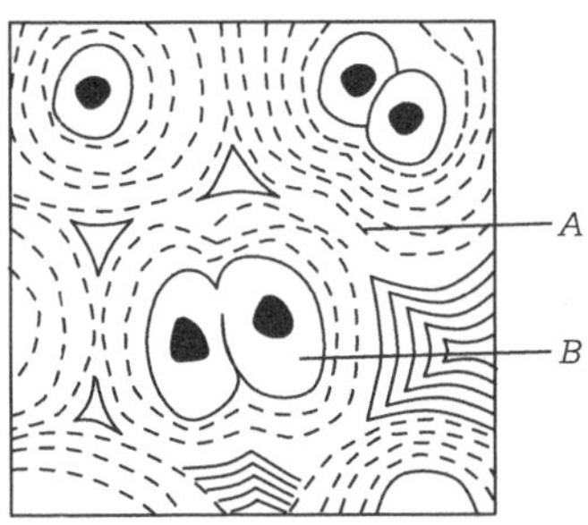

A. ______________________

B. ______________________

41. Name a fluid connective tissue which helps in transport of various substances.

____

42. Match the following columns.

	Column I (Connective tissues)		Column II (Functions)
A.	Tendons	1.	Attach one bone to another
B.	Ligaments	2.	Weight bearing function
C.	Bones	3.	Attach skeletal muscle to bones
D.	Blood	4.	Contain formed elements

A. __________ B. __________

C. __________ D. __________

Topic 3 : Muscular Tissues

43. Each muscle is made of many long cylindrical fibres arranged in parallel arrays. These fibres are composed of numerous fine fibrils, called __________.

44. Match the following columns.

	Column I (Figures)		Column II (Names)		Column III (Locations)
A.		1.	Striated muscle	K.	Wall of blood vessels
B.		2.	Smooth muscle	L.	Nodal tissue
C.	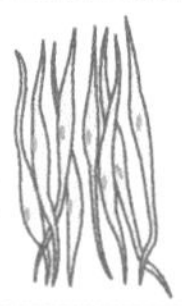	3.	Cardiac muscle	M.	Biceps

A. __________ B. __________

C. __________

45. Write the scientific names for the following.

(i) Cytoplasm of muscle cell __________

(ii) Membrane covering the muscle cell __________

(iii) Give two special properties shown by muscular tissue.

____

46. Complete the following table.

Skeletal muscles	Smooth muscles	Cardiac muscles
(A)	Uninucleate	(B)
Striated and cylindrical	(C)	(D)
(E)	(F)	Involuntary
(G)	Slow contraction and do not get fatigued	(H)
(I)	(J)	Present only in the heart

A. ______________________ B. ______________________
C. ______________________ D. ______________________
E. ______________________ F. ______________________
G. ______________________ H. ______________________
I. ______________________ J. ______________________

47. Communication junctions or intercalated discs allow the skeletal muscle to contract as a unit.

True ☐ False ☐

Topic 4 : Neural Tissues

48. Which animal tissue exerts the greatest control over the body's responsiveness to changing conditions?

____

49. Name the excitable cells of the body which are known as the unit of neural system.

____

50. Which cell of neural system protect and support neurons?

____

51. In human body, neuroglia cells occur in the ______________________

52. Write down the correct labels *A* to *D* for the given neural tissue.

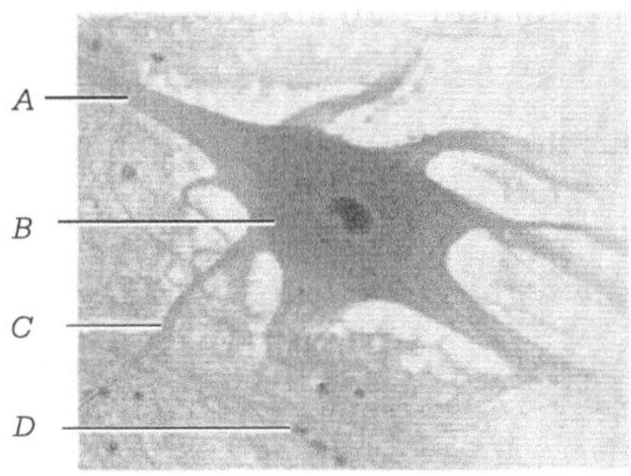

A. ______________________
B. ______________________
C. ______________________
D. ______________________

53. Name an organ which consists of all the four types of animal tissues.

____

54. The word anatomy is conventionally used for the study of morphology of internal organs in animals.

True ☐ False ☐

Topic 5 : Earthworm

55. ____________ (Earthworm/Cockroach) is a reddish-brown terrestrial invertebrate that inhabits the upper layer of the moist soil.

56. During the daytime, earthworm lives in burrows made by ____________ and ____________ the soil.

57. In gardens, the earthworm can be traced by their faecal deposits known as ____________.

58. Name two common Indian earthworms.

____

59. Write the approximate total number of segments (metamers) in *Pheretima.*

____

60. The dorsal surface and ventral surface of the earthworm is marked by ____________ and ____________, respectively.

61. Earthworm does not have a distinct head.

True ☐ False ☐

62. Identify the structure in earthworm based on the characters listed below.
- (i) Present on the anterior end
- (ii) Sensory in function
- (iii) Serves to open cracks in the soil during crawling.
- (iv) Is a fleshy lobe which covers the mouth.

____

63. Give the scientific term for the first segment of earthworm bearing mouth.

____

64. In a mature worm, segments 14th, 15th and 16th are covered by a prominent dark band of glandular tissue called ____________.

65. Name the three prominent regions into which the body of the mature earthworm is divided.

____

66. Identify the figure '*X*' and write the names of its parts *A* to *D*.

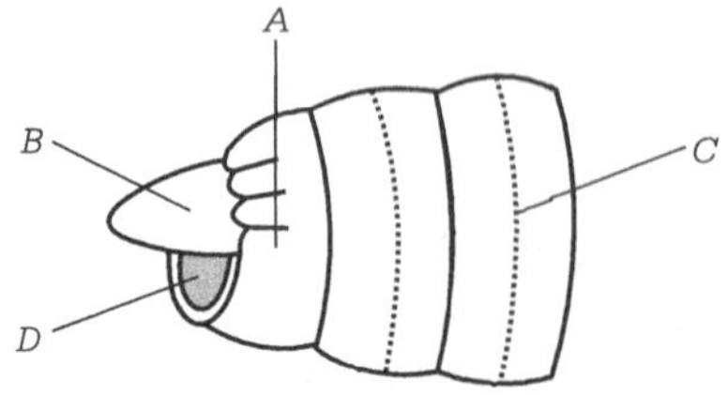

X

X. ____________

A. ____________

B. ____________

C. ____________

D. ____________

67. Identify the figure '*X*' and write the correct names of its parts *A* to *E*.

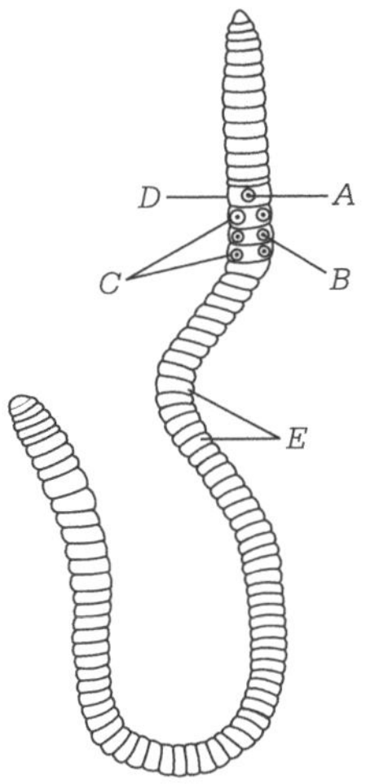

X. ______________________

A. ______________________

B. ______________________

C. ______________________

D. ______________________

E. ______________________

68. Match the following columns.

	Column I		Column II		Column III
A.	Male genital pore(s)	1.	10th and 11th segments	K.	Single
B.	Female genital pore(s)	2.	14th segment	L.	1 Pair
C.	Spermathecal apertures	3.	Intersegmental grooves, i.e. 5th-9th segments	M.	4 Pairs

A. ______________ B. ______________

C. ______________

69. State True or False for the given statements.

(i) Setae are present in all body segments except first, last and clitellum.

True ☐ False ☐

(ii) Nephridia are present in all body segments except first two segments of body.

True ☐ False ☐

(iii) Intestine is present only in the post-clitellar segments.

True ☐ False ☐

70. Identify the structure in earthworm based on the characters listed given below.

(i) S-shaped chitinous structures.

(ii) Help in locomotion by extension and retraction.

(iii) Embedded in the epidermal pits in the middle of each segment.

____

71. Write the correct sequence of layer of the body wall of earthworm from outside to inside.

Epidermis, Longitudinal muscle, Circular muscle, Coelomic epithelium, Non-cellular cuticle

____

72. Refer to the figure and write its parts labelled as *A* to *G* in diagram of alimentary canal of earthworm.

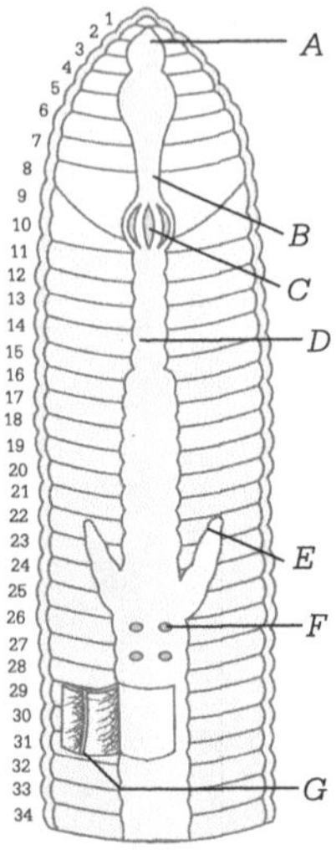

A. ______________________

B. ______________________

C. ______________________

D. ______________________

E. ______________________

F. ______________________

G. ______________________

73. Alimentary canal in earthworm runs from 1st segment to the last comprising of various parts. Write the segments in which the listed parts of alimentary canal are present.

(i) Gizzard__________ (ii) Intestine__________

(iii) Stomach__________ (iv) Buccal cavity__________

(v) Oesophagus__________ (vi) Pharynx__________

(vii) Typhlosole__________

74. Give one role of each of the parts of earthworm.

(i) Typhlosole______________________

(ii) Gizzard______________________

(iii) Calciferous glands in stomach______________________

75. What is the food source of earthworm?

____

76. *Pheretima* exhibits a ______________ (closed/opened) blood vascular system.

77. How many features are correct for blood vascular system in earthworm.

(i) Blood contains RBCs.

(ii) Haemoglobin dissolved in blood plasma.

(iii) Blood confined to heart and blood vessels.

____

78. Blood glands produce respiratory pigment, ______________ and blood cells which are ______________ in nature. These glands are present on ______________ segments.

79. Earthworm has four pairs of lateral hearts present in the segments ______________.

80. Complete the following analogy in earthworm.

Nephridia : Excretory organ : : ________________ : Respiratory organ.

81. Go through the blood vascular system of earthworm and write names of the missing parts *A* to *E*.

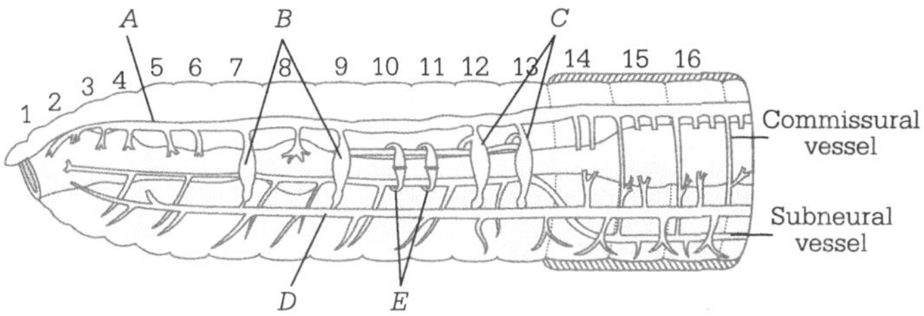

A. ________________

B. ________________

C. ________________

D. ________________

E. ________________

82. Match the following columns for the three different types of nephridia in earthworm.

Column I	Column II	Column III
A. Integumentary nephridia	1. 4th to 6th segment	K. Exonephric
B. Septal nephridia	2. 15th to last segment	L. Enteronephric
C. Pharyngeal nephridia	3. 3rd to last segment	

A. ________________ B. ________________

C. ________________

83. Rewrite the statements by changing the underlined word.

(i) A nephridium starts out as a funnel that collects excess fluid from <u>intestinal lumen</u>.

____

(ii) Earthworm is mainly <u>uricotelic</u>.

____

(iii) <u>Blood glands</u> regulate the volume and composition of body fluids.

____

84. In the given figure of nephridial system in earthworm, some labels are misleading. You have to identify that how many labels are correct in the figure?

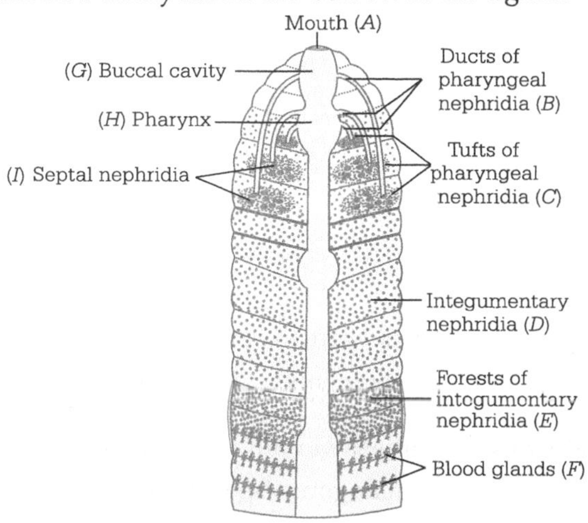

____

85. Nerve cord of *Pheretima* is
(i) single / paired__________ (ii) solid / hollow__________
(iii) dorsal / ventral__________

86. In the earthworm, the nerve cord in the anterior region (3rd and 4th segments) bifurcates laterally encircling the pharynx and joins the cerebral ganglia dorsally to form a

____

87. (i) Sensory system of earthworm does not have eyes.

True ☐ False ☐

(ii) How many senses are found in earthworm?

Touch, Taste, Sound, Light, Vibration

____

88. *Pheretima* is ____________________ (unisexual/bisexual).

89. Refer to the figure of reproductive system of earthworm and write names of its missing parts labelled as *A* to *F*.

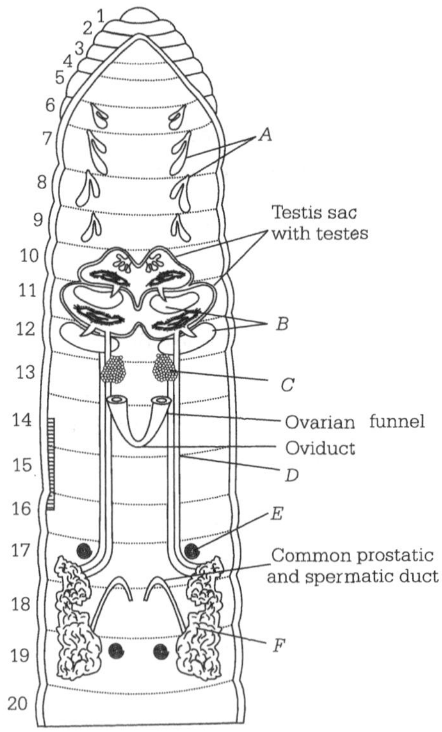

A. ______________________
B. ______________________
C. ______________________
D. ______________________
E. ______________________
F. ______________________

90. Complete the following analogy in earthworm.
Seminal vessicle : 11th and 12th segments : : Accessory gland : _______ segments.

91. _________ (2/4) pairs of spermathecae in 6th to 9th segments in earthworm function to receive ______________ (spermatogonia/spermatozoa) during copulation for storage.

92. One pair of ovaries is attached at the intersegmental septum of 13th and 14th segments.
(*Rewrite the statement by changing the underlined word.*)

____

93. Ovarian funnels are present beneath the ____________ which continue into oviduct, join together and open on ventral side as single median ____________ genital pore on the 14th segment.

94. Write the names of the described structure in earthworm.

(i) Packets of sperms formed during mating. ____________

(ii) Site of fertilisation and development. ____________

95. Development of earthworms is indirect, i.e. there is no larva formed.
(*Rewrite the statement by changing the underlined word.*)

____

96. Earthworms are called farmer's friend because
(i) it burrows in the soil.
(ii) it is a hermaphrodite.

____

97. Write the scientific term for the process of increasing fertility of soil by the earthworms.

____

Topic 6 : Cockroach

98. *Periplaneta* belongs to class ____________ of phylum ____________.

99. Choose the correct option for cockroach.

(i) Omnivorous/Herbivorous ____________

(ii) Monoecious/Dioecious ____________

(iii) Coelomate/Pseudocoelomate ____________

(iv) Economically important/Serious pest and vector of several diseases

____

100. The body of cockroach is segmented and divisible into ____________

101. Exoskeleton of cockroach is formed of ____________ (keratin/chitin).

102. In each segment of cockroach, the exoskeleton has hardened plates called

____________.

103. Complete the following analogy related to cockroach.

Dorsal sclerites : Tergites : : Ventral sclerites : ____________.

104. Flexibility of the cockroach body is due to arthrodial membrane or articular membrane.

True ☐ False ☐

105. Write the correct names for the labels *A* to *E* in the figure of cockroach.

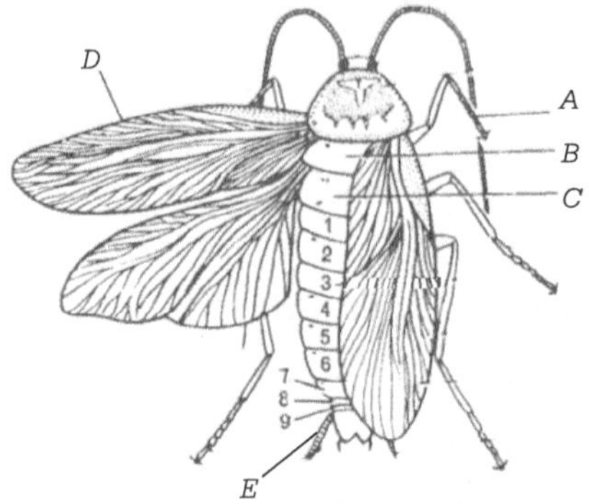

A. ____________________

B. ____________________

C. ____________________

D. ____________________

E. ____________________

106. How many segments of cockroach body are fused to form its head?____________

107. Triangular head of cockroach shows great mobility in all directions due to

____

108. Correctly identify and label only the mouth parts of cockroach in the given diagram of head of cockroach.

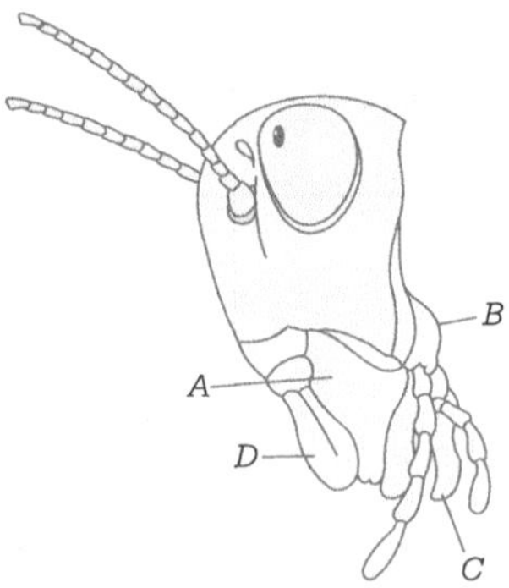

A. ____________________

B. ____________________

C. ____________________

D. ____________________

109. Hypopharynx of cockroach acts as a ____________________ .

110. The mouth parts of cockroach are ______________ and ______________ type.

111. Identify the parts labelled as *A, B, C, D* and *E* in the given figure related with mouthparts of the cockroach.

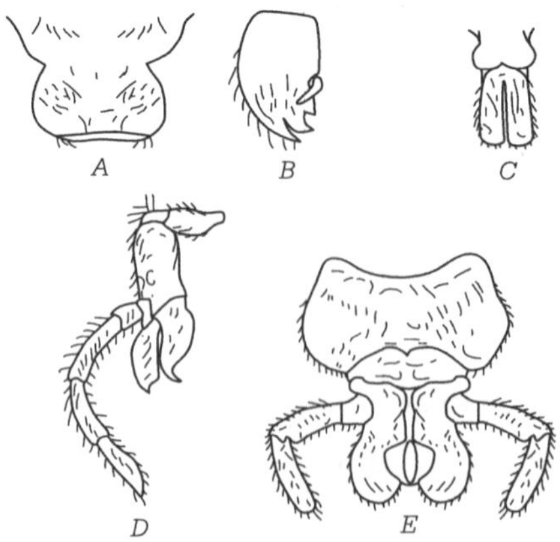

A. ____________________

B. ____________________

C. ____________________

D. ____________________

E. ____________________

112. The number of thoracic segments in cockroach are

____

113. In the mouth parts of cockroach, the organ of mastication is ____________.

114. In cockroach,

(i) pair of walking legs arises from ____________ .

(ii) 1st pair of wings arises from ____________ and 2nd pair of wings arises from ____________ .

115. The abdomen in male and female consists of ____________ segments.

116. How many statements are correct for forewings of cockroach?

(i) Also called mesothoracic wings or tegmina.
(ii) Opaque, dark and leathering.
(iii) Cover the hindwings when at rest.
(iv) Used in flight.

____

117. In female cockroach, shape of the 7th sternum is ____________.

118. In female cockroach, the 7th sternum together with the 8th and 9th sterna forms a

____

119. In female cockroach, anterior part of the genital pouch contains ____________.

120. In both the sexes of cockroaches, the 10th segment bears a pair of jointed filamentous structure called ____________ .

121. Among the structures given in the box, which one is used for storing food?

Crop, Gastric caeca, Gizzard, Oesophagus

____

122. In the digestive system of cockroach, gastric caecum is present at the junction of midgut and hindgut which secrete digestive juice.
(*Rewrite the statement by changing the underlined word.*)

____

123. How many Malpighian tubules are present in cockroach?

____

124. Identify the missing parts labelled as *A* to *E* in the given figure of alimentary canal of cockroach.

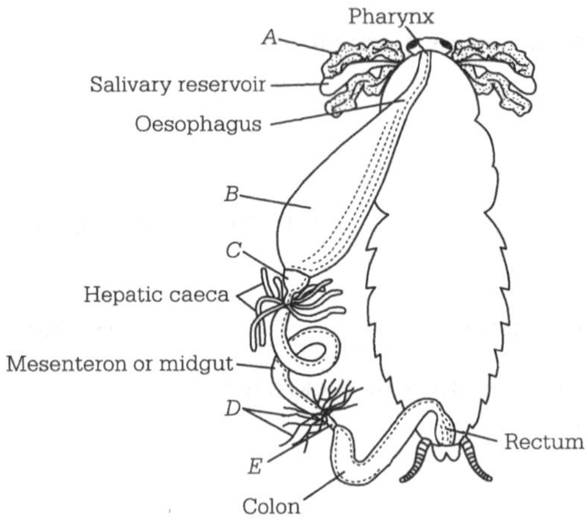

A. ____________________

B. ____________________

C. ____________________

D. ____________________

E. ____________________

125. Write the correct sequence of atleast 7 structures in the alimentary canal of cockroach starting from mouth.

____

126. Thin Malpighian tubules in cockroach are present at the junction of

____

127. Which part of the cockroach helps in the removal of excretory products from the haemolymph?

____

128. Identify the parts labelled as *A*, *B* and *C* in the given figure of open circulatory system of cockroach.

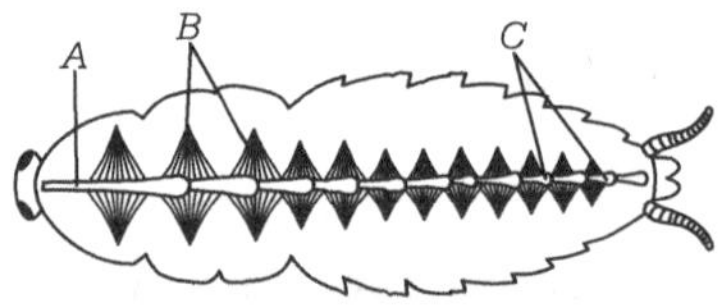

A. ____________________

B. ____________________

C. ____________________

129. Blood of a cockroach contains erythrocytes and plasma.

True ☐ False ☐

130. Heart of cockroach consists of elongated muscular tube lying along mid dorsal line and ____________________ (caecum/abdomen).

131. Complete the flowchart of blood circulation in cockroach.

Blood from sinus ⟶ *A* ⟶ Pumped anteriorly ⟶ *B*

A. ____________________ B. ____________________

132. The respiratory system of the cockroach consists of a pair of lungs that open through 10 pairs of small holes called spiracles present on the lateral side of the body.
(*Rewrite the statement by changing the underlined word.*)

____

133. Tracheal tubes in cockroach are sub-divided into tracheoles which carry ________ from the air to all the parts. The opening of spiracles is regulated by the ______________ .

134. Exchange of gases takes place in cockroaches by the process of ______________.

135. Malpighian tubules in the cockroach are lined by

Glandular and ciliated cells, Cuboidal and ciliated cells, Columnar and glandular cells, Glandular and cuboidal cells

____

136. In addition to the Malpighian tubules, excretion of waste products in cockroach occurs by three other structures *viz,*______________________________

137. The nervous system of cockroach consists of a series of fused, segmentally arranged ganglia joined by paired longitudinal connectives on the ventral side.

True ☐ False ☐

138. In the head region of the cockroach, brain is represented by ______________

139. Choose the odd one out related to sense organ pair in cockroach.

Antennae and eyes, Maxillary palp and labial palps, Antennae and anal cerci, Mandible

____

140. The position of compound eyes of cockroaches with respect to head is ____________ .
Each eye consists of about ____________ ommatidia.

141. In cockroach, with the help of ommatidia it can receive several images of an object. This kind of vision is known as nocturnal vision.
(*Rewrite the statement by changing the underlined word.*)

____

142. The vision of cockroach is
(i) less sensitive with high resolution
(ii) more sensitive with less resolution

____

143. Name the accessory gland in 6th-7th abdominal segments of male cockroach.

____

144. Match the following columns with reference to cockroach and choose the correct option.

	Column I (Body parts of cockroach)		Column II (Positions in the body)
A.	Pair of testes	1.	6th segment
B.	Pair of spermatheca	2.	2nd-6th abdominal segment
C.	Two ovaries	3.	4th-6th abdominal segment
D.	Total abdominal segments	4.	10th segment

A. ____________ B. ____________

C. ____________ D. ____________

145. Answer the following.

(i) The external genitalia of cockroach is called ____________.

(ii) Chitinous asymmetrical structure surrounding the male gonopore is ____________.

146. Answer the following.

(i) In male cockroaches, sperms are stored in which part of the reproductive system?

____

(ii) The sperms are transferred through ____________

147. The characteristic of female cockroach is one pair of spermatheca present in the 6th segment and open in genital chamber.

True ☐ False ☐

148. (i) In cockroach, 14-16 fertilised eggs are stored in ____________

(ii) The number of ovarioles in each ovary is ____________

149. The development of *Periplaneta americana* is <u>holometabolous</u>.
(*Rewrite the statement by changing the underlined word.*)

____

150. The number of moultings through which the nymphs of cockroaches reach the adult form is ____________.

151. What external changes are visible after the last moult of a cockroach nymph?

____

152. Refer to the diagram of male cockroach's reproductive system and identify the missing labelled parts as *A* to *E*.

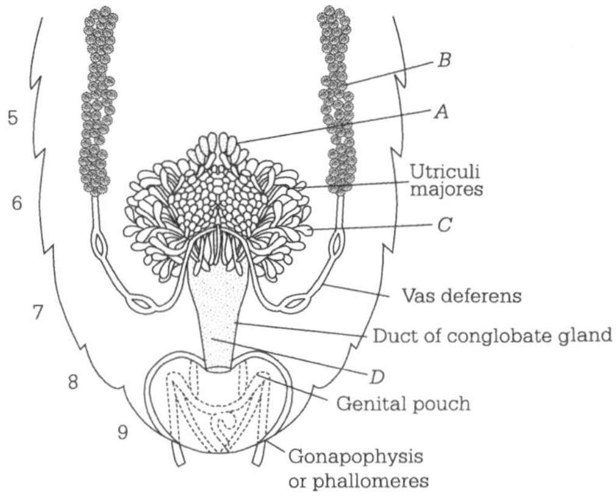

A. ______________________

B. ______________________

C. ______________________

D. ______________________

153. Identify the missing parts labelled as *A* to *E* in the given diagram of female reproductive system of cockroach.

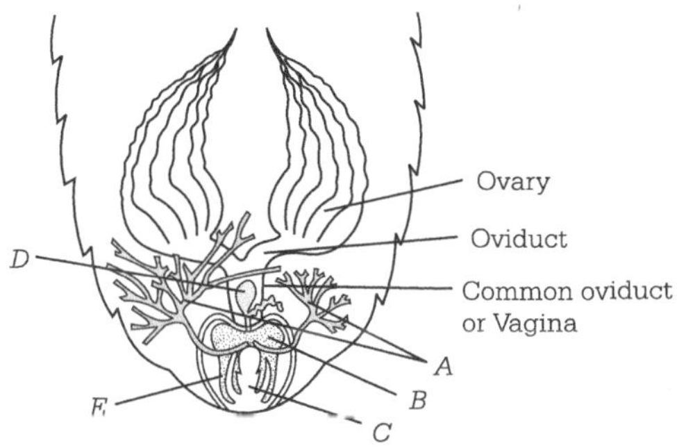

A. ______________________

B. ______________________

C. ______________________

D. ______________________

E. ______________________

Topic 7 : Frog

154. Choose the odd one out related to *Rana tigrina* habits.

Camouflage, Aestivation, Hibernation, Endothermy

____

155. The frogs have the ability to change its colour to hide them from their enemies. This protective colouration is called ______________________ .

156. The skin of frog is slippery and smooth due to the presence of gelatin.

True ☐ False ☐

157. Body of a frog is divisible into __________ and __________ .

158. Refer to the given figure and answer the questions that follows.

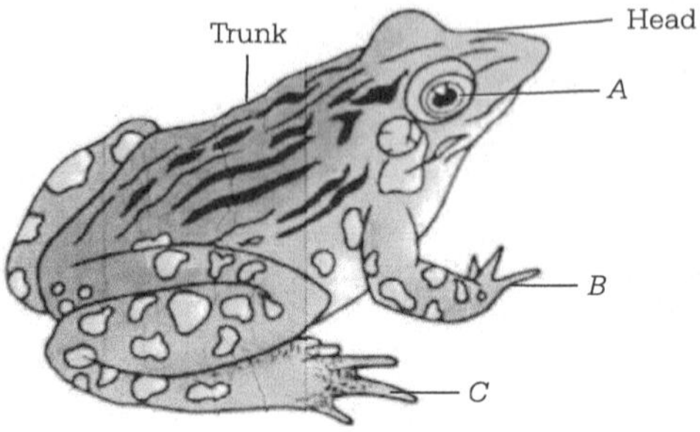

(i) Name the membrane that is found at site *A* and help to protect eyes under water. ______

(ii) Number of digits at site *B* is ______

(iii) Number of digits at site *C* is ______

(iv) Functions of *B* and *C* is ______

159. Male frog can be distinguished from female frog by the presence of
(i) hindlimb ends in the five digits.
(ii) vocal sacs and copulatory pad on first digit of forelimb.

______

160. The alimentary canal of frog is short because frogs are ______

161. State whether the statement True or False.

(i) In frog, excess of the bile juice secreted by the liver is stored in intestine.

True ☐ False ☐

(ii) In frog, for the digestion of food, walls of the stomach secrete pepsin and rennin.

True ☐ False ☐

(iii) In frogs, digested food is absorbed by villi and microvilli in intestine.

True ☐ False ☐

162. Given below is the diagram of internal organs of frog. Identify the missing parts labelled as *A* to *F*.

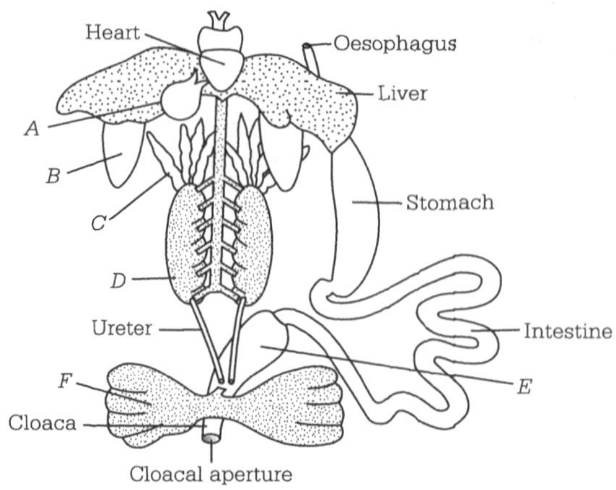

A. ______

B. ______

C. ______

D. ______

E. ______

F. ______

163. Give biological term of the following sentences.

(i) The respiration through lungs in frog. ____________________

(ii) The membrane that covers the heart of the frog. ____________________

164. State, whether the statement True or False.

(i) Frogs respire on land and in the water by two different methods. In water, lungs act as aquatic respiratory organ (cutaneous respiration).

True ☐ False ☐

(ii) In frog, dissolved oxygen in the water is exchanged through the skin by diffusion.

True ☐ False ☐

(iii) On land, the buccal cavity, skin and lungs act as a respiratory organs.

True ☐ False ☐

165. During aestivation and hibernation in frogs, gaseous exchange takes place through

____

166. The blood vascular system of the frog is ____________________ (open type/closed type).

167. How many of the listed components are found in blood vascular system of frog?

Arteries, Veins, Capillaries, Heart, Blood without RBC and WBC

____

168. Three-chambered heart of the frog contains ____________________

169. In frog, a triangular structure called sinus venosus joins the <u>left atrium</u>. It receives blood through major veins called vena cava.
(*Rewrite the statement by changing the underlined word.*)

____

170. Right atrium receives blood through the major veins called ____________________.

171. In frog, ventricle opens into sac-like conus arteriosus on the ____________________ side of heart.

172. In frog, blood from the heart is carried to all parts of body by arteries. The ____________________ collect blood from different parts of the body to the heart.

173. In frog, a special venous connection is present between liver and ____________________

174. Complete the following analogy given below.

Liver and intestine : Hepatic portal system : : Kidney : ____________________

175. The lymph of frog lacks ____________________ .

176. Excretory system of the frog consists of ____________________

177. Choose the right word.

(i) In male frog, ureters act as ____________________ (urinogenital ducts/cloaca).

(ii) Structural and functional unit of kidney in the frog ____________________ (nephrons/bidder's canal).

(iii) In frogs, cloaca is an opening of ____________ (excretory ducts only/excretory and reproductive ducts)

(iv) The frog is ____________ (ureotelic animal/uricotelic animal).

178. The system for control and coordination in frogs comprises

____

179. The number of pairs of cranial nerves arising from the brain of frog is ____________

180. Fill the blanks for different parts of the brain in frog.

(i) The three major components of the forebrain of

____

(ii) The midbrain of the frog is characterised by a pair of

____

(iii) Hindbrain of a frog consists of____________

181. Which of the following pairs is not correctly matched?
(i) Hearing – Tympanum with external ears
(ii) Touch – Sensory papillae
(iii) Balancing – Ears
(iv) Vision – Simple eyes

____

182. In frogs, sensory papillae, taste buds and nasal epithelium are ____________, while eyes and internal ears are well-organised structures.

183. Match the following columns with reference to frog.

	Column I		Column II
A.	Touch	1.	Nasal epithelium
B.	Smell	2.	Foramen magnum
C.	Cranial nerves	3.	Sensory papillae
D.	Medulla oblongata	4.	Peripheral nervous system

A. ____________ B. ____________
C. ____________ D. ____________

184. The number of vasa efferentia that arises from testes in frog's male reproductive system is

____

185. How many of the listed products in male frogs are passed through cloaca?

Sperms, Urine, Faecal matter, Blood

____

186. Fill the correct route for the passage of sperms in male frogs.

Testes → *A* → Kidney → *B* → Urinogenital duct → *C*

A. ____________ B. ____________

C. ____________

187. Identify the missing parts labelled as *A*, *B*, *C*, *D* and *E* in the given figure of male reproductive system of frog.

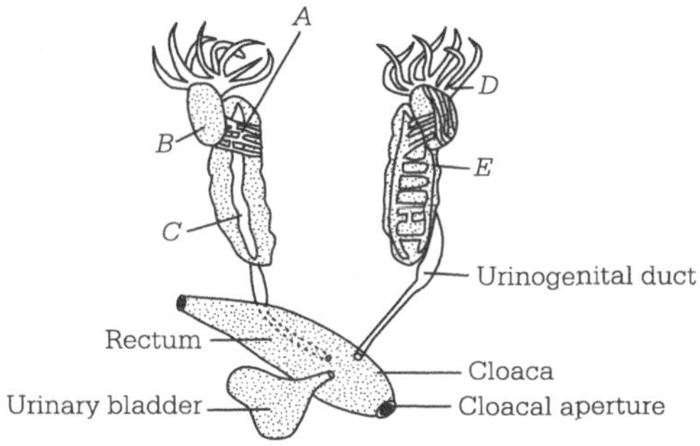

A. ____________

B. ____________

C. ____________

D. ____________

E. ____________

188. Identify the missing parts labelled as *A* to *E* in the following figure of female reproductive system of frog.

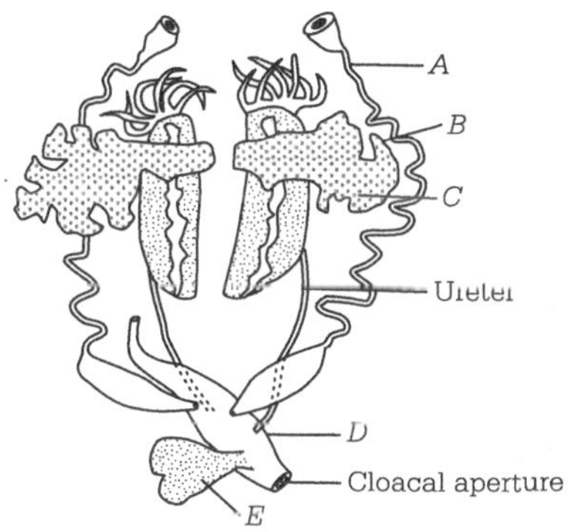

A. ____________

B. ____________

C. ____________

D. ____________

E. ____________

189. In female frogs, ovaries are not connected with kidney functionally.

True ☐ False ☐

190. A mature female egg lays how much ova at a time? ____________

191. Fill the blanks for frog.

(i) Fertilisation is ____________ (external/internal).

(ii) Development involves larval stage called ____________

(iii) Larva undergoes ____________ to form the adult.

(iv) Edible part of frog body in some countries is ____________

Answers

1. Tissues
2. 4, 2
3. Structure and function
4. *A* – Connective tissue, *B* – Muscular tissue, *C* – Neural tissue, *D* – Ectoderm or endoderm, mesoderm
5. Epithelial tissue
6. (i) Compound epithelium, (ii) Simple epithelium
7. Squamous, cuboidal and columnar epithelia.
8. *A* - Squamous epithelium, *B* - Simple cuboidal epithelium, *C* - Simple columnar epithelium, *D* - Columnar cells bearing cilia
9. A-2-M, B-3-N, C-4-K, D-1-L
10. (i) Simple squamous epithelium
    (ii) Simple cuboidal epithelium and columnar epithelium,
    (iii) Ciliated epithelium
11. (i) Cuboidal epithelium,
    (ii) Ciliated columnar epithelium
    (iii) Squamous epithelium
    (iv) Columnar epithelium
12. Glandular, unicellular and multicellular
13. *A*– (i) Multicellular, (ii) Cluster of cells, (iii) Cells of salivary gland.
    B– (i) Unicellular, (ii) Isolated cells, (iii) Goblet cells of alimentary canal.
14. Ciliated epithelium
15. Hormones
16. Oil and milk glands of humans
17. 3 statements (i, iii and iv) are related to compound epithelium. It has very limited role in secretion and absorption.
18. Exocrine and endocrine glands.
19. Endocrine glands
20. 6 (All of them are secretions of exocrine glands except hormones).
21. (i) True, (ii) True
22. Cell junctions
23. A-2, B-3, C-1
24. Blood
25. (i) Fibres
    (ii) Bone, adipose and cartilage
26. Loose connective tissue
27. Areolar tissue
28. Adipose tissue
29. (i) Dense connective tissue
    (ii) Dense regular tissue
    (iii) Dense irregular tissue
30. A-2, B-4, C-1, D-3
31. *X*-Areolar tissue, *A*-Macrophages, *B*-Fibroblast, *C*-Collagen fibres, *D*-Mast cells
32. *X*-Adipose tissue, *A*-Fat storage area, *B*-Nucleus, *C*-Plasma membrane
33. *A*-Dense irregular connective tissue, *B*-Dense regular connective tissue
34. Cartilage
35. Tip of nose, outer ear, joints between adjacent bones of vertebral column.
36. Bone
37. Osteoclast
38. Osteocytes, lacunae
39. Bone marrow
40. *A*–Collagen fibres, *B*–Cartilage cell (chondrocytes)
41. Blood
42. A–3, B–1, C–2, D–4
43. Myofibrils
44. A-1-M, B-3- L, C-2-K
45. (i) Sarcoplasm
    (ii) Sarcolemma
    (iii) Contractility, electrical excitability
46. *A* – Multinucleate, *B* – Uninucleate, *C* – Non-striated and spindle-shaped, *D* – Striated, Cylindrical with intercalated discs, *E* – Voluntary, *F* – Involuntary, *G* – Very rapid contraction and soon get fatigued, *H* – Rapid contraction and never get fatigued, *I* – Present attached to skeletal bones, *J* – Wall of internal organs such as the blood vessel, stomach and intestine
47. True
48. Neural tissue
49. Neuron
50. Neuroglial cell
51. Brain and spinal cord
52. *A*-Axon, *B*-Cell body with nucleus, *C*-Dendrite, *D*-Neuroglea
53. Heart
54. True
55. Earthworm
56. Boring, swallowing
57. Worm castings
58. *Pheretima* and *Lumbricus*
59. 100-120 segments
60. A dark, median mid dorsal line (dorsal blood vessel), presence of genital pores.
61. True
62. Prostomium
63. Peristomium
64. Clitellum
65. Preclitellar, clitellar and post-clitellar.
66. Figure '*X*' is of lateral view of the body showing mouth opening. *A*-Peristomium, *B*-Pro-stomium, *C*-Ring of setae, *D*-Mouth opening
67. The figure '*X*' is of ventral view of body of earthworm. *A*-Female genital aperture, *B*-Male genital aperture, *C*-Genital papillae, *D*-Clitellum, *E*-Annuli
68. A-3- L, B-2- K, C-1-M,
69. (i) True
    (ii) True
    (iii) False. Intestine is present in the clitellar and post-clitellar segments.
70. Setae
71. Non-cellular cuticle → Epidermis → Circular muscle layer → Longitudinal muscle layer → Coelomic epithelium

72. *A* – Mouth, *B* – Oesophagus, *C* – Gizzard, *D* – Stomach, *E* – Caecum, *F* – Lymph gland, *G* – Typhlosole

73. (i) Gizzard – 8th and 9th
(ii) Intestine – 15th to last
(iii) Stomach – 9th-14th
(iv) Buccal cavity 1st-3rd
(v) Oesophagus – 5th-7th
(vi) Pharynx – 3rd and 4th
(vii) Typhlosole – 26th-35th

74. (i) Increases the effective area of absorption in the intestine.
(ii) Grinding soil particles and decaying leaves
(iii) Neutralise the humic acid present in humus.

75. Decaying leaves and organic matter mixed in the soil.

76. Closed

77. 2, i.e. (ii) and (iii). Earthworm's blood does not contain RBCs

78. Haemoglobin, phagocytic, 4th-6th.

79. 7th, 9th, 12th and 13th

80. General moist body surface/skin

81. *A*-Dorsal vessel, *B*-Lateral hearts, *C*-Lateral oesophageal hearts, *D*-Ventral vessel, *E*-Anterior loops

82. A-3-K, B-2-L, C-1-L

83. (i) Coelomic chamber
(ii) Ureotelic
(iii) Nephridia

84. 7 labels are correct. I and F are incorrect because.
I is blood glands and F is septal nephridia.

85. (i) Paired
(ii) Solid
(iii) Ventral

86. Nerve ring

87. (i) True, (ii) Touch, taste, light and vibration

88. Hermaphrodite (bisexual)

89. *A*-Spermathecae, *B*-Seminal vesicles, *C*-Ovary, *D*-Vas deferens, *E*-Accessory gland, *F*-Prostate gland

90. 17th and 19th segments

91. 4, spermatozoa

92. 12th and 13th segments.

93. Ovaries, female

94. (i) Spermatophores
(ii) Cocoon

95. Direct

96. (i) Earthworms make burrows in the soil and make it porous which helps in respiration and penetration of the developing plant roots.

97. Vermicomposting

98. Insecta, Arthropoda

99. (i) Omnivorous, (ii) Dioecious, (iii) Coelomate, (iv) Serious pest and vector of several diseases.

100. Head, thorax and abdomen

101. Chitin

102. Sclerites

103. Sternites

104. True

105. *A* – Prothoracic leg, *B* – Mesothorax, *C* – Metathorax, *D* – Tegmina, *E* – Anal cerci

106. 6 segments

107. Flexible neck

108. *A* – Mandible, *B* – Maxilla, *C* – Labium, *D* – Labrum

109. Tongue

110. Chewing and biting

111. *A*–Labrum, *B*–Mandible, *C*–Hypopharynx, *D*–Maxilla, *E*–Labium

112. 3 (Prothorax, mesothorax and metathorax).

113. Mandibles

114. (i) All the three thoracic segments
(ii) Mesothorax, metathorax

115. 10

116. 3 statements, i.e. (i), (ii) and (iii) are correct. Forewings are not used in flight.

117. Boat-shaped

118. Genital pouch

119. Gonopore

120. Anal cerci

121. Crop

122. Foregut and midgut

123. 100-150

124. *A*–Salivary gland, *B*–Crop *C*–Gizzard, *D*–Malpighian tubules, *E*–Ileum

125. Pharynx → Oesophagus → Crop → Gizzard → Ileum → Colon → Rectum

126. Midgut and hindgut

127. Malpighian tubules

128. *A*–Anterior aorta, *B*–Alary muscles, *C*–Chambers of heart

129. False. Blood of a cockroach contains plasma and haemocytes.

130. Abdomen

131. *A*–Heart, *B*–To sinuses again

132. A network of trachea

133. Oxygen, sphincters

134. Diffusion

135. Glandular and ciliated cells

136. Fat bodies, nephrocytes, urecose glands

137. True

138. Supraoesophageal ganglion

139. Mandible is odd one out.

140. Dorsal, 2000 hexagonal

141. Mosaic vision

142. (ii) More sensitive with less resolution

143. Mushroom gland

144. A–3, B–1, C–2, D–4

145. (i) Gonapophysis,
(ii) phallomere

146. (i) Seminal vesicles
(ii) Spermatophores

147. True

148. (i) Oothecae, (ii) 8

149. Paurometabolous

150. 13

151. Both forewings and hindwings develop

152. *A*–Phallic gland
B–Testis

C–Seminal vesicle
D–Ejaculatory duct

153. *A*–Collaterial glands, *B*–Genital chamber, *C*–Vestibulum, *D*–Spermatheca, *E*–Gonapophyses

154. Endothermy is odd one out because frogs do not have constant body temperature.

155. Mimicry

156. False. The skin of frog is slippery and smooth due to the presence of mucus.

157. Head, trunk

158. (i) Nictitating membrane
(ii) 4
(iii) 5
(iv) Swimming, leaping burrowing

159. (ii) Male frog can be distinguished from female frog be the presence of vocal sacs and copulatory pad on the first digit of the forelimb

160. Carnivores

161. (i) False. In frog, excess of the bile juice secreted by the liver is stored in gall bladder.
(ii) False. In frog, for the digestion of food, walls of the stomach secrete HCl and gastric juice.
(iii) True

162. *A*–Gall bladder, *B*–Lung, *C*–Fat bodies, *D*–Kidney, *E*–Rectum, *F*–Urinary bladder

163. (i) Pulmonary respiration
(ii) Pericardium

164. (i) False. Frogs respire on land and in the water by two different methods. In water, skin act as aquatic respiratory organ (cutaneous respiration.)
(ii) True
(iii) True

165. Skin

166. Closed type

167. 4(All of them except blood without RBC and WBC). Frog blood contains RBCs and WBCs.

168. Two atria and one ventricle

169. Right atrium

170. Vena cava

171. Ventral

172. Veins

173. Intestine

174. Renal portal system

175. RBCs and few proteins

176. Paired kidney, ureters, urinary bladder, cloaca

177. (i) Urinogenital ducts
(ii) Nephrons
(iii) Both of them
(iv) Ureotelic animal

178. Highly evolved neural system and endocrine glands

179. 10 pairs

180. (i) Olfactory lobes, unpaired diencephalon and paired cerebral hemispheres
(ii) Optic lobes,
(iii) Cerebellum and medulla oblongata

181. (i) is not correct match because Hearing–Tympanum with internal ears.

182. Cellular aggregations around nerve endings

183. A–3, B–1, C–4, D–2

184. 10-12

185. 3(Urine, sperm, faecal matter).

186. Testes → Vasa efferentia (*A*) → Kidney → Bidder's canal (*B*) → Urinogenital duct → Cloaca (*C*)

187. *A*–Vasa efferentia, *B*–Testis, *C*–Adrenal gland, *D*–Fat body, *E*–Kidney

188. *A*–Oviduct, *B*–Ovary, *C*–Ova, *D*–Cloaca, *E*–Urinary bladder

189. True

190. 2500-3000 ova.

191. (i) External
(ii) Tadpole
(iii) Metamorphosis
(iv) Muscular legs

CHAPTER EIGHT

CELL : THE UNIT OF LIFE

Topic 1 : Introduction and Cell Theory

1. Tissue is the fundamental structural and functional basic unit of all living organisms. (*Rewrite the statement by changing the underlined word.*)

   ____

2. Living organisms composed of a single cell are called as ____________ and organisms composed of many cells are called as ____________ .

3. Identify thc odd onc out from the following series with respect to their cellularity.

Amoeba, Euglena, Pandorina, Chlorella, Streptobacillus

   ____

4. Name the scientist who first saw and described a living cell.

   ____

5. Who is known as Father of Cytology? ____________

6. Discovery of ____________ revealed the minute structural details of the cell.

7. ____________ (Von Mohl/Matthias Schleiden) examined a large number of plants, observed that all plants are composed of different kinds of plant cells.

8. Theodore Schwann studied different types of fungal cells. (*Rewrite the statement by changing the underlined word.*)

   ____

9. What is the unique character of the plant cells according to Schwann?

____

10. The bodies of animal and plants are composed of __________ and __________.

11. Match the following columns.

	Column I (Scientists)		Column II (Discoveries made by)
A.	Schleiden and Schwann	1.	*Omnis cellula-e-cellula*
B.	Rudolf Virchow	2.	Protoplasm
C.	W Flemming	3.	Cell theory
D.	Purkinje	4.	Chromatin

A.__________ B. __________ C. __________. D__________

12. What is the meaning of *omnis cellula-e-cellula*?

____

13. Schleiden and Schwann gave final shape of cell theory, according to which all living organisms are composed of cells and arises from pre-existing cells.

True ☐ False ☐

14. Cell theory also known as cell kinase theory.

True ☐ False ☐

15. Regarding cell theory what is the common in between the following ?

RBCs, Seive tubes, Viruses, Some coenocytic organism

____

16. __________ (Nucleus/Cytoplasm) is a dense membrane bound structure which lies inside the cell.

17. A cell with membrane bound nuclei is known as __________

18. In both prokaryotic and eukaryotic cells, a semifluid matrix called mesosomes occupies the maximum volume of the cell.

(*Rewrite the statement by changing the underlined word.)*

____

19. Identify the part of cell on the basis of given characters.
   (i) Main arena of cellular activities in all living cells.
   (ii) Various reactions take place in it.

____

20. How many of these organelles is/are present in prokaryotic cells?

Endoplasmic reticulum, Golgi complex, Lysosomes, Mitochondria, Ribosomes, Microbodies

____

21. Identify the odd one out from the given series.

Chloroplast, Mitochondria, Lysosomes, Rough ER.

____

22. ______________ (Centrosomes/Mesosomes) is a non-membrane bound organelle present in animal cells, which helps in cell division.

23. The largest isolated single cell is the egg of kiwi.

(*Rewrite the statement by changing the underlined word.*)

____

24. Match the following columns.

	Column I (Types of cell)		Column II (Approximately size of cells)
A.	*Mycoplasma*	1.	3 – 5 μm
B.	Nerve cell	2.	7.0 μm
C.	Bacteria	3.	0.3 μm
D.	Red blood cell	4.	30– 90 cm

A. ____________________ B. ____________________

C. ____________________ D. ____________________

25. Match the following analogy with respect to size.

Smallest cell : *Mycoplasma* : : Longest cell : ________________ .

26. ____________________ is the smallest human cell.

27. Write the shape of the cells from given diagram.

A. 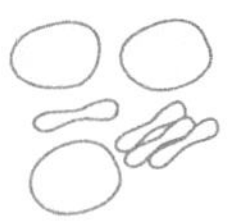 (Red blood cells)

____

B.  (White blood cells)

____

C. 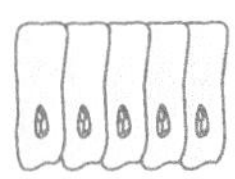 (Columnar epithelial cells)

____

D. 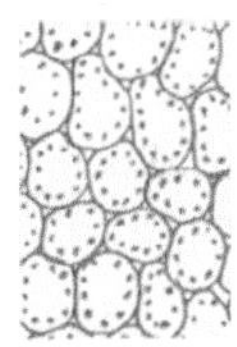 (Mesophyll cells)

____

E. 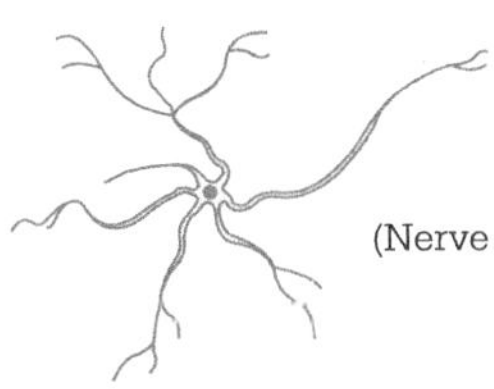 (Nerve cell)

____

F.  (A tracheid)

____

28. ________ (Pleomorphic/Homomorphic) cells change their shape constantly.

29. The size of the cells are affected by ________, i.e. the ratio between the volume of nucleus and the cytoplasm.

30. Match the following columns.

	Column I (Shapes of the cells)		**Column II** (Examples)
A.	Flattened	1.	Secretory cell
B.	Cubical	2.	Epidermis
C.	Filament	3.	*Spirogyra*
D.	Comma	4.	Smooth muscle cell

A. ________ B. ________

C. ________ D. ________

Topic 2 : Prokaryotic Cell

31. How many of the following characters are related to prokaryotic cell?

(i) Smaller
(ii) Multiply more rapidly
(iii) PPLO is a prokaryotic
(iv) Have defined nucleus
(v) Contains peroxisomes

____

32. Complete the following diagrammatic analogy with respect to structure of bacteria.

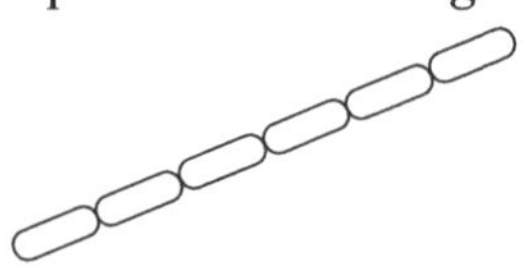 : Rod shape : :  : ________

33. Find the odd one out from the following series.

Staphylococcus, Mycoplasma, Bacillus, Spirillum

____

34. In prokaryotes, genetic material is ________, i.e. not enveloped by a ________ membrane.

35. Many bacterial cells contain small circular DNA outside the genomic DNA known as carboxysomes.
(*Rewrite the statement by changing the underlined word.*)

____

36. How many of the given characters are related to plasmid?

(i) Resistance to antibiotic
(ii) Monitor bacterial transformation
(iii) Known as extrachromosomal
(iv) Only one plasmid is found

____

37. Answer the following questions, considering the given below.

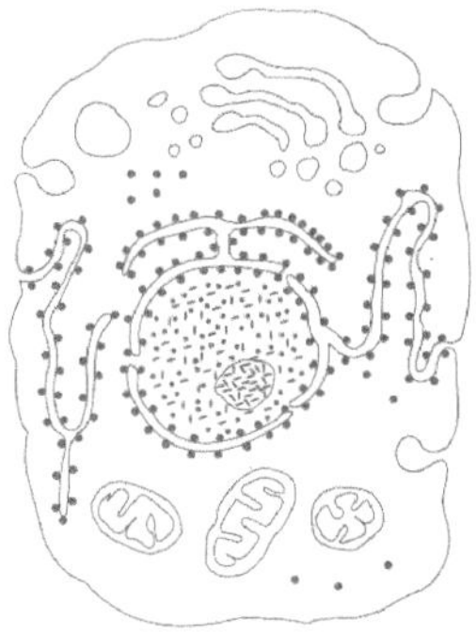

(i) The cell represents a ____________________ (eukaryotic/prokaryotic) cell.

(ii) The size of this cell range between 1 µm - 9 µm.

True ☐ False ☐

38. The bacterial cell envelope consists of a tightly bound ____________________ (two/three) layered structure.

39. Rearrange the following sequence of layer of bacterial cell envelope from outside to inside.

Cell wall → Plasma membrane → Glycocalyx.

____

40. Bacteria can be classified into ____________ groups on the basis of the differences in the cell envelope.

41. The presence of <u>mucoglycan</u> in the cell envelope is the basis of the classification of bacteria.
(*Rewrite the statement by changing the underlined word.*)

____

42. Name the scientist who developed staining procedure of bacteria.

____

43. Bacteria that take up the Gram stain are known as ____________________ others that do not are called ____________________.

44. Complete the following analogy with respect to Gram staining.

Gram positive : ____________________ : : Gram negative : *Escherichia coli*

45. The composition and thickness of glycocalyx is same among different bacteria.

True ☐ False ☐

46. ____________________ is the loose sheath of glycocalyx present on bacterial cell.

47. When glycocalyx is tough and thick it is called ____________ (chondrions/capsule).

48. How many of the following characters are not related to the bacterial cell wall?

(i) Determines the shapes of the cell (ii) Provides structural support
(iii) Helps in bursting of cell

____

49. The ________________________ (ribosome/plasma membrane) of prokaryotes is structurally similar to the eukaryotes.

50. ____________________(Basal body/Mesosome) is formed by the extensions of plasma membrane into the cell.

51. How many of these given structures are the extensions of mesosomes?

Hook, Pili, Lamellae, Tubules, Filament, Form of vesicles

____

52. How many of the following processes are carried out by mesosome?
   (i) Cell wall formation
   (ii) DNA replication
   (iii) Respiration
   (iv) Secretion processes
   (v) Increase surface area of plasma membrane
   (vi) Enzymatic activity
   (vii) Reproduction

____

53. In PPLO, chromatophores contain pigments which are membranous extensions into the cytoplasm.
(*Rewrite the statement by changing the underlined word.*)

____

54. Bacterial cell contains one or more ________________ for cellular locomotion.

55. Flagellum is a thin filamentous extenstion that arise from the cytoplasm and penetrate cell wall.

True ☐ False ☐

56. In bacteria, flagellum is composed of three parts, i.e. ________________, ____________ and ______________________________.

57. Which is the longest structural component of the flagella? ________________________

58. Flagellum is formed by which protein? ____

59. Match the following columns.

	Column I (Parts of bacterial cell)		Column II (Characteristics)
A.	Mesosome	1.	Bristle-like fibres
B.	Pili	2.	Elongated tubular structure
C.	Fimbriae	3.	Chondriod

A. ____________________ B. ____________________

C. ____________________

60. In some bacteria, ______________ (fimbriae/pili) attach the bacteria to rocks and also to the host tissues.

61. Pili is made up of protein <u>leukocidin.</u>
(*Rewrite the statement by changing the underlined word.*)

____

62. In prokaryotes, ribosomes are associated with the ____________________ (plasmid/ plasma membrane) of the cell.

63. Write the size of ribosomes in prokaryotes.______________________________

64. In prokaryotes, the 70S ribosome is made up of two subunits ______________ and ____________________ .

65. The prokaryotic ribosome is made up of 40% *r*RNA and 60% ribosomal protein.

True ☐ False ☐

66. Several ribosomes may attach to a single *m*RNA and form a chain called

____.

67. Name the structure which store reserve material in the cytoplasm of prokaryotic cell.

____

68. Give a common biological term for the following series.

Phosphate granules, Cyanophycean granules, Glycogen granules

____

69. Name two bacteria in which gas vacuoles are found? ______________________

70. Match the following columns.

	Column I		Column II
A.	Polyribosome	1.	Inclusion body
B.	Carboxysome	2.	Murein
C.	Nucleoid	3.	Translate *m*RNA into protein
D.	Peptidoglycan	4.	Teichoic acid
E.	Gram (+ ve) bacteria	5.	Nuclear material

A. ____________________ B. ____________________

C. ____________________ D. ____________________

E. ____________________

71. <u>Phycobilisomes</u> contain enzymes involved in carbon dioxide fixation.
(*Rewrite the statement by changing the underlined word.*)

____

Topic 3 : Eukaryotic Cell

72. Which type of cell is found in all protists, plants, animals and fungi?

____

73. You have given a cell, which shows following properties.

(i) compartmentalisation of cytoplasm (ii) membrane bound organelle
(iii) nucleus with nuclear envelope (iv) cytoskeletal structure

Identify the cell.

____

74. Eukaryotic genetic material is organised into chromosomes.

True ☐ False ☐

75. The _______________ possesses cell walls, large central vacuole and plastids.

76. Name the organelle which is present in animal cell, but absent in plant cell.

____

77. Consider the following figure of plant cell, identify the organelles *A*, *B*, *C* and *D*.

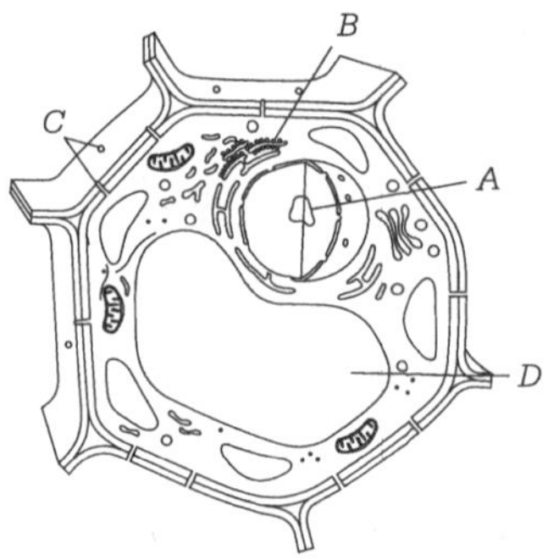

A. _______________

B. _______________

C. _______________

D. _______________

78. Consider the following figure of animal cell, identify which alphabets represent the organelle centriole, microvilli and ribosomes?

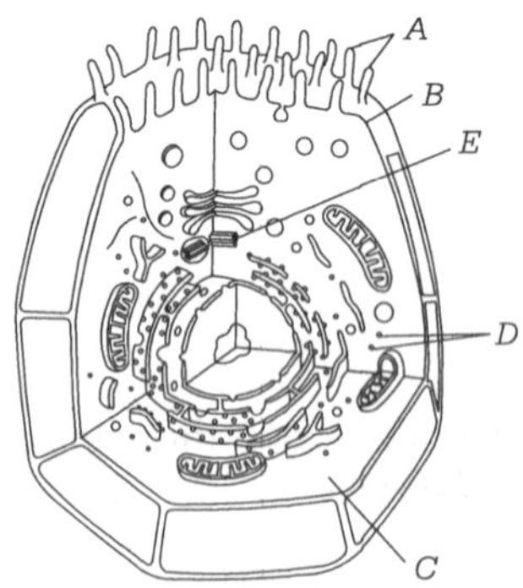

____

79. The detailed structure of the cell membrane was studied only after the discovery of the compound microscope.
(*Rewrite the statement by changing the underlined word.*)

____

80. Match the following columns.

	Column I (Components of cell membrane)		Column II (Compositions of cell membrane)
A.	Protein	1.	1 – 2%
B.	Lipid	2.	58 – 59%
C.	Carbohydrate	3.	40%

A. ______________ B. ______________
C. ______________

81. ______________ are the major lipids that are arranged in a bilayer membrane of eukaryotic cell.

82. In bilayer membrane, the lipids polar head lies towards the inner side and tails towards the outer side.

True ☐ False ☐

83. Complete the following analogy with respect to membrane structure.

Hydrophobic : Tail of lipid : : Hydrophilic : ______________

84. Lipid layer in plasma membrane protects the non-polar tail ______________ hydrocarbon from the aqueous environment.

85. The lipid component of the membrane mainly consists of stearic acid.
(*Rewrite the statement by changing the underlined word.*)

____

86. The membrane proteins are of two types, i.e. ______________ and ______________ protein.

87. Name the protein which lie on the surface of the membrane.

____

88. The ______________ proteins are partially or totally buried in the membrane.

89. Integral membrane proteins are easy to extract.

True ☐ False ☐

90. The peripheral membrane constitutes about ________ % of total protein content of the membrane.

91. Name the main carbohydrate group present in the cell membrane.

____

92. Singer and Nicolson proposed a cell membrane structure model known as unit membrane model.

True ☐ False ☐

93. According to the Fluid Mosaic Model (FMM), the quasi-fluid nature of lipid enables the ______________ movement of protein within the overall bilayer.

94. The <u>rigidity</u> of membrane is measured by the ability of protein movement within the bilayer.
(*Rewrite the statement by changing the underlined word.*)

____

95. Consider the following diagram of structure of plasma membrane, answer the given questions.
(i) Identify parts *A*, *B*, *C* and *D*.

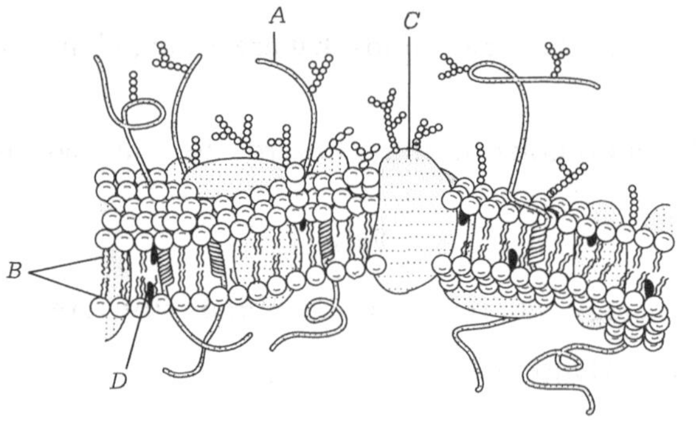

A. ______________
B. ______________
C. ______________
D. ______________

(ii) Which model is represented by this figure? ______________

96. ______________ is a cell membrane protein which allows facilitated diffusion between cell and extracellular fluid.

97. Glycolipids and cholesterol are usually present in the membrane of the plant cells.

True ☐ False ☐

98. How many of the given functions are related to plasma membrane?

(i) Formation of ribosomes
(ii) Formation of intercellular junctions
(iii) Secretion
(iv) DNA replication
(v) Endocytosis
(vi) Cell division
(vii) Cell growth

____

99. The plasma membrane is <u>permeable</u> to all molecules present on either side of it.
(*Rewrite the statement by changing the underlined word.*)

____

100. In ______________ (active/passive) transport, many molecules can move briefly across the membrane without any requirement of energy.

101. Name the solutes which can move across the membrane *via* simple diffusion.

____

102. In _______________ process, solutes move from higher concentration to the lower.

103. Movement of water from higher concentration to the lower *via* membrane is called _______________ (active transport/osmosis).

104. The polar molecules require a carrier <u>lipid</u> of the membrane to facilitate their transport across the non-polar lipid bilayer.
(*Rewrite the statement by changing the underlined word.*)

____

105. The process of movement of molecules with the help of carrier protein is called _______________ (active transport/facilitated diffusion).

106. Give one example of facilitated diffusion in human body.

____

107. In _______________ transport, molecules and ions move across the membrane against the concentration gradient.

108. A process of transport which require energy in the form of ATP is known as facilitated diffusion.

True ☐ False ☐

109. Give one example through which active transport operates in human body.

____

110. Match the following columns.

	Column I (Functions of plasma membrane)		**Column II** (Features)
A.	Endocytosis	1.	Transported out of cells
B.	Phagocytosis	2.	Intake of liquid particles
C.	Pinocytosis	3.	Intake of solid particles
D.	Exocytosis	4.	Materials are transported into cells

A. _______________ B. _______________

C. _______________ D. _______________

111. The cell wall is a _______________(living/non-living) rigid structure.

112. Name two organisms in which the cell wall functions as an outer covering of plasma membrane. _______________

113. The plasma membrane gives shape to the cell and protects the cell from mechanical damage and infection.

True ☐ False ☐

114. The cell wall helps in cell to cell interaction and transport macromolecules.

True ☐ False ☐

115. How many of the given components are found involved in algal cell wall?

Pectin, Cellulose, Chitin, Galactans, Mannans, Calcium carbonate, Silicic acid

____

116. Match the following columns.

	Column I (Cell walls)		Column II (Chemical constituents)
A.	Fungal cell wall	1.	Alginic acid
B.	Plant cell wall	2.	Cellulose, hemicellulose, pectins
C.	Algal cell wall	3.	Glucosamine, chitin

A. ________________ B. ________________

C. ________________

117. Rearrange the sequence of cell wall layers from inside to outside.

Primary wall, Middle lamella, Secondary wall

____

118. The cell wall of a young plant cell contains secondary wall.
(*Rewrite the statement by changing the underlined word.*)

____

119. The middle lamella is made up of ________________ (cellulose/calcium pectate).

120. The middle lamella ________________ the different neighbouring cells together.

121. The plasmodesmata cannot cross the cell wall and middle lamella.

True ☐ False ☐

122. Name the structure which connects the cytoplasm of neighbouring cells.

____

123. The ripening of fruits occurs due to dissolution of middle lamella. Name the enzyme which is responsible for it. ________________

124. Find the odd one out with respect to organelles related to endomembrane system.

Endoplasmic reticulum, Peroxisomes, Golgi complex, Lysosomes

____

125. The endomembrane systems are considered together because their functions are

________________________________.

126. Name the cell organelle which is present in network or reticulum of tiny tubular structures. ______________________

127. Endoplasmic reticulum divides the intracellular space into two distinct compartments, i.e. ______________ and ______________.

128. Name the organelle which is attached to the outer surface of Endoplasmic Reticulum (ER).______________________

129. Complete the following analogy with respect to cell organelles.

Rough Endoplasmic Reticulum (RER) : Ribosome present : : Smooth Endoplasmic Reticulum (SER) : ______________

130. RER is frequently observed in the cells actively involved in <u>lipid</u> synthesis and secretion. (*Rewrite the statement by changing the underlined word.*)

____

131. The ________ (RER/SER) are extensive and continuous with the outer membrane of the nucleus.

132. Which molecule synthesise in smooth endoplasmic reticulum?

____

133. Answer the following questions considering the following diagram of endoplasmic reticulum.

(i) Identify the labelled parts *A* and *B*.

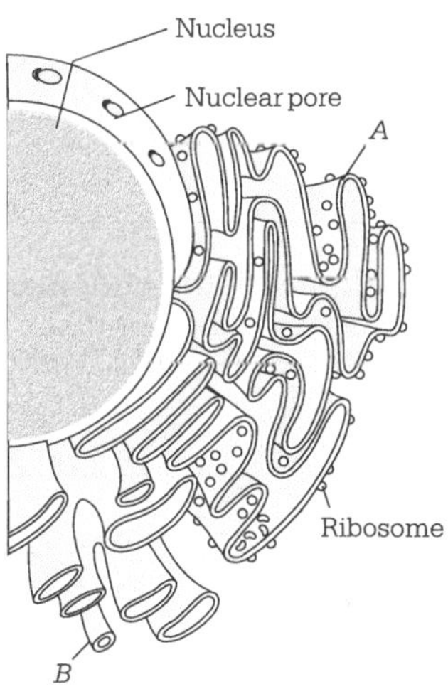

A. ______________________

B. ______________________

(ii) In which part calcium storage and protein pakaging takes place? ______________

134. The ______________ are densely stained reticular structure near the nucleus.

135. The Golgi apparatus consist of many elongated, round or cisternae.

True ☐ False ☐

136. The diameter of cisternae range from ________ to ________ .

137. Who first observed the Golgi apparatus?______________________

138. The cisternae of Golgi apparatus remain stacked perpendicular to each other and present in varied number.
(*Rewrite the statement by changing the underlined word.*)

____

139. Identify the given figure and labelled part.

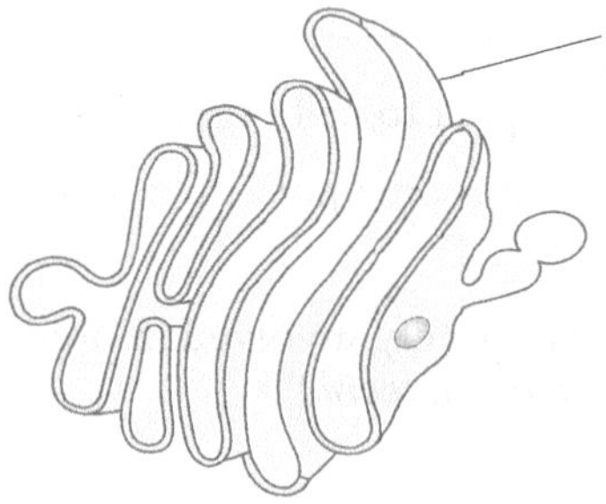

A. _______________

140. The convex cisternae of Golgi apparatus lies towards the _______________ and concave cisternae lies towards the _______________.

141. Complete the following analogy related with cisternae of Golgi apparatus.
Cis : _______________ : : *Trans* : Maturing

142. The *cis* and *trans* faces of the organelle are entirely
(i) different but interconnected
(ii) similar but separated

____

143. The principle function of Golgi apparatus is _______________ of material.

144. The Golgi apparatus remains in close association with the endoplasmic reticulum for
(i) protein and lipid synthesis
(ii) packaging of material

____

145. Materials are packed in the form of tubules in Golgi apparatus.
(*Rewrite the statement by changing the underlined word.*)

____

146. The proteins on the endoplasmic reticulum are modified in the cisternae of the Golgi apparatus.

True ☐ False ☐

147. The modified protein of Golgi apparatus released from _______________ (*cis*/ *trans*) face.

148. How many of the following materials are packed by Golgi apparatus?

Zymogen in pancreatic cells, Lactoprotein in mammary gland cells, Mucus in Goblet, Hormones in endocrine cells

____

149. Golgi apparatus is the important site of formation of ____________________ and ____________________.

150. Lysosomes are formed by the process of packaging in the endoplasmic reticulum.

True ☐ False ☐

151. Lysosomes are membrane bound ____________________ structure.

152. Name the enzyme which is present in isolated lysosomal vesicles. ____________________

153. Give a common term for the following enzymes.

Lipases, Proteases, Carbohydrases

____

154. Lysosomal enzymes are capable of digesting carbohydrate, protein, lipids and nucleic acids.

True ☐ False ☐

155. Lysosomes are also known as suicidal bags because

(i) they cause autoimmune diseases.

(ii) they are capable of digesting the substances of their own cell.

____

156. Hydrolytic enzymes of lysosomes are optimally active at __________(basic/acidic) pH.

157. In electron microscope, you see a membrane bound space in the cytoplasm. What is it?

____

158. How many of the following materials are present in vacuoles?

Glycogen, Water, Sap, Carbon dioxide, Anthocyanin, Permeases, Lipoprotein, Excretory product

____

159. The vacuole is bound by a single membrane called ____________ (cytoplast/tonoplast).

160. The vacuoles can occupy up to <u>20%</u> of the volume of plant cell.
(*Rewrite the statement by changing the underlined word.*)

____

161. The tonoplast facilitates the transport of a number of ions and other materials from higher concentration to the lower concentration.

True ☐ False ☐

162. Mark the sign of less than/greater than/equal (>/</=) in given box.

Concentration of cytoplasm ☐ Concentration of vacuole

163. In *Amoeba*, the ______________________ is important for excretion.

164. The ______________________ contains digestive enzymes in protists.

165. Match the following columns.

	Column I (Types of vacuoles)		Column II (Features)
A.	Food vacuole	1.	Helps in osmoregulation
B.	Contractile vacuole	2.	Contains hydrolases enzyme
C.	Sap vacuole	3.	Found in green bacteria
D.	Gas vacuole	4.	Contains dissolved salts and pigments

A. ______________ B. ______________

C. ______________ D. ______________

166. Mitochondria are easily visible under the microscope without staining.

True ☐ False ☐

167. A ______________________ (senescent cell/meristematic cell) contains high number of mitochondria.

168. The ______________ and ______________ of mitochondria varied from cell to cell.

169. What is the typical size and shape of a mitochondria?

____

170. Each mitochondrion is a ______________ (single/double) membrane bound structure.

171. The outer membrane and the inner membrane of mitochondria divide its lumen distinctly into ______________ aqueous compartments.

172. Answer the following questions, considering the LS of mitochondria given below.

(i) Identify the labelled parts *A*, *B* and *C*.

A. ______________________

B. ______________________

C. ______________________

(ii) Which of the labelled part contains oxysomes? ______________________

173. The inner compartment of mitochondria is filled with a dense homogeneous substance called <u>cristae</u>.
(*Rewrite the statement by changing the underlined word.*)

____

174. The continuous limiting boundary of mitochondria is formed by ________________ (outer membrane/inner membrane).

175. Name the mitochondrial membrane which form number of infoldings called cristae towards the matrix.________________

176. (i) The ________________ (cristae/matrix) increase the surface area of mitochondria.

(ii) The inner mitochondrial membrane bears minute regularly spaced tennis racket-shaped structure known as __________ it is also known as ______________ complex.

(iii) An oxysome consists of three parts, *viz*, ________________ , ________________ and ______________ .

177. Name the organelle which is the site of aerobic respiration. ________________

178. Mitochondria produce cellular energy in the form of __________ , hence they are called of the cell.

179. Name the part of mitochondria which contain single circular DNA, RNA, ribosomes, and the components required for the protein synthesis. ________________

180. Name the type of ribosome found in mitochondria. ________________

181. The mitochondria divide by <u>budding</u>.
(*Rewrite the statement by changing the underlined word.*)

____

182. ____________ are found in all plant cells and in euglenoids. These are easily observed under microscope.

183. The plants possess different colour due to the ______________ present in plastids.

184. Plastids can be of three types, i.e. ________________ , ________________ and ________________ .

185. Match the following columns.

	Column I (Types of plastids)		Column II (Features)
A.	Chloroplasts	1.	Carotenoid
B.	Chromoplasts	2.	Chlorophyll
C.	Leucoplasts	3.	Colourless

A. ________________ B. ________________ C. ________________

186. Identify the odd one out from the following series.

Elaioplast, Rhodoplast, Amyloplast, Aleuroplast

____

187. Complete the following analogy.
Elaioplast : Oils : : ______________ : Protein

188. Give example of two carotenoid pigments that give the parts of the plant a yellow, orange or red colour.________________

189. Majority of the chloroplasts of the green plants are found in the ________________ (epidermis/mesophyll) cells of the leaves.

190. The number of chloroplast in *Chlamydomonas* and mesophyll cells are ____________ and ____________, respectively.

191. Chloroplasts are double membrane bound organelles.

True ☐ False ☐

192. In chloroplast the ________________________ (inner membrane/outer membrane) is relatively less permeable.

193. The space enclosed by the inner membrane of the chloroplast is called <u>thylakoids</u>. (*Rewrite the statement by changing the underlined word.*)

____

194. The thylakoids are the organised flattened membranous sacs present in the ________________ of chloroplast.

195. By considering the given figure of chloroplast, answer the following questions.

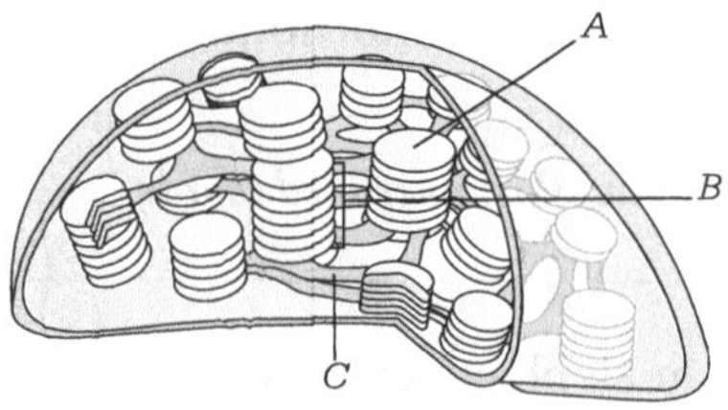

(i) Part *B* are arranged in stacks like piles of coins. What is part *B* ? ________________

(ii) The single unit of part *B* is part *A*. What is part *A*? ________________

(iii) Name the labelled part *C*. ________________

196. The flat membranous tubules connecting the thylakoids of the different grana is called

____.

197. The membrane of the thylakoids encloses a space called a lumen.

True ☐ False ☐

198. The ______________ (thylakoid/stroma) contains enzymes required for the synthesis of carbohydrates and proteins.

199. The chloroplast contains small, double-stranderd circular DNA molecules and <u>peroxisomes</u>. *(Rewrite the statement by changing the underlined word.)*

____

200. Complete the following analogy with respect to ribosome.

Chloroplast ribosome : 70S : : ________________ : 80S

201. ______________ are the granular structures that were first observed under the electron microscope as dense particles.

202. Name the scientist who first observed ribosomes. ______________________

203. Name the organelle which is not surrounded by any membrane and composed of RNA and proteins. ______________________

204. Consider the following diagrams of ribosomes and write the names of cells in which these ribosomes are found?

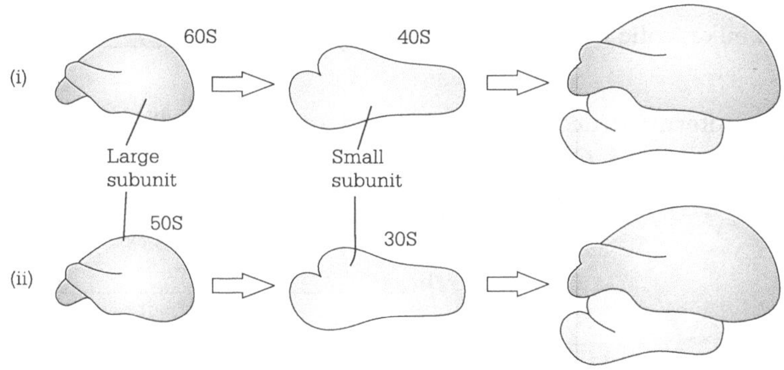

(i) ______________________

(ii) ______________________

205. What is the final coefficient of ribosome shown in figure, in Q 204.

(i) ______________________ (ii) ______________________

206. The S (Svedberg's unit) indirectly measures ________ and __________ of ribosomal units.

207. Ribosomes attached to ER through the specialised protein called________________ .

208. ______________________ (Centrosome/Cytoskeleton) is an elaborate network of filamentous proteinaceous structures present in the cytoplasm.

209. How many given functions are related to cytoskeleton?

Mechanical support, Protein synthesis, Motility, Shape of the cell

____

210. Give a collecting term for the following series.

Microtubules, Microfilaments, Intermediate filaments

____

211. Match the following columns.

	Column I (Cytoskeletons)		**Column II** (Features)
A.	Microtubules	1.	Composed of acidic proteins
B.	Microfilaments	2.	Constituents of spindle fibres, cilia
C.	Intermediate filament	3.	Consists of actin and myosin proteins
D.	Desmin filament	4.	Present in muscle cells

A. ______________________ B. ______________________

C. ______________________ D. ______________________

212. Name the hair-like outgrowths of the cell membrane. ____________

213. Cilia are long structures which work like oars, causing the movement of either the cell or the surrounding fluid.

True ☐ False ☐

214. Flagella are comparatively ____________ and help in cell ____________.

215. The prokaryotic and eukaryotic flagella are structurally ____________ (similar/different).

216. Consider the figure of internal structure of flagella and answer the following questions.

(i) Identify the labelled parts *A, B, C, D* and *E*.

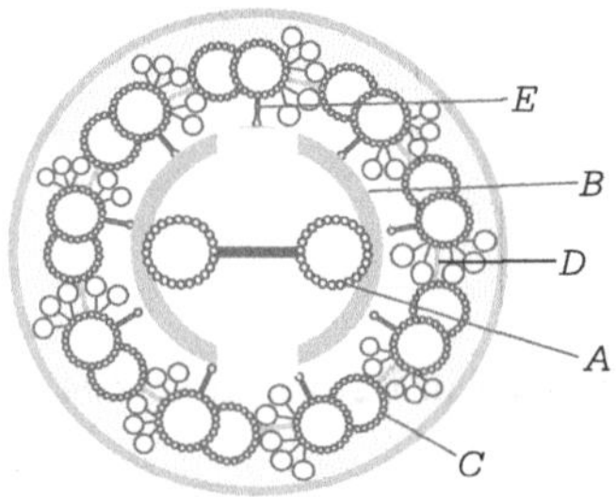

A. ____________

B. ____________

C. ____________

D. ____________

E. ____________

(ii) How many pairs of microtubules are found in this figure? ____________

217. Cilium and flagellum are covered with ____________ (plasma membrane/cell wall)

218. The core of flagella/cilia is called the ____________. It possesses a number of ____________ running parallel to the long axis.

219. Write the arrangement of axonemal microtubules. ____________

220. The central microtubules are connected by ____________ (connective/bridges) and is enclosed by a central sheath.

221. The central sheath is connected to one of the tubules of each peripheral doubles by a interdoublet bridge.
(*Rewrite the statement by changing the underlined word.*)

____

222. How many radial spokes are found in cilia/flagella? ____________

223. The cilium and flagellum emerge from lomasome like structure called the hook bodies.
(*Rewrite the statement by changing the underlined words.*)

____

224. The two cylindrical structures of a centrosome are called axoneme.

True ☐ False ☐

225. What is the surrounding material of centrioles? ____________

226. The centriole in a centrosome lie ____________ (parallel/perpendicular) to each other to form a structure like cartwheel.

227. Centrioles made up of nine evenly spaced peripheral fibrils of ____________ protein. These peripheral fibrils form a ____________ .

228. Identify the structure of centriole that is being described in statements (i) and (ii).

(i) This is central part of the centriole, it is proteinaceous in nature. ____________

(ii) It connects the hub to the tubules of the peripheral triplets. ____________

229. The centrioles form the basal body of ____________ and ____________ .

230. The centrioles form the ____________ during cell division in animals.

231. What is the name proposed by Flemming to the material of the nucleus stained by basic dye?

____

232. The interphase nucleus has highly extended and elaborated ____________ fibres called chromatin, nuclear matrix and one or more spherical bodies called ____________ .

233. Nucleus is enveloped by nuclear envelope, it consists of two parallel membranes with a space between them called the ____________ (paranuclear/perinuclear) space.

234. The nuclear envelope forms a barrier between the materials present inside the nucleus and that of the cytoplasm.

True ☐ False ☐

235. The outer membrane of nucleus attaches with endoplasmic reticulum and bears centrosome on it.
(*Rewrite the statement by changing the underlined word.*)

____

236. Consider the following image of nucleus and answer the following questions

(i) Identify the labelled parts *A*, *B*, *C* and *D*.

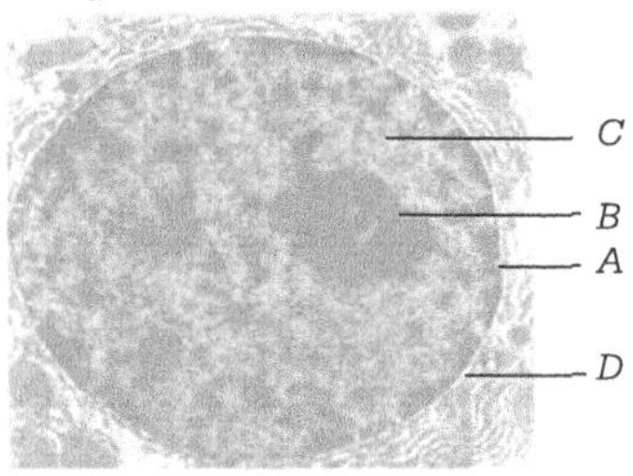

A. ____________

B. ____________

C. ____________

D. ____________

(ii) Which of the labelled parts provide passage for the movement of RNA and protein in both direction from nucleus and cytoplasm? ____________

(iii) Which of the labelled part contains chromatin? ____________

237. Nuclear pores are formed by the fusion of ____________ (2/3) membranes of nuclear envelop.

238. What raw materials are contained in nucleoplasm for the synthesis of DNA and RNA?

____

239. How many of the following enzymes are present in nucleoplasm?

DNA polymerase, RNA polymerase, NAD synthetase, PEPCase, Adenosine deaminase, Pyruvate kinase

____

240. Nucleoplasm contains histone and non-histone proteins for combination with DNA.

True ☐ False ☐

241. How many nucleus are present in erythrocytes of mammals and sieve tube cells of vascular plants?______________________________

242. Find the odd one out from the following on the basis of number of nucleus.

Osteoclasts, Tapetal cell, Parenchymatous cell, Liver cell, Muscle cells

____

243. Name the spherical structure present in the nucleoplasm.______________________________

244. The content of nucleolus is not continuous with rest of the nucleoplasm because it is membrane bound structure.

True ☐ False ☐

245. When nucleus undergoes protein synthesis, it contains more and large ______________.

246. Match the following columns.

	Column I (Structures of a nucleus)		Column II (Characteristics)
A.	Nuclear membrane	1.	It is dark staining granular structure
B.	Nucleoplasm	2.	It is octagonal
C.	Nucleolus	3.	It is also known as karyotheca
D.	Nuclear pore	4.	It is non-staining part

A. ______________ B.______________ C.______________ D. ______________

247. Complete the following analogy with respect to nucleus.

Uninucleate : *Amoeba* : : Binucleate : ______________________________

248. In the interphase, nucleus contains ______________________________, but during the different stages of cell division, it gets condensed into ______________________________ .

249. What does a chromatin contain ?______________________________

250. A single human cell has approximately ______________ (2/3) metre long thread of DNA distributed among its ____________ (46/48) chromosomes.

251. Complete the following analogy with respect to nuclear structure.

Heterochromatin : ______________________ : : Euchromatin : Loose

252. The primary constriction of a chromosome is termed as <u>satellite</u>.
(*Rewrite the statement by changing the underlined word.*)

____

253. Considering the following diagram, answer the following questions.

(i) '*A*' is a disc-shaped structure present on the side of centromere.
What is '*A*'? ______________

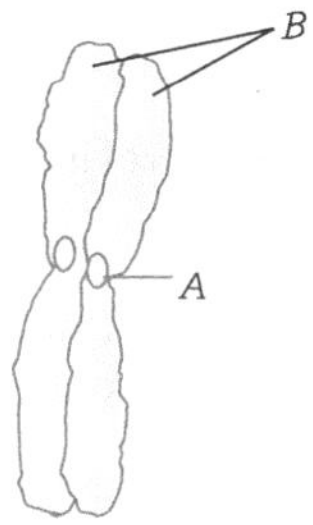

(ii) Identify the labelled part *B*. ______________________________

254. Chromatids of a chromosome held together by ______________________

255. Match are following columns.

	Column I		Column II
A.		1.	Telocentric
B.		2.	Sub-metacentric
C.		3.	Metacentric
D.		4.	Acrocentric

A. ______________ B. ______________
C. ______________ D. ______________

256. Identify types of chromosomes from the following statements.

(i) The chromosome has middle centromere that form two equal arms of the chromosome.

____

(ii) The chromosome has terminal centromere. ______________________

257. The ____________________ (telocentric/sub-metacentric) chromosome has centromere slightly away from the middle of the chromosome resulting into one shorter arm and one longer arm.

258. The <u>telocentric</u> chromosome has centromere situated close to its ends forming one extremely short and one very long arm.
(*Rewrite the statement by changing the underlined word.*)

____

259. ____________________ is the characteristic pattern of chromosomes present in a somatic cell of any organisms.

260. A few chromosomes have non-staining secondary constrictions at a constant location, it gives the appearance of a small fragement called kinetochore.

True ☐ False ☐

261. ____________________ are membrane bound minute vesicles that contain various enzymes.

262. Give a common collecting term for the following.

Peroxisomes, Glyoxysomes, Spherosomes

____

263. Match the following columns.

	Column-I (Types of microbodies)		Column-II (Characteristics)
A.	Glyoxysomes	1.	It contains peroxins protein
B.	Peroxisomes	2.	It contains hydrolases, proteases
C.	Spherosomes	3.	Found in endosperm of germinating seeds

A. ____________________ B. ____________________ C. ____________________

Answers

1. Cell
2. Unicellular, multicellular
3. *Pandorina* is odd one out. Rest all are unicellular organisms.
4. Antonie van Leeuwenhoek
5. Robert Hooke
6. Electron microscope
7. Matthias Schleiden
8. Animal
9. Presence of cell wall
10. Cells, its product
11. A-3, B-1, C-4, D-2
12. New cells are derived from pre-existing cells.
13. True
14. False. Cell theory also known as cell lineage theory.
15. These are considered to be exceptions to cell theory.
16. Nucleus
17. Eukaryotic cell
18. Cytoplasm
19. Cytoplasm
20. 1, All of them absent in prokaryotes except ribosomes.
21. Lysosomes is odd one out as all of them contain ribosomes except lysosomes.
22. Centrosome
23. An ostrich
24. A-3, B.-4, C-1, D- 2
25. Nerve cell
26. Red blood cell
27. *A*–Round and biconcave, *B*–Amoeboid, *C*–Long and narrow, *D*–Round and oval, *E*–Branched and long *F*–Elongated
28. Pleomorphic
29. Nucleocytoplasmic index
30. A-2, B-1, C-4, D-3
31. 3 (i, ii, iii). Prokaryotic cell do not contain defined nucleus and peroxisomes.
32. Comma-shaped
33. All of them possess cell wall except *Mycoplasma.*
34. Naked, nuclear
35. Plasmid
36. 3(i, ii, iii). Plasmids are of different types thus, (iv) is not correct.
37. (i) Eukaryotic
    (ii) False. The size of this cell range between 10-20 μm.
38. Three
39. Glycocalyx → Cell wall → Plasma membrane
40. 2
41. Peptidoglycan
42. Hans Christian Gram
43. Gram positive, Gram negative
44. *Bacillus subtilis*
45. False. The composition and thickness of glycocalyx is different among different bacteria.
46. Slime layer
47. Capsule
48. 1 (iii). As cell wall prevent bursting of cell.
49. Plasma membrane
50. Mesosome
51. 3 (Form of vesicles, Tubules, Lamellae)
52. 6(i, ii, iii, iv, v, vi)
53. Cyanobacteria
54. Flagella
55. True
56. Filament, Hook, Basal body
57. Filament
58. Flagellin
59. A-3, B-2, C-1
60. Fimbriae
61. Pilin
62. Plasma membrane
63. 15nm to 20nm
64. 50S, 30S
65. False. The prokaryotic ribosome is made up of 60% *r*RNA and 40% ribosomal protein.
66. Polyribosomes or Polysome
67. Cell inclusion bodies
68. Inclusion bodies
69. Blue-green, purple and green photosynthetic bacteria.
70. A-3, B-1, C-5, D-2, E-4
71. Carboxysomes
72. Eukaryotic cell
73. Eukaryotic cell
74. True
75. Plant cell
76. Centrioles
77. *A*–Nucleolus, *B*–Rough endoplasmic reticulum, *C*–Plasmodesmata, *D*–Vacuole
78. *E*–Centriole, *A*–Microvilli, *D*–Ribosome
79. Electron microscope
80. A-2, B-3, C-1
81. Phospholipids
82. False. In bilayer membrane, the lipids polar head lies towards the outer side and tails towards the inner side.
83. Polar head of lipid bilayer
84. Saturated
85. Phosphoglycerides
86. Integral, peripheral
87. Peripheral protein
88. Integral
89. False. Integral membrane proteins are difficult to extract.
90. 30%
91. Oligosaccharide
92. False. Singer and Nicolson proposed a cell membrane structure model known as fluid mosaic model.
93. Lateral
94. Fluidity
95. (i) *A*–Peripheral protein, *B*–Phospholipid bilayer, *C*–Integral protein, *D*–Cholesterol,
    (ii) Fluid mosaic model
96. Permease
97. False. Glycolipids and cholesterol are usually present in the membrane of the animal cells.
98. 5(ii, iii, v, vi, vii)
99. Selectively permeable
100. Passive
101. Neutral solutes
102. Diffusion

103. Osmosis

104. Protein

105. Facilitated diffusion

106. The entry of glucose in erythrocytes takes place *via* facilitated diffusion.

107. Active

108. False. A process of transport which require energy in the form of ATP is known as active transport.

109. The transport of Na^+/K^+ takes place by active transport.

110. A-4, B-3, C-2, D-1

111. Non-living

112. Fungi, plants

113. False. The cell wall gives shape to the cell and protects the cell from mechanical damage and infection.

114. False. The cell wall helps in cell to cell interaction and provides barrier to undesirable macromolecules.

115. 4(cellulose, galactans, mannans, calcium carbonate).

116. A-3, B-2, C-1

117. Secondary wall → Primary wall → Middle lamella

118. Primary wall

119. Calcium pectate

120. Holds or Glues

121. False. The plasmodesmata pass through the cell wall and middle lamella.

122. Plasmodesmata

123. An enzyme pectinase is responsible for it.

124. All of these are the part of endomembrane system except peroxisomes.

125. Coordinated

126. Endoplasmic reticulum

127. Luminal (inside ER), Extraluminal (Cytoplasm)

128. Ribosomes

129. No ribosome

130. Protein

131. RER

132. Lipid molecule and steroid hormones.

133. (i) *A*–Rough Endoplasmic Reticulum (RER), *B*-Smooth Endoplasmic Reticulum (SER)
(ii) The calcium storage takes place at SER and protein packaging takes place at RER.

134. Golgi apparatus

135. False. The Golgi apparatus consist of many flat, disc-shaped sacs or cisternae.

136. 0.5μm, 1.0 μm

137. Camillo Golgi (1898)

138. Parallel

139. This is Golgi apparatus. *A*–Cisternae

140. Cell membrane, nucleus

141. Forming

142. (i), Different but interconnected

143. Packaging

144. (ii) The ER fuses with the *cis* face of the Golgi apparatus and move towards the maturing face for the packaging of material.

145. Vesicles

146. True

147. *Trans*

148. All of them packed by Golgi apparatus.

149. Glycoproteins, glycolipids

150. False. Lysosomes are formed by the process of packaging in the Golgi apparatus.

151. Vesicular

152. Hydrolytic enzyme

153. All of these are hydrolytic enyzmes, isolated from lysosomes.

154. True

155. (ii), Lysosomes contain hydrolytic enzyme, these enzymes are capable of digesting the own substances of cell.

156. Acidic

157. Vacuole

158. 5 (Water, sap, carbon dioxide, anthocyanin and excretory product)

159. Tonoplast

160. 80%

161. False. The tonoplast facilitates the transport of a number of ions and other materials against concentration gradient.

162. Concentration of cytoplasm $<$ Concentration of vacuole

163. Contractile vacuole

164. Food vacuole

165. A-2, B-1, C-4, D-3

166. False. Mitochondria, unless specifically stained are not easily visible under the microscope.

167. Meristematic cell

168. Shape, size

169. Shape–sausage or cylindrical shaped and size–diameter (0.2-1.0 μm) and length (1.0-4.1 μm)

170. Double

171. Two

172. (i) *A*–Matrix, *B*–Crista, *C*–Inner membrane, (ii) *B*–Cristae

173. Matrix

174. Outer membrane

175. Inner membrane

176. (i) Cristae, (ii) Oxysomes, $F_0 - F_1$ (iii) F_1-subunit, stalk, F_0-subunit

177. Mitochondria

178. ATP, powerhouse

179. Matrix

180. 70S

181. Fission

182. Plastids

183. Pigments

184. Chloroplasts, chromoplasts, leucoplasts

185. A-2, B-1, C-3

186. All of them are leucoplast except rhodoplast.

187. Aleuroplasts

188. Carotene, xanthophyll

189. Mesophyll

190. 1, 20-40

191. True

192. Inner membrane

193. Stroma

194. Stroma

195. (i) *B*–Granum or integral thylakoid, (ii) *A*–Thylakoid, (iii) Stroma lamellae

196. Stroma lamellae

197. True

198. Stroma

199. Ribosomes

200. Cytoplasmic ribosome
201. Ribosomes
202. George Palade (1953)
203. Ribosomes
204. (i) Eukaryotic cell, (ii) Prokaryotic cell
205. (i) 80S (ii) 70S
206. Density, size
207. Ribophorins
208. Cytoskeleton
209. 3 (Mechanical support, motility, shape of the cell).
210. Cytoskeleton elements
211. A-2, B-3, C-1, D-4
212. Cilia and flagella
213. False. Cilia are small structures which work like oars, causing the movement of either the cell or the surrounding fluid.
214. Longer, movement
215. Different
216. (i) *A*–Central microtubule *B*–Central sheath, *C*–Peripheral microtubules (doublets), *D*–Interdoublet bridge, *E*–Radial spoke
(ii) Nine pairs at periphery and 1 pair at centre.
217. Plasma membrane
218. Axonema, microtubules
219. 9 + 2 arrangement
220. Bridges
221. Radial spoke
222. Nine
223. Centriole, basal
224. False. The two cylindrical structures of a centrosome is called centrioles.
225. Amorphous pericentriolar material.
226. Perpendicular
227. Tubulin, triplet
228. (i) Hub, (ii) Radial spoke
229. Cilia, flagella
230. Spindle fibres
231. Chromatin
232. Nucleoprotein, nucleoli
233. Perinuclear
234. True
235. Ribosomes
236. (i) *A*–Nuclear pore, *B*–Nucleolus, *C*–Nucleoplasm, *D*–Nuclear membrane
(ii) *A*–Nuclear pore
(iii) *C*–Nucleoplasm.
237. 2
238. Nucleotides, enzymes, metal ions.
239. 5(DNA polymerase, RNA polymerase, NAD synthetase, Adenosine deaminase, Pyruvate kinase).
240. True
241. Zero. Both of these lack nucleus.
242. All of them contain many nuclei per cell except parenchymatous cell, which contain only one nucleus.
243. Nucleoli
244. False. The content of nucleolus is continuous with rest of the nucleoplasm because it is not a membrane bound structure.
245. Nucleoli
246. A-3, B-4, C-1, D-2
247. *Paramecium*
248. Chromatin, chromosomes
249. DNA, histones, non-histones, RNA
250. 2, 46
251. Compact
252. Centromere
253. (i) *A*–Kinetochore (ii) *B*–Chromatids
254. Centromere
255. A-3, B-2, C-4, D-1
256. (i) Metacentric, (ii) Telocentric
257. Sub-metacentric
258. Acrocentric
259. Karyotype
260. False. A few chromosomes have non-staining secondary constrictions at a constant location, it gives the appearance of a small fragment called satellite.
261. Microbodies
262. These are microbodies.
263. A-3, B-1, C-2

CHAPTER NINE

BIOMOLECULES

Topic 1 : Biomicromolecules

1. Name the four major elements obtain by the elemental analysis of plant tissue, animal tissue or a microbial paste? ____________________

2. The relative abundance of carbon and hydrogen with respect to other elements is higher in living organisms than in earth's crust.

   True ☐ False ☐

3. Complete the following table of comparison of elements present in non-living and living matter.

	Elements	% weight of earth's crust	% weight of human body
(i)	Hydrogen	(a) 0.14	(1) (A)
(ii)	Carbon	(b) 0.03	(2) (B)
(iii)	Oxygen	(c) (C)	(3) 65.0
(iv)	Nitrogen	(d) Very little	(4) (D)
(v)	Sodium	(e) (E)	(5) 0.2

   A. ____________ B. ____________

   C. ____________ D. ____________

   E. ____________

4. For chemical analysis of living tissue, it is grind in ____________ using a mortar and a pestle.

5. Consider the following flowchart of process of chemical analysis of living tissue and answer the questions that follows.

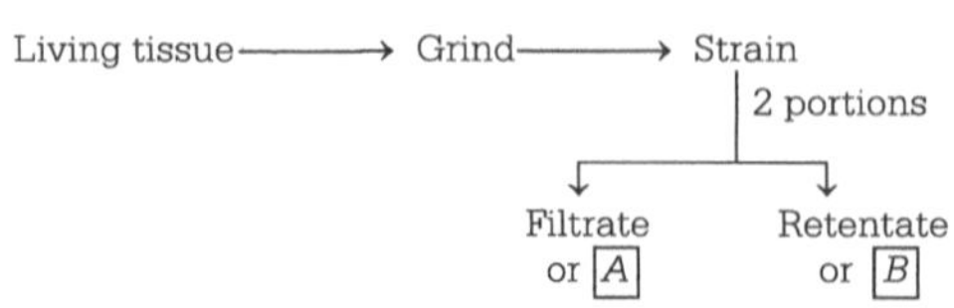

(i) Name the material used for straining. ______________________

(ii) Identify *A* and *B* portion.

A.______________________ B. ______________________

(iii) Which of the following portion contain organic compounds?

____

6. After isolating and purifing a compound from tissues, the extracted material undergoes various ______________ techniques.

7. What information (any two) is provided by analytical technique of organic compounds?

____

8. ______________ (Biomolecules/Micromolecules) are carbon compound obtained from living tissues.

9. The inorganic materials of living organisms are obtained from ______________.

10. Consider the following process and answer the given questions.

$$\text{Living tissue} \longrightarrow \text{Dry it} \longrightarrow \boxed{A} \xrightarrow[\text{water vapour}]{CO_2,} \text{Ash}$$

(i) Identify process '*A*' ______________.

(ii) Name any two inorganic elements present in ash. ______________.

11. The weight of living tissue is called ______________ .

12. Name any two inorganic compounds which are present in the acid soluble fraction of living tissue.______________________

13. Elemental analysis gives elemental composition of living tissues in the form of __________, __________, __________, __________ etc.

14. The analysis for compounds gives an idea of the kind of ______________ and ______________ constituents.

15. Two students, one from chemistry and other one from biology did an experiment on analysis of compound. Write the results of these experiments on the basis of chemistry and biology.

(i)______________________

(ii)______________________

16. Proteins are organic compounds containing an amino group and an acidic group as substituents on the same carbon, i.e. the β-carbon.
(*Rewrite the statement by changing the underlined word.*)

____

17. Amino acids are also known as ______________ because it contains an amino group and an acidic group on the α-carbon.

18. Amino acids are substituted methanes because
(i) amino acids are derived from methane.
(ii) four valency position in amino acids is occupied by four substituted groups similar to hydrogen and carbon atoms in methane.

____

19. The α-carbon of amino acid has four substituted groups *viz*, ____________, ____________, ____________ and a ____________ designated as *R*-group.

20. On the basis of nature of carboxyl group, amino acids are of several type.

True ☐ False ☐

21. How many amino acids are found in nature which form various proteins? ____________

22. Complete the following analogy with respect to amino acid structure.

Hydrogen : Methyl group : : ____________ : Alanine

23. Match the following columns.

Column I (Types of amino acid)		Column II (Structures of amino acid)
A. Serine	1.	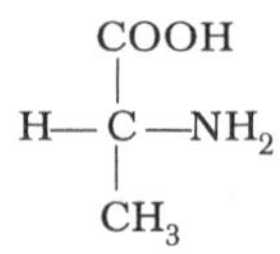
B. Alanine	2.	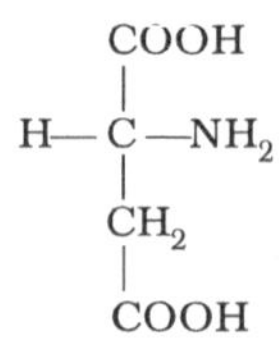
C. Glycine	3.	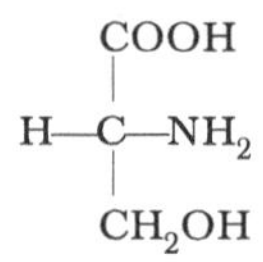
D. Aspartic acid	4.	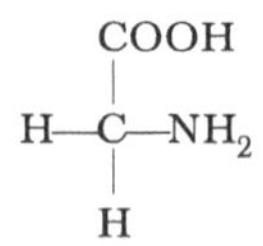

A. ____________ B. ____________
C. ____________ D. ____________

24. The physical and chemical properties of amino acids are only dependent on the *R*, i.e. functional group.

True ☐ False ☐

25. The number of ______________ (amino/*R*-group) and carboxyl groups differentiate the amino acids into acidic and basic.

26. Identify the odd one out with respect to nature of amino acids.

Glycine, Valine, Leucine, Arginine, Alanine

____

27. The amino acids having one amino group and one carboxylic group is ______________ (alcoholic/neutral) in nature.

28. Give a common collective term for the following amino acids on the basis of their nature.

Phenylalanine, Tyrosine, Tryptophan

____

29. Complete the following analogy with respect to nature of amino acids.

Lysine : Glutamic acid : : ______________ : Acidic

30. What is the nature of amino acid having an extra carboxylic group? ______________

31. The specific property of amino acids is the ionisable nature of ______________ and ______________ group.

32. Answer the following questions considering the given equation.

$$\underset{(A)}{H_3\overset{+}{N}-\overset{\overset{R}{|}}{C}H-COOH} \rightleftharpoons \underset{(B)}{H_3\overset{+}{N}-\overset{\overset{R}{|}}{C}H-COO^-} \rightleftharpoons \underset{(C)}{H_2N-\overset{\overset{R}{|}}{C}H-COO^-}$$

(i) Which of the following structure is zwitter ion form? ______________

(ii) Which of the following structure found in acidic pH? ______________

(iii) Which of the following structure is anion? ______________

33. In amino acid serine, methyl group is present at the *R*-group position. (*Rewrite the statement by changing the underlined word.*)

____

34. Amino acids are crystalline solids that dissolve in ______________, but are insoluble in ______________ .

35. The 20 types of amino acids that occur in proteins are called as

____

36. ______________ amino acids occur in both free as well as in combined state, but not proteins.

37. Lipids are water ______________ (soluble/insoluble) and they could be simple ______________ (fatty acids/glycerols).

38. A fatty acid has a ______________ group attached to an *R*-group.

39. The *R*-group in lipids could be a methyl or ethyl or higher number of —CH_2 range between 1 carbon to 40 carbon.

True ☐ False ☐

40. Complete the following analogy.
Palmitic acid : ________ : : Arachidonic acid : 20

41. How many double bonds are present in the fatty acids given below?

(i) Saturated ________________________________

(ii) Unsaturated________________________________

42. Glycerol is the form of simple lipid called ________________________

43. Many lipids have both glycerol and fatty acids, these are esterified with glycerol.

True ☐ False ☐

44. Esterified lipids are of three types, *viz,* ______________, ________________ and ______________

45. Oils have higher melting point and hence remain as oil in winters.
(*Rewrite the statement by changing the underlined word.*)

____

46. Phosphlipids have phosphorus and a ______________ (carbonated/ phosphorylated) organic compound. These are found in ____________________ .

47. Lipids consist of a hydrophilic head and one to three hydrophobic tails, hence are called

____

48. Complete the following analogy with respect to organic acid.

Saturated fatty acid : ______________________ : : Unsaturated fatty acid : Oleic acid

49. Identify the missing boxes of the lipids from following structure.

(i) CH_3 —Ⓐ— COOH A.________________________
Palmitic acid

(ii)

CH_2 —Ⓐ A.________________________
|
CH — OH
|
CH_2 —Ⓑ B.________________________
Glycerol

50. Give a common collective term for the following lipids.

Phospholipids, Glycolipids, Sphingolipids, Lipoproteins

____

51. Name the tissue which stores fat in animal and plants, respectively.

____

52. (i) Identify the compound *A* whose chemical structure is shown below.

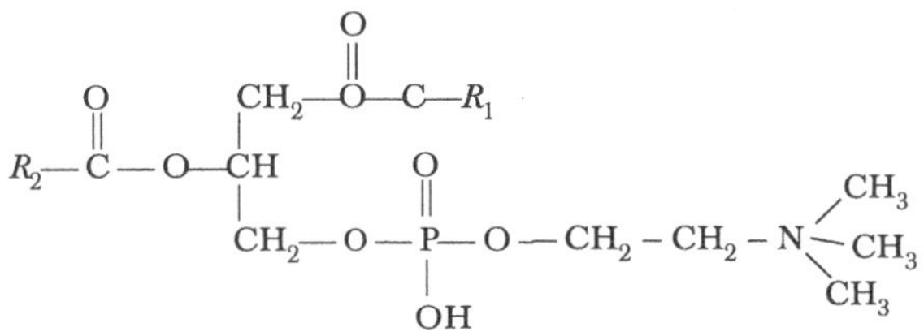

A. ______________________________

(ii) The given compound *A* is ______________ (phospholipid/glycolipid).

53. In the structure of triglycerides, how many oxygen atoms are bounded by double bond?

____

54. ______________ (Glycolipids/Lipoproteins) occur in myelin sheath of nerve fibres and on outer surfaces of nerve cell and of chloroplast membranes.

55. ______________ (Cholesterol/Prostaglandin) is the most abundant steroid in the animal tissues.

56. Answer the following questions considering the given chemical structure of a lipid.

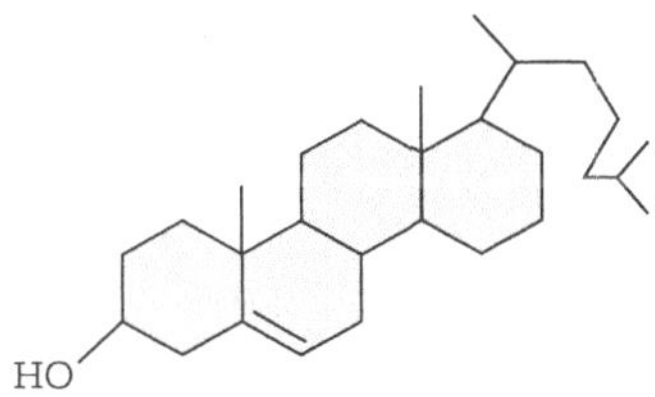

(i) Identify the structure ______________.

(ii) How many of the listed hormones are synthesised by this lipid?

Testosterone, Estradiol, Aldosterone, Ecdysone, Thyroxine

____

57. Give a collective term for the following sequence.

Adenine, Guanine, Cytosine, Uracil, Thymine

____

58. Complete the following analogy.

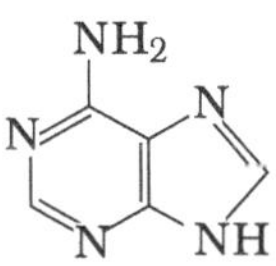

: Adenine : : ______________ : Uracil

59. The nitrogen base molecule is joined to the sugar molecule by a glycosidic bond to form ______________ (nucleosides/nucleotides).

60. Give a common collective term for the following sequence.

Adenosine, Guanosine, Thymidine, Uridine, Cytidine

____

61. The nucleoside combines with a phosphate group at 5' position by an ester bond to form a _______________ (nucleosides/nucleotide).

62. Give a common collective term for the following sequence.

Adenylic acid, Thymidylic acid, Guanylic acid, Uridylic acid, Cytidylic acid

____

63. Complete the following analogy with respect to nucleic acid.

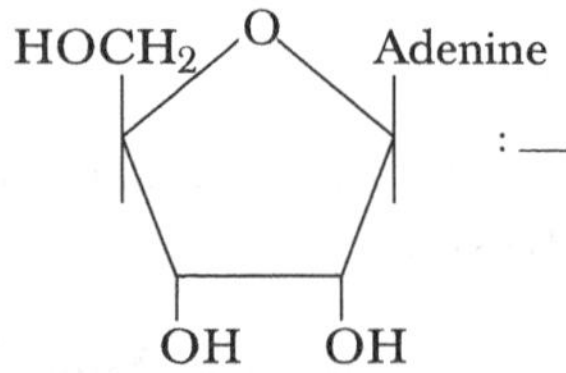

: _______________ : : Adenosine : Uridine

64. The structure of a nucleotide is given below, what is formed if phosphate group is removed from this structure. Draw the new structure.

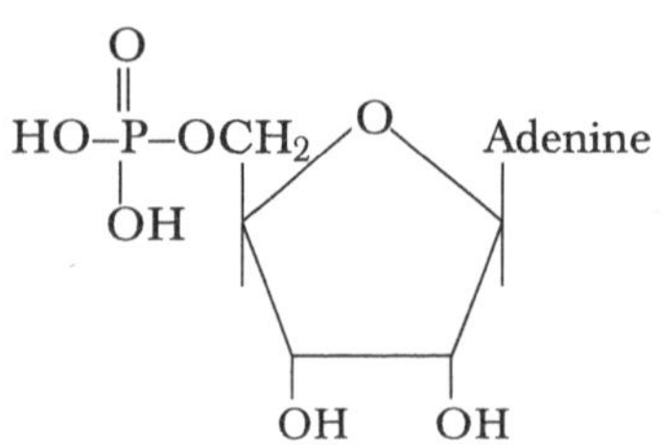

____

65. Nucleic acid, e.g. DNA and RNA consist of <u>nucleosides</u> only.
(*Rewrite the statement by changing the underlined word.*)

____

66. Identify the products *A*, *B* and *C* in the following reaction.

$$\text{Adenine} + \text{Ribose} + H_3PO_4 \rightarrow \boxed{A} + H_3PO_4 \rightarrow \boxed{B} + H_3PO_4$$
$$\boxed{C} \text{ (linked below } B\text{)}$$

A. _______________ B. _______________ C. _______________

Topic 2 : Primary and Secondary Metabolites

67. ________________ are the intermediate products produce during metabolism, catalysed by various enzymes that occur naturally within cells.

68. The metabolites which are required for the growth and maintenance of cellular functions are called secondary metabolites.

    True ☐ False ☐

69. How many of the following are primary metabolites?

    Alanine, Thiamine, Morphine, Ricin, Uridine

    ____

70. The primary metabolites consist of vitamins, amino acids, nucleosides and organic acids.

    True ☐ False ☐

71. All the compounds other than primary metabolites in plant, fungal and microbial cells are called ____

72. Match the following columns.

	Column I (Secondary metabolites)		Column II (Types)
A.	Pigments	1.	Morphine, codenine
B.	Alkaloids	2.	Abrin, ricin
C.	Toxins	3.	Vinblastin, curcumin
D.	Lectins	4.	Carotenoids, anthocyanin
E.	Drug	5.	Rubber, gum, cellulose
F.	Terpenoids	6.	Concanavalin-A
G.	Polymeric substances	7.	Monoterpenes

A. ____________ B. ____________ C. ____________

D. ____________ E. ____________ F. ____________

G. ____________

73. Name any five secondary metabolites which are useful for humans.

    ____

Topic 3 : Biomacromolecules

74. The compounds found in acid soluble pool have molecular weight ranging from ______ to ______ Daltons (Da).

75. The molecular weight of compounds found in acid insoluble pool is approximately 10 thousand Daltons and above excluding lipids.

    True ☐ False ☐

76. ________________________ (Biomolecules/Biomacromolecules) have molecular weights less than one thousand Dalton.

77. Complete the following analogy.

Biomolecules : ________________ : : Biomacromolecule : Acid insoluble

78. (i) The molecular weight of lipids do not exceed ______________ Da.

(ii) Lipids are extracted from the acid ________________ (soluble /insoluble) fraction.

(iii) Lipids are often found in macromolecular fraction during separation because

(a) they itself are macromolecules.

(b) they are found in cell membranes which get broken into pieces and form water insoluble vesicles.

____

79. The biomacromolecules are formed by ______________ of the micromolecules.

80. The acid soluble pool roughly represents the ________________________ (cytoplasmic/mitochondrial) composition and acid insoluble pool represents macromolecules from cytoplasm and organelles.

81. Match the following columns.

	Column I (Compositions of cells)		Column II (% of the total cellular mass)
A.	Water	1.	2
B.	Protein	2.	1
C.	Lipids	3.	5-7
D.	Carbohydrate	4.	70-90
E.	Nucleic acid	5.	10-15
F.	Ions	6.	3

A.____________ B. ____________ C. ____________

D.____________ E. ____________ F. ____________

82. Rearrange the following compounds on the basis of abundance in living organisms in decreasing order.

Protein, Water, Lipid, Carbohydrate

____

Topic 4 : Proteins

83. Proteins are linear chains of ______________ linked by ______________ bonds.

84. Name the polymer which have only one type of monomer repeating '*n*' number of times.

____

85. ______________ (Essential/Non-essential) amino acids are not synthesised in our body and have to be supplied through diet.

86. What is the source of essential amino acids?________________________

87. Complete the following analogy with respect to category of amino acids.

Essential amino acid : Valine : : Non-essential amino acid : ________________

88. Essential amino acids are synthesised by our body.

True ☐ False ☐

89. Match the following columns.

	Column I (Protein)		Column II (Functions)
A.	Collagen	1.	Enzyme
B.	Trypsin	2.	Sensory reception
C.	Insulin	3.	Intercellular ground substance
D.	Antibody	4.	Enables glucose transport into cells
E.	Receptor	5.	Hormone
F.	GLUT4	6.	Fights infectious agents

A.________ B. ________ C. ________

D. ________ E. ________ F. ________

90. Complete the following analogy.

Collagen : Animal : : ________ : Plant

91. Identify the types of proteins that are being described in following statements.

(i) These proteins are made up of only amino acid. ________

(ii) These are obtained from protein through denaturation, coagulation.

____

(iii) These proteins are thread-like. ________

92. The ________ (conjugated/globular) proteins have non-amino prosthetic groups.

93. Collagen is the most abundant protein in the whole of the biosphere.

True ☐ False ☐

Topic 5 : Polysaccharides

94. Complete the following analogy.

Carbohydrate : ________ : : Protein : Amino acid

95. Polypeptides are threads containing different monosaccharides as building blocks. (*Rewrite the statement by changing the underlined word.*)

____

96. Match the following columns.

	Column I (Types of carbohydrate)		Column II (Examples)
A.	Monosaccharides	1.	Cellulose, starch
B.	Disaccharides	2.	Raffinose, stachyose
C.	Oligosaccharides	3.	Maltose, sucrose
D.	Polysaccharides	4.	Glucose, fructose

A.________ B. ________ C. ________ D. ________

97. Complete the following analogy with respect to structure of molecule.

Homopolymer : Cellulose : : Heteropolymer : ____________________

98. Complete the following analogy with respect to polymer of molecules.

Glycogen : Glucose : : ________ : Fructose

99. The right end of polysaccharide chain is called ____________ end and the left end is known as ____________________ end.

100. Answer the following questions considering the diagrammatic representation of a polysaccharides.

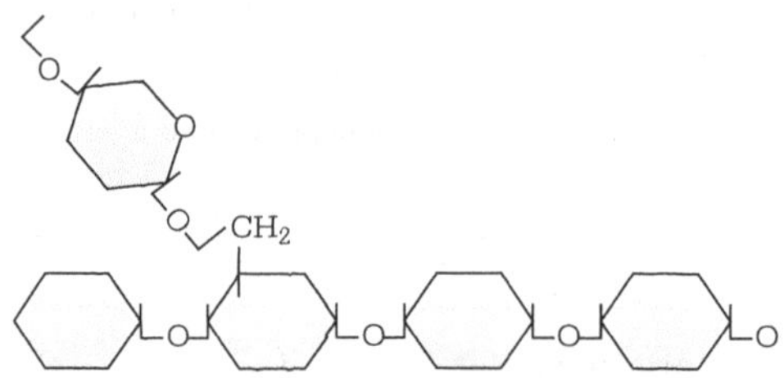

(i) It is the branched structure of ________________.

(ii) It consists of β-D glucose units.

True ☐ False ☐

101. Starch form helical secondary structure and can hold ________ molecules in the helical portion.

102. Complete the following analogy with respect to reaction of iodine with carbohydrates.

Glycogen : ________________ : : Starch : Blue

103. Cellulose possesses complex helices structure and its helices hold I_2.

True ☐ False ☐

104. Write the type of carbohydrate found in the following structures /products.

(i) Muscles and liver of animal ____________________

(ii) Plant cell wall, paper, cotton fibre ____________________

(iii) Immunostimulant ____________________

(iv) Exoskeleton of arthropods ____________________

105. Reducing sugar contain free ____________ or __________ group while non-reducing sugar lacks it.

106. Which of the following sugar is suitable for the reaction?

____________________ (Reducing/Non-reducing sugar) $\xrightarrow{\text{Benedict's test}}$ Green, yellowish, orange

107. Identify the odd one out from the following compounds.

Glucose, Fructose, Galactose, Sucrose

____

108. Name any two modified sugars used as building blocks in amino sugar.

____

109. Name the product which is formed in the following reaction.

Glucose + Glutamine → ______________

110. Glucosamine and N-acetylglucosamine are the ______________ (homopolymer/heteropolymer) and monomer of chitosan and chitin.

Topic 6 : Nucleic Acids

111. A nucleotide has three chemically distinct components, *viz*, ______________, ______________ and ______________.

112. The nucleotide functions as the buildings block of <u>polysaccharides.</u>

(*Rewrite the statement by changing the underlined word.*)

____

113. Match the following columns.

Column I (Nitrogen bases)		**Column II** (Types of nitrogen base)
A. Purine	1.	Guanine
B. Pyrimidine	2.	Cytosine
	3.	Thymine
	4.	Adenine
	5.	Uracil

A. ______________ B. ______________

114. Purines and pyrimidines are the homocyclic rings in nucleic acids.

True ☐ False ☐

115. A nucleotide having ribose sugar is called ______________ and a nucleotide containing deoxyribose sugar is termed ______________.

116. Polysaccharide, polypeptide and polynucleotide comprise the true macromolecular fraction of any living tissue.

True ☐ False ☐

Topic 7 : Structure of Proteins

117. In inorganic chemistry, the structure invariable refers to the ______________.

118. Up to which level, the structure of a molecule is described in the following fields.

(i) Organic chemistry ______________

(ii) Physics ______________

(iii) Biology ______________

119. The sequence of amino acid in a protein structure is known by <u>sequential</u> information.
(*Rewrite the statement by changing the underlined word.*)

____

120. Which protein structure is being studied to determine the first and last amino acid of a protein?

____

121. If a protein is imagined as a line, the left end represented by first amino acid is known as C-terminal amino acid.

True ☐ False ☐

122. The right hand of protein is represented by the last amino acid and it is known as ______________ (N-terminal/C-terminal) amino acid.

123. The portion of protein thread is folded in the form of a ______________.

124. The most common types of helix found in protein is the ______________ (right/left) handed.

125. The linear polypeptide chain folded into a specific coiled structure is known as <u>quaternary</u> structure.
(*Rewrite the statement by changing the underlined word.*)

____

126. Refer to the given figure of structure of proteins and answer the questions that follows.

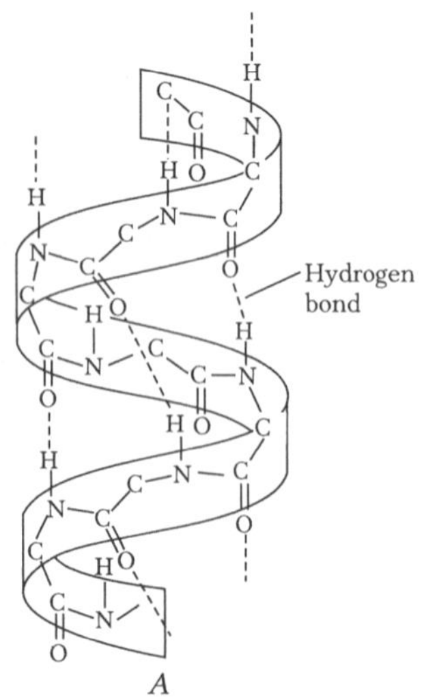

A

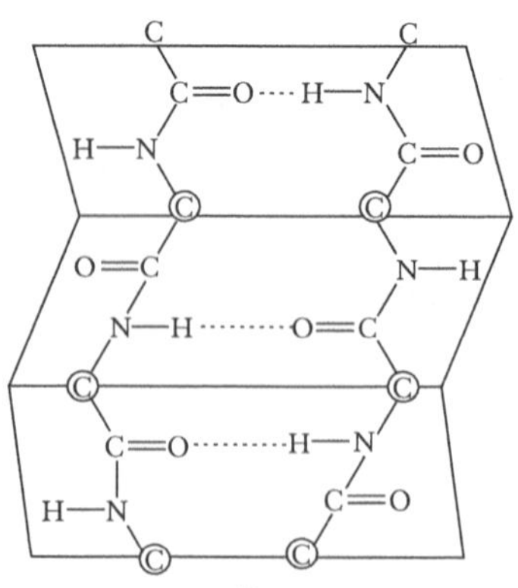

B

(i) Identify the level of protein structure of figure *A* and *B*.

A. ____________ B. ____________

(ii) Figure *A* is coiled in ____________ (right/left) handed manner.

(iii) Which of the following structure is found in myosin and tropomyosin? ____________

127. Complete the following analogy with respect to structure of protein.

Parallel β-sheet : β-keratin : : Antiparalle β-sheet : ____________

128. The long protein chain is folded into a hollow woolen ball to form ____________ structure of protein.

129. ____________ (Tertiary/Secondary) structure gives three dimensional structure, and it is necessary for many biological activities of proteins.

130. Consider the given structure of protein and answer the questions that follows.

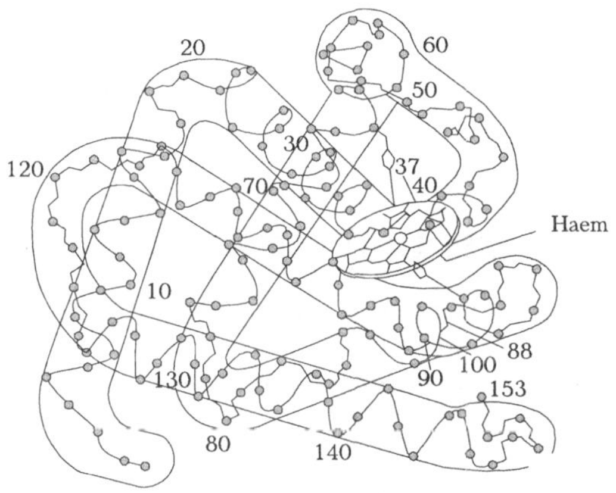

(i) Identify the level of this protein structure. ____________

(ii) Name any three bonds which stabilised this structure. ____________

131. Identify the level of protein structures shown in the diagrams *A* and *B*.

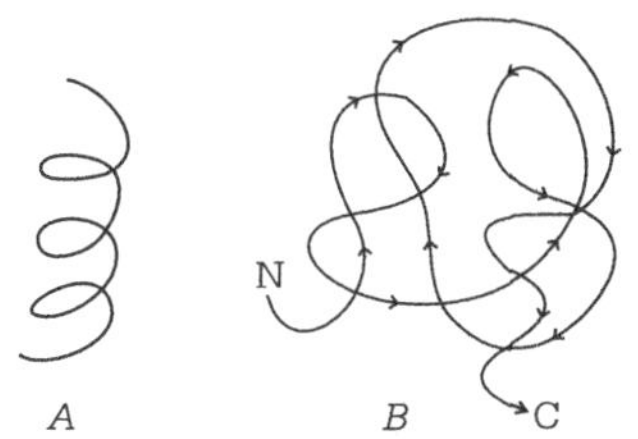

A. ____________

B. ____________

132. ____________ structure of proteins are an assembly of more than one polypeptide or subunits.

133. Answer the following questions based on the figure given below.

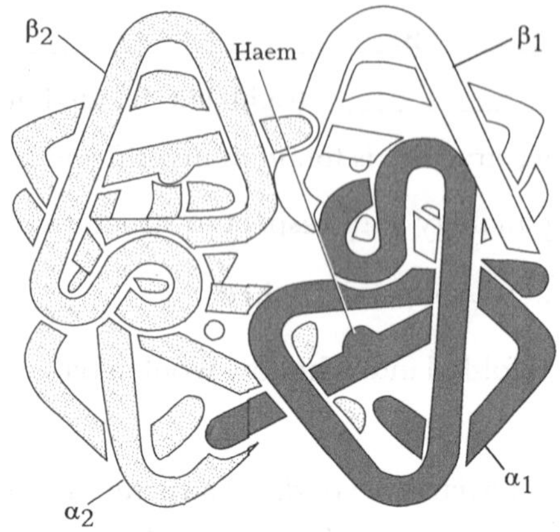

(i) Identify the level of structure of the protein. ____________________

(ii) The protein subunits are arranged in linear string of spheres.

True ☐ False ☐

134. The adult haemoglobin represents ____________________ structure of protein and it consists of ______________ subunits.

Topic 8 : Nature of Bonds Linking Monomers in a Polymer

135. The carboxyl group of one amino acid reacts with amino group of the next amino acid with the help of hydrogen bond.
(*Rewrite the statement by changing the underlined word.*)

____

136. The permanent straightening treatment of hair, affect the disulphide linkage of hair proteins.

True ☐ False ☐

137. Identify the types of bonds *A*, *B* and *C* shown in the following polypeptide backbone.

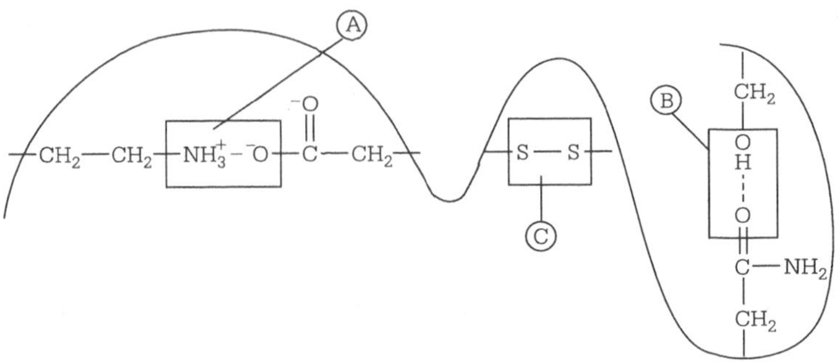

A. ____________________

B. ____________________

C. ____________________

138. The monosaccharides of a polysaccharides are linked by a ________________ bond.

139. The glycosidic bond and peptide bond are formed by the process of ________________ (hydrolysis/dehydration).

140. Which of the following box (*A, B* and *C*) represent the glycosidic bond is in the given structure ?

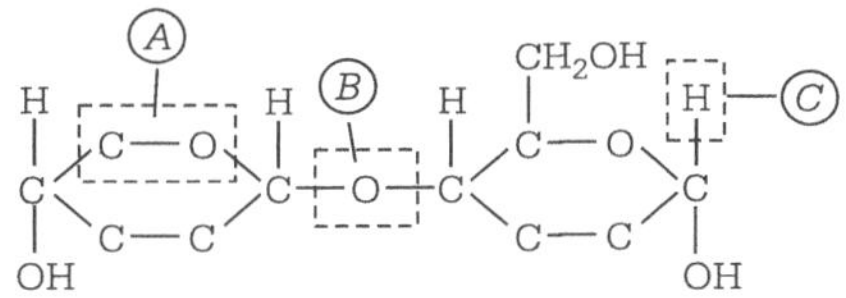

____

141. The phosphate group of nucleic acid links the ________ carbon of sugar of one nucleotide to the ______ carbon of the sugar of the succeeding nucleotide thus, forming ________________ bond.

142. The bond between the phosphate and hydroxyl groups of sugar is an __________ bond.

143. In the following figure of structure of DNA, phosphodiester bond is wrongly draw. Redraw the structure after correcting the phosphodiester bond.

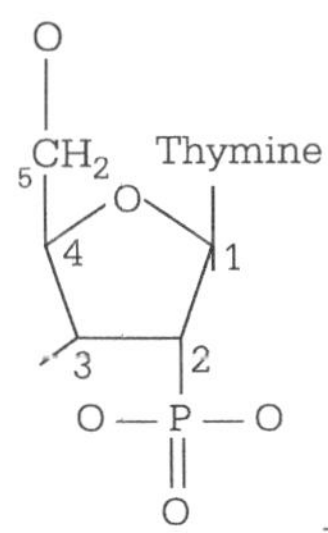

____

144. The famous Watson-Crick model of DNA represent its tertiary structure. (*Rewrite the statement by changing the underlined word.*)

____

145. How many of the following statements are correct for DNA structure?

(i) Double helix

(ii) Two strands of polynucleotide are antiparallel.

(iii) Polynucleotides run in same direction.

(iv) Backbone of nucleic acid is formed by sugar-phosphate sugar chain.

____

146. In a DNA molecule, the nitrogenous bases are projected ________________ (parallel/perpendicular) to the polynucleotide chain and face ________________ (outside/ inside).

147. Consider the given diagram and answer the following questions.

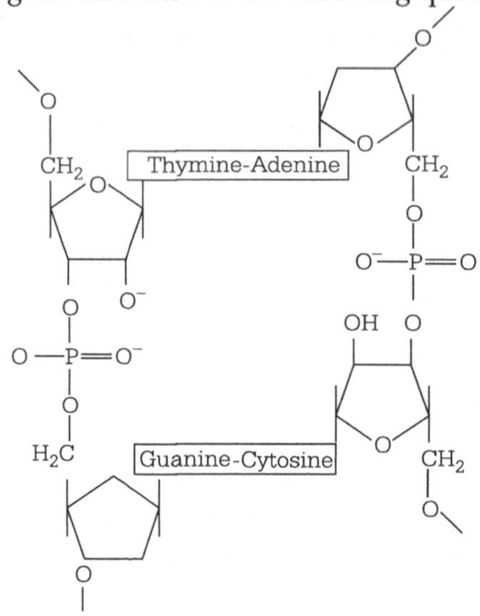

(i) How many hydrogen bond are formed between AT and GC, respectively?

(ii) What happens to the nucleic acid structure when T is replace with G?

---

148. How many of the following statements are incorrect for the given figure?

(i) At each step of acent the strand turns 36°.

(ii) One full turn of helical structure contain 10 base pairs.

(iii) The pitch of DNA helix is 36 Å.

(iv) The rise per base pair is 3.6 Å.

Topic 9 : Metabolism

149. The biomolecules are found in certain concentrations and thus, expressed as ________ or ____________

150. ________________ include all chemical reactions involved in maintaining the living state of cells, and results in the transformation of biomolecules .

151. How many of the following reactions are the example of metabolic transformation?

(i) Carboxylation

(ii) Removal of CO_2 from amino acid.

(iii) Removal of amino group in a nucleotide base.

(iv) Hydrolysis of a glycosidic bond in a disaccharide.

____

152. The metabolites are inconvertible in a series of linked reaction called metabolic pathways.

True ☐ False ☐

153. Different metabolic pathways occur in isolation and do not overlap each other.

True ☐ False ☐

154. The flow of metabolite in metabolic pathway has

(i) definite rate and direction

(ii) indefinate rate and direction less

____

155. The metabolite flow is called the ________________ (dynamic/constant) state of body constituents.

156. How many of the following statement(s) is/are incorrect for the metabolic pathway?

(i) Metabolic reactions are catalysed.

(ii) Some metabolic reactions are uncatalysed.

(iii) These reactions are very smooth and without any mishap.

____

157. __________ (Enzymes/Hormones) are proteins with catalytic power which _________ (hasten/slow) the metabolic reaction.

158. Identify whether the given pathway are anabolic or catabolic?

(i) Conversion of simple substances into complex form. ____________________

(ii) Degradation of complex substances into simple form. ____________________

159. Which of the following column represents anabolic reactions?

Column A	Column B
Conversion of acetic acid into cholesterol.	Glucose becomes lactic acid.
Conversion of protein into glycogen.	Conversion of fatty acid into acetyl Co-A.
Conversion of glucose into glycogen.	Glucose becomes pyruvate.

____

160. Which of the following column in above question (Q. 159) represent the energy releasing reactions?

____

161. The metabolic pathway from glucose to lactic acid occurs in 12 metabolic steps and is called gluconeogenesis.
(*Rewrite the statement by changing the underlined words.*)

____

162. The process of glucose degradation into lactic acid in human skeletal muscle, requires energy.

True ☐ False ☐

163. How many of the following reactions require energy input?
(i) Assembly of protein from amino acid.
(ii) Breakdown of stored fat.
(iii) Degradation of protein into amino acid.
(iv) Lipid synthesis.

____

164. The energy liberated during the metabolic process is stored in the form of ______________. Which is then utilised for biosynthetic and osmotic work.

165. The most important form of energy currency in living system is the bond energy in a chemical called ______________.

166. The blood concentration of glucose in a healthy individual is 4.5-5.0 m mol/L, while that of hormones would be present in litres.

True ☐ False ☐

167. The steady state of living organisms is affected by the ______________ (concentration of biomolecules/size of biomolecules).

168. ______________ refers to the amount of a metabolite processed by one or more catalytic steps per unit time.

169. The chemical and physical process move spontaneously to ______________ and steady state is a non-equilibrium state.

170. How many of the following are the character of living state?

(i) Possess equilibrium state.

(ii) Non-equilibrium steady state to be able to perform work.

(iii) Prevent the onset of equilibrium state through energy input.

____

171. The living state is constantly maintained by ______________________________

Topic 10 : Enzymes

172. A non-protein enzyme is called as <u>exoenzymes</u>. These are nucleic acid in nature. (*Rewrite the statement by changing the underlined word.)*

____

173. During the tertiary structure formation, the protein chain folds and *criss-crosses* itself to form many ____________ sites.

174. Where does the substrate fits into the enzyme?

____

175. Inorganic catalysts and enzymes differ in some context. Identify the effect of the given conditions on both inorganic catalysts I and enzymes II.

(i) High temperature ______________________________

(ii) High pressure ______________________________

176. Some enzymes isolated from thermophilic organisms are stable and retain their catalytic power at high temperature up to 80-90° C.

True ☐ False ☐

177. ____________________ are secreted externally from the cell and function outside the cells.

178. How many of the following transformations depicts the chemical reaction?

(i) Change in shape of a molecule.

(ii) Formation of new bonds.

(iii) Ice melts into water.

(iv) Water becomes vapour.

(v) Hydrolysis of starch into glucose.

____

179. The ______________ of a physical or chemical process refers to the amount of product formed per unit time.

180. Answer the following questions, considering the given expression.

$$\text{Rate} = \frac{\delta P}{\delta t}$$

(i) It is not affected by temperature.

True ☐ False ☐

(ii) The rate ________________________ (doubls only /double or decrease) by half for every 10°C change in either direction.

181. How does the following conditions affect the rate of the reaction?

(i) Catalysed reaction ______________________________

(ii) Uncatalysed reaction ______________________________

182. Answer the following questions considering the given equation.

$$CO_2 + H_2O \underset{}{\overset{A}{\rightleftharpoons}} H_2CO_3$$

(i) Identify the enzyme A. ____________________

(ii) Which of the following graph is correct for the enzyme A? ____________________

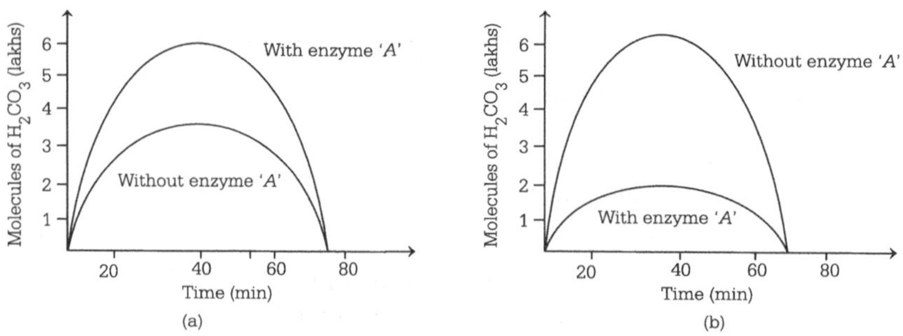

(iii) The enzyme has accelerated the reaction rate by 10 million times as about 600,000 molecules of H_2CO_3 are formed instead of 200.

True ☐ False ☐

183. Answer the following questions, considering the given metabolic pathway.

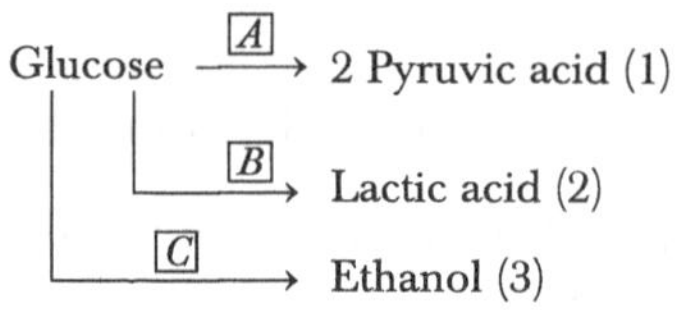

(i) Identify the conditions A and B in which these reaction occur.

A. ____________________ B. ____________________

(ii) Identify the process C by which ethanol is produced from glucose. C. ____________________

(iii) How many enzymes are used in the metabolic pathway 1?

____

(iv) The metabolic pathway 3 occurs in ____________________ .

(v) The metabolic pathway 2 occurs in ____________________ .

184. The enzyme with three dimensional structures including an active site convert a ____________ into a product.

185. The substrate changes the conformation of active site during the binding process.

True ☐ False ☐

186. The substrate binds to the enzyme's active site to form a complex called enzyme product complex. This complex remain in <u>non-transition</u> state.
(*Rewrite the statement by changing the underlined word.*)

____

187. The ______________ is released from the active site after the expected bond formation or dissociation is completed.

188. In an enzymatic pathway the structure of ______________ (enzyme/substrate) gets converted into the structure of protein.

189. The enzyme produce many altered ______________ between the stable substrate and the product which are unstable.

190. All the molecules of substrate possess energy, which is utilised to collide with other molecules, so as to reach between the reactant and product. This energy is called ______________ .

191. Answer the following questions considering the given graph.

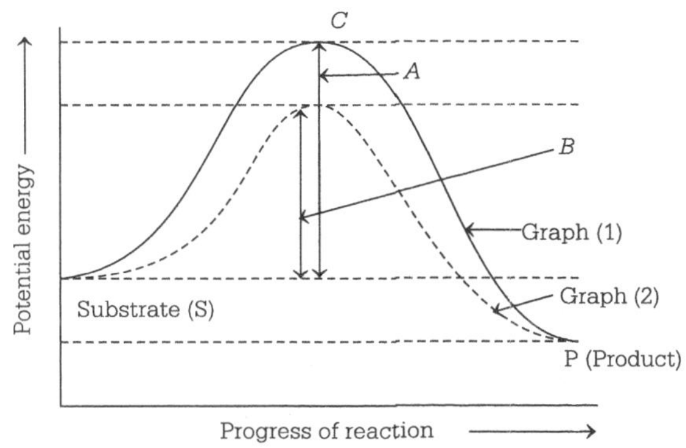

(i) Identify *A*, *B* and *C*.

A.______________ B. ______________

C. ______________

(ii) Which of the following graph represents lower activation energy? ______________

(iii) If product is at lower level than substrate, the reaction is an ______________ reaction.

(iv) The rate of reaction in graph 1 is low because of

____

192. The activation energy of any reaction is ______________ (lowered/increased) by enzyme.

193. The energy required to lower the activation energy is called ______________ .
It is obtained from the ______________ interaction between enzyme and substrate.

194. The peak of energy required for the reaction represents the <u>activation state</u> in which high intermediate is formed during the conversion of reactant to product.
(*Rewrite the statement by changing the underlined word.*)

____

195. Answer the following questions, considering the given equation.

$$E+S \rightleftharpoons \boxed{A} \longrightarrow EP \longrightarrow E+\boxed{B}$$

(i) Identify *A* and *B*. A.________________ B. ________________

(ii) Complex *A* is less reactive and stable.

True ☐ False ☐

(iii) The released enzyme *E* at the and of reaction is reused.

True ☐ False ☐

(iv) *A* and EP are ________________

(v) Which of the following complex is essential for catalysis? ________________

196. Complete the following flowchart.

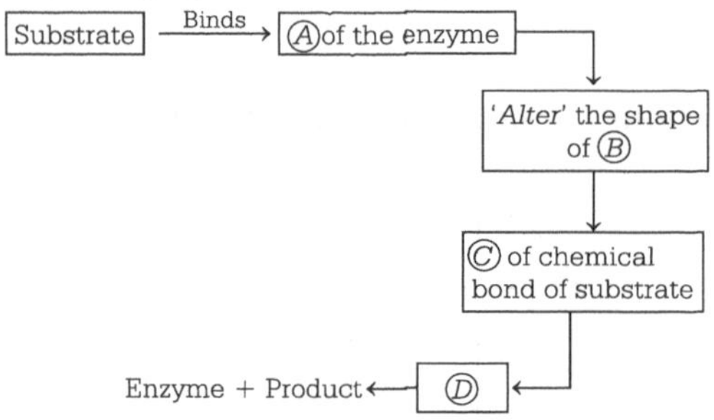

A. ________________

B. ________________

C. ________________

D. ________________

197. Name the hypothesis which depicts that both enzyme and substrate molecules have specific geometrical shapes.________________

198. According to the hypothesis given by Koshland, just like a lock can be opened by its specific key, a substrate molecule can be acted upon by a particular enzyme. It also explains the efficiency of enzyme action.
(*Rewrite the statement by changing the underlined words.*)

____

199. Enzymes are used in very small concentration because

(i) these are toxic chemical compounds.

(ii) these remain unaltered and thus, reused.

____

200. According to induced fit hypothesis, the active site of the enzyme contains two groups, ________________ and ________________ .

201. Which of the following figure represent the lock and key hypothesis?

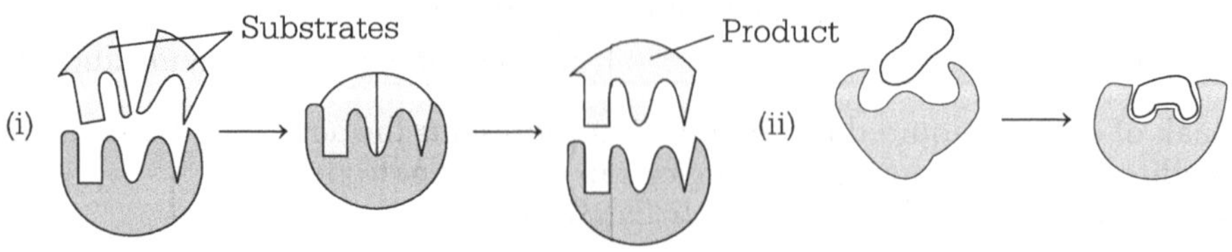

____

202. The activity of an enzyme can be affected by many factors which can alter the ______________ (secondary/tertiary) structure of the protein.

203. Each enzyme shows its highest activity at an optimum temperature and pH.

True ☐ False ☐

204. Answer the following questions by considering the given graph.

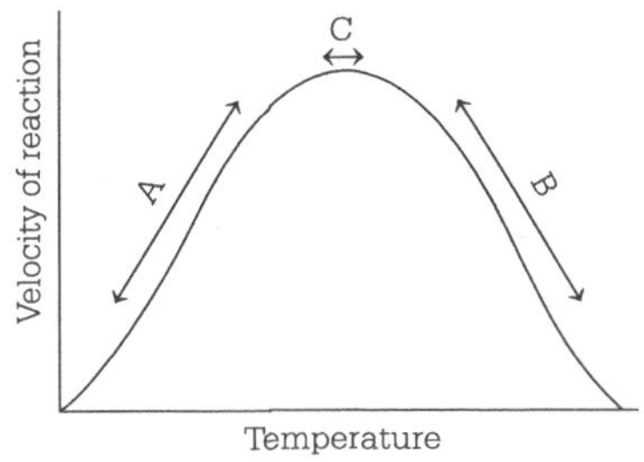

(i) The enzyme remain in inactive state at the slope ______________ .

(ii) Which of the following slope represents optimum temperature? ______________

(iii) During slope ______________ (*A/B*), proteins are denatured.

205. The rate of enzymatic activity ______________ with increase in substrate concentration until all the ______________ are saturated with substrate molecules.

206. The maximum velocity of reaction is denoted by ______________, which is not exceeded by any further rise in concentration of the substrate.

207. Answer the following questions by considering the given graph.

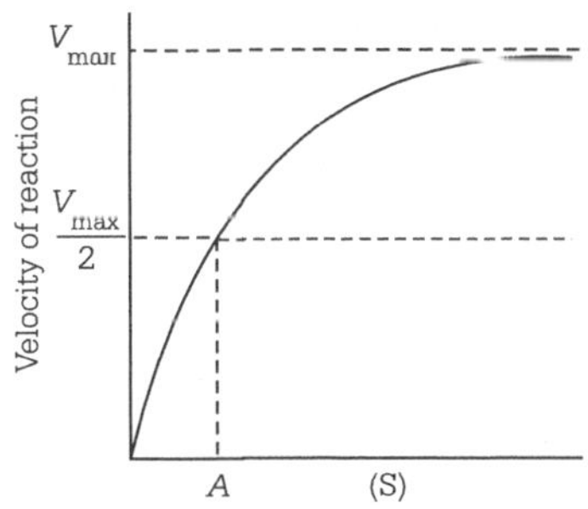

(i) Give relation between velocity of reaction and initial substrate concentration.

____

(ii) At what concentration of substrate, the velocity of reaction becomes independent?

____

(iii) What is '*A*' in this graph? ______________

208. The rate of reaction becomes constant at V_{max} because all the enzyme molecules are used by substrate molecule.

True ☐ False ☐

209. Represent the relation between V_1, V_{max}, K_m, S. ______________

210. The higher the K_m value ________ (lower/higher) is the affinity of the enzyme for the substrate and *vice versa.*

211. The binding of any chemical which shuts off the enzyme activity is called ______________ and the chemical is called ______________ .

212. When the inhibitor closely resembles the substrate in its molecular structure and inhibits the activity of the enzyme, it is known as ____________ (non-competitive/competitive) inhibition.

213. The ________________ (competitive/non-competitive) inhibitor competes with substrate for the substrate binding site of the enzyme due to similar molecular structure.

214. ____________ closely resembles the substrate succinate in structure due to which it can inhibit succinic dehydrogenase.

215. The competitive inhibitors are used in the control of herbicides.
(*Rewrite the statement by changing the underlined word.*)

____

216. Name the enzyme which catalyse the oxidoreduction of the following reaction.
S(reduced) + S(oxidised) $\rightarrow$ S (oxidised) + S(reduced)

____

217. __________________ enzymes catalyse the transfer of a group (other than hydrogen) between a pair of substrate.

218. Hydrolases catalyse the hydrolysis of ester, ether, peptide, glycosidic, C—C, C — halide or P — N bonds.

True ☐ False ☐

219. Enzymes that catalyse the removal of groups from substrates by the mechanisms other than hydrolysis thus, leaving double bonds is known as ____________ (ligases/lyases).

220. Ligases include all enzymes that catalyse the interconversion of optical, geometric or positional isomers.

True ☐ False ☐

221. Hydrolases catalyse the joining of bonds like C — O, C — S, C — N, etc.
(*Rewrite the statement by changing the underlined word.*)

____

222. Match the following columns.

	Column I (Enzyme classes)		Column II (Types of enzymes)
A.	Oxidoreductase	1.	Glycosidases
B.	Trasferases	2.	Aldolases
C.	Hydrolases	3.	Racemases
D.	Lyases	4.	Synthetases
E.	Isomerases	5.	Oxygenases
F.	Ligases	6.	Acyltransferase

A. ____________ B. ____________ C. ____________

D. ____________ E. ____________ F. ____________

223. Give a collecting term for the following.

Dehydrogenases, Oxidases, Deaminases, Peroxidases

____

224. Name the non-protein constituent of enzyme which binds to the enzyme to make it catalytically active. ____________________________

225. The protein portion of the enzyme is called <u>isozymes</u>.
(*Rewrite the statement by changing the underlined word.*)

____

226. Name the all three types of cofactors. __________________________

227. ________________ (Prosthetic group/Zymogen) are organic compounds which are tightly bound to the ____________ (cofactor/apoenzyme).

228. Name the prosthetic group of peroxidases and catalase enzyme, which catalyse the breakdown of H_2O_2 to H_2O and O_2. __________________________

229. The association of ________________ with the apoenzyme is only transient and it usually occur during the course of catalysis.

230. The essential chemical components of many coenzymes are ________________

231. Many enzymes require <u>vitamin</u> for their activity, which form covalent bonds with active site as well as with the substrate.
(*Rewrite the statement by changing the underlined word.*)

____

232. ________________ is a cofactor for the proteolytic enzyme carboxypeptidase.

233. Match the following columns.

	Column I (Cofactors)		Column II (Examples)
A.	Prosthetic group	1.	Molybdenum
B.	Coenzyme	2.	Flavin Adenine Dinucleotide
C.	Metal ions	3.	NAD and NADP (Nicotinamide Adenine Dinucleotide)

A. __________ B. __________ C. __________

234. Enzyme with cofactor is known as ________________________

235. Catalytic activity is lost when the cofactor is removed form the enzyme.

True ☐ False ☐

Answers

1. Carbon, hydrogen, oxygen, nitrogen.
2. True
3. *A*–0.5, *B*–18.5, *C*–46.6, *D*–3.3, *E*–2.8
4. Trichloroacetic acid (Cl_3CCOOH)
5. (i) Cheesecloth or cotton
(ii) *A*–Acid soluble pool, *B*–Acid insoluble pool,
(iii) Acid soluble pool (*A*)
6. Separation
7. Analytical technique gives an idea of molecular formula and structure of the organic compound.
8. Biomolecules
9. Ash
10. (i) *A*–Burn, (ii) Calcium, magnesium
11. Wet weight
12. Sulphate, phosphate
13. Hydrogen, oxygen, chlorine, carbon
14. Organic, inorganic
15. (i) Chemistry's student can identify functional groups like aldehydes, ketones, aromatic compounds, etc.
(ii) Biology's student can identify amino acid, nucleotide bases, fatty acids, etc.
16. Amino acids
17. α-amino acids
18. (ii) The chemical formula of methane is CH_4 in which four valency position of carbon is occupied by hydrogen. Similarly in amino acids, the four valency position occupied by four substituent groups.
19. Hydrogen, carboxyl group, amino group, variable
20. False. On the basis of the nature of *R*-group there are many amino acids.
21. Twenty
22. Glycine
23. A-3, B-1, C-4, D-2
24. False. The physical and chemical properties of amino acids are dependent on amino, carboxyl and *R*-functional groups.
25. Amino
26. Arginine is odd one out, because all of them are neutral amino acids except arginine.
27. Neutral
28. All of them are aromatic amino acid.
29. Basic
30. Acidic
31. NH_2, COOH
32. (i) *B*, (ii) *A*, (iii) *C*
33. Alanine
34. Water, organic solvents
35. Protein amino acids
36. Non-protein
37. Insoluble, Fatty acids
38. Carboxyl
39. False. The *R*-group could be a methyl or ethyl or higher number of $—CH_2$ range between 1 carbon to 19 carbon.
40. 16
41. (i) Without double bond,
(ii) With one or more double bonds
42. Trihydroxy propane
43. True
44. Monoglycerides, diglycerides, triglycerides
45. Lower
46. Phosphorylated, cell membrane
47. Amphipathic
48. Palmitic acid
49. (i) *A*- $(CH_2)_{14}$, (ii) *A*–OH, *B*–OH
50. All of these are conjugate lipids.
51. Animals–Adipose tissue, Plants–Seed
52. (i) *A*–Lecithin
(ii) Phospholipid
53. 3, as shown below

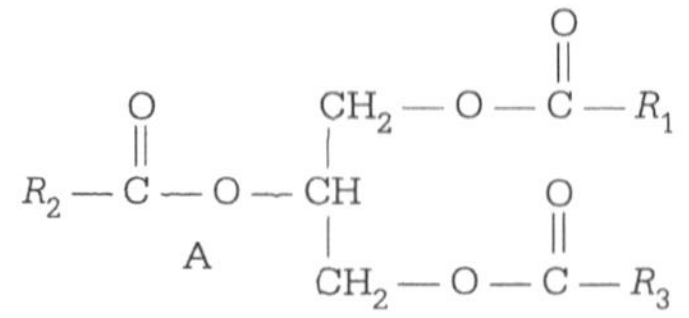

54. Glycolipids
55. Cholesterol
56. (i) Cholesterol,
(ii) 4 (Testosterone, estradiol, aldosterone, ecdysone)
57. Nitrogen base
58.

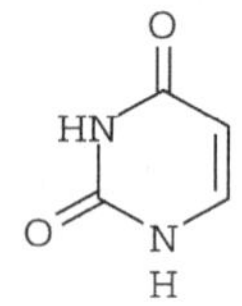

59. Nucleosides
60. Nucleoside
61. Nucleotide
62. Nucleotide
63.

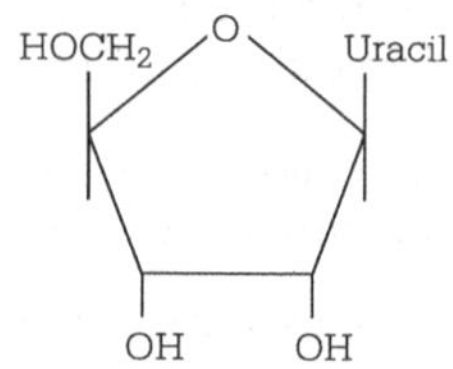

64.

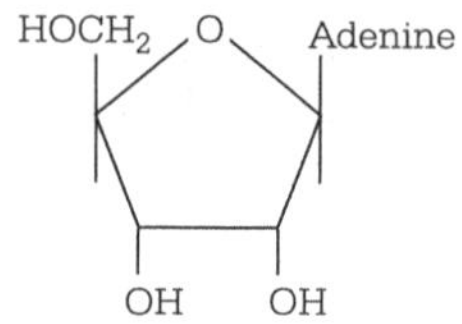

65. Nucleotides
66. *A*–Adenosine monophosphate, *B*–Adenosine diphosphate, *C*–Adenosine triphosphate.
67. Metabolites
68. False. The metabolites which are required for the growth and maintenance of cellular functions are called primary metabolites.
69. 3 (Alanine, thiamine, uridine)
70. True
71. Secondary metabolites
72. A-4, B-1, C-2, D-6, E-3 F-7, G-5
73. Rubber, drugs, spices, scents, pigments.
74. 18 to 800
75. True
76. Biomolecules
77. Acid soluble
78. (i) 800, (ii) Insoluble, (iii) (b)
79. Polymerisation
80. Cytoplasmic

81. A-4, B-5, C-1, D-6, E-3, F-2
82. Water > Protein > Carbohydrate > Lipid
83. Amino acids, peptide
84. Homopolymer
85. Essential
86. Dietary
87. Serine
88. False. Non-essential amino acids are synthesised by our body.
89. A-3, B-1, C-5, D-6, E-2, F-4
90. RuBisCO (Ribulose Bisphosphate Carboxylase Oxygenase)
91. (i) Simple protein, (ii) Derived protein, (iii) Fibrous protein
92. Conjugated
93. False. RuBisCO is the most abundant protein in the whole of the biosphere.
94. Monosaccharides
95. Polysaccharides
96. A-4, B-3, C-2, D-1
97. Arabinogalactan
98. Inulin
99. Reducing, non-reducing
100. (i) Glycogen, (ii) False. It consists of α-D glucose units.
101. Iodine
102. Red
103. False. Cellulose does not contain complex helices and hence cannot hold I_2.
104. (i) Glycogen, (ii) Cellulose, (iii) Inulin, (iv) Chitin
105. Aldehyde, keto
106. Reducing
107. All of them are monosaccharides except sucrose.
108. Glucosamine, N-acetyl galactosamine
109. Glucosamine
110. Homopolymer
111. Monosaccharide, heterocyclic compound, phosphoric acid
112. Nucleic acids
113. A-1, 4; B-2, 3, 5
114. False. Purine and pyrimidines posses heterocyclic ring in nucleic acids.
115. Ribonucleotide, deoxyribonucleotide
116. True
117. Molecular formula
118. (i) Two dimensions, (ii) Three dimensions, (iii) Four levels
119. Positional
120. Primary structure
121. False. If a protein is imagined as a line, the left end represented by first amino acid is known as N-terminal amino acid.
122. C-terminal
123. Helix
124. Right
125. Secondary
126. (i) *A*–α-helix, *B*–β-pleated sheet, (ii) right, (iii) Figure *A*
127. Fibroin of silk
128. Tertiary
129. Tertiary
130. (i) Tertiary structure, (ii) van der Waals' forces, electrostatic bond and hydrogen bond.
131. *A* – Secondary structure, B – Tertiary structure
132. Quaternary
133. (i) Quaternary structure, (ii) False. The protein subunits are arranged one upon each other in the form of a cube.
134. Quaternary structure, 4
135. Peptide
136. True
137. *A*–Ionic bond, *B*–Hydrogen bond, *C*–Disulphide bond
138. Glycosidic
139. Dehydration
140. *B*
141. 3', 5', phosphodiester
142. Ester
143.

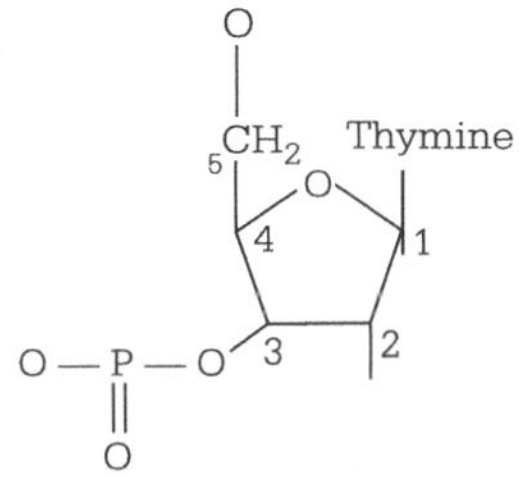

144. Secondary
145. 3 (i, ii, iv) is correct for DNA structure. (iii) is incorrect as polynucleotide run in antiparallal direction.
146. Perpendicular, inside
147. (i) There are two hydrogen bonds between A and T and three hydrogen bonds between G and C.
(ii) It will effect the double helical structure of nucleic acid. G and T will not pair, hence helical structure would not form.
148. 2 (iii, iv) is incorrect statement. The pitch is 34Å and the rise per base pair is 3.4 Å.
149. mol/cell, mol/litre
150. Metabolism,
151. 4 (All of them).
152. True
153. False. The metabolic pathways are either linear or circular and also *criss-cross* each other.
154. (i) Definite rate and direction
155. Dynamic
156. Only one, i.e. (ii).
157. Enzymes, hasten
158. (i) Anabolic pathway, (ii) Catabolic pathway
159. Column A
160. Column B (catabolic reaction)
161. 10, glycolysis
162. False. The process of glucose degradation into lactic acid in human skeletal muscle, liberate energy.
163. 2 (i, iv)
164. Chemical bond
165. Adenosine Triphosphate (ATP)
166. False. The blood concentration of glucose in a healthy individual is 4.2 – 6.1 m mol/L, while that of hormones would be present in nanograms/mL.
167. Concentration of biomolecules
168. Metabolic flux
169. Equilibrium
170. 2 (ii, iii)
171. Metabolism
172. Ribozymes.
173. Active

174. The active site of an enzyme is a crevice or pocket into which substrate fits.

175. (i) Inorganic catalyst works efficiently at high temperature while enzyme gets denatured.
(ii) High pressure is favourable for inorganic catalyst while it affects the efficiency of enzyme.

176. True

177. Exoenzymes

178. 2 (ii, v)

179. Rate

180. (i) False. It is affected by temperature.
(ii) Doubles or decreases

181. (i) The rate of the reaction will increase due to the catalysis process of enzymes.
(ii) The rate of the reaction would not affected due to the absence of enzymes.

182. (i) Carbonic anhydrase, (ii) (a), (iii) True

183. (i) *A*–Aerobic condition, *B*–Anaerobic condition,
(ii) *C*–Fermentation,
(iii) 10 different enzymes (during glycolysis)
(iv) Yeast, (v) Skeletal muscles of human.

184. Substrate (S)

185. False. The substrate does not change the conformation of active site during binding process.

186. Transition

187. Product

188. Substrate

189. Structural state

190. Activation energy

191. (i) *A*–Activation energy without enzyme, *B*–Activation energy with enzyme, *C*–Transition state,
(ii) Graph (2),
(iii) Exothermic, (iv) The rate of graph 1 is low because of large activation energy.

192. Lowered

193. Binding energy, non-covalent

194. Transition state

195. (i) *A*–ES, *B*–P, (ii) False. Complex *A* is short lived and reactive
(iii) True,
(iv) Intermediate complex,
(v) E + S

196. *A*–Active site, *B*–Enzyme, *C*–Breakdown, *D*–Enzyme-Product complex

197. Lock and key hypothesis

198. Emil Fischer, specificity

199. (ii) The chemical reactants do not cause any alteration in the composition or physiology of the enzyme. The same enzyme molecule can be used again and again. Hence, it is used in very small concentration.

200. Buttressing, catalytic

201. (i) represent lock and key hypothesis.

202. Tertiary

203. True

204. (i) Slope *A*, (ii) *C*, (iii) *B*

205. Increases, active sites

206. V_{max}

207. (i) Velocity of reaction α initial substrate, (ii) High concentration,
(iii) K_m (Michalis - Menten's constant)

208. True

209. $V_1 = \dfrac{V_{max}[S]}{K_m + [S]}$

210. Lower

211. Inhibition, inhibitor

212. Competitive

213. Competitive

214. Malonate

215. Bacterial pathogens.

216. Oxidoreductases/ dehydrogenases

217. Transferases

218. True

219. Lyases

220. False. Isomerases include all enzymes that catalyse the interconversion of optical, geometric or positional isomers.

221. Ligases

222. A-5, B-6, C-1, D-2, E-3, F-4

223. All of them are the oxidoreductase enzymes.

224. Cofactor

225. Apoenzymes.

226. Prosthetic group, coenzyme, metal ions.

227. Prosthetic group, apoenzyme

228. Haem

229. Coenzyme

230. Vitamin

231. Metal ions

232. Zinc

233. A-2, B-3, C-1

234. Holoenzyme

235. True

CHAPTER TEN

CELL CYCLE AND CELL DIVISION

Topic 1 : Cell Cycle

1. In all living organisms, growth and reproduction are the main characteristics of the cell.

   True ☐ False ☐

2. Such cycles of growth and division allows a ____________ to form a structure consisting of millions of cells.

3. Name the two process which also takes place during cell division.

   (i) ____________ (ii) ____________

4. Write the scientific word for the process given in the statement.

   The sequence of events by which a cell duplicates its genome, synthesises the other constituents of the cell and eventually divides itself in two cells. ____________

5. Cell growth is a continuous process in terms of ____________

6. During cell cycle, the DNA synthesis occurs how many times? ____________

7. The events of cell cycle are under the control of ____________ (cell/genes).

8. The replicated chromosomes get distributed into the ____________ of daughter cells.

9. Typical eukaryotic cells (human cell) divide once every ________ hours (12/24).

10. The duration of cell cycle in the cells of all organisms is same.

    True ☐ False ☐

11. *Saccharomyces cerevisiae* divides in every ____________ (45 minutes/ 90 minutes).

12. Name two basic stages of cell cycle.

    (i) ____________ (ii) ____________

13. Which phase of cell cycle is the actual phase of cell division? ____________

14. The phase between two successive M-phase is called ____________

15. What is the approximate percentage duration of cell cycle that comes under interphase in humans? ____________

16. Identify the following events of cell cycle on the basis of given statements.

(i) Separation of daughter chromosomes. ____________

(ii) Division of cytoplasm. ____________

17. <u>M-phase</u> is also called as resting phase.
(*Rewrite the statement by changing the underlined word.*)

____

18. Choose the odd one out from the following phases.

G_1-phase, S-phase, G_2-phase, M-phase

____

19. The G_1-phase corresponds to the ____________ between mitosis and initiation of DNA replication.

20. The enzymes needed for DNA replication in synthesis phase are synthesised in G_2-phase.

True ☐ False ☐

21. Match the following columns.

Column I (Phases of interphase)	Column II (Events occur)
A. G_0-phase	1. Synthesis of proteins and enzymes
B. G_1-phase	2. Duplication of DNA
C. G_2-phase	3. Inactive stage of the cell cycle
D. S-phase	4. Cell metabolically active and continuously grow

A. ____________ B. ____________ C. ____________ D. ____________

22. If the initial amount of DNA in a cell is 2C, what is the content of DNA after the synthesis phase gets over in the cell cycle? ____________

23. The amount of DNA per cell reduces during synthesis phase of interphase.

True ☐ False ☐

24. The number of chromosome increases during S-phase of cell cycle.

True ☐ False ☐

25. In which phase of cell cycle, the centriole duplicates in the cytoplasm? ____________

26. Name the cell phase which synthesise protein to cause condensation of chromosomes to initiate mitosis. ____________

27. The heart cells enter the G_0-stage and occassionally divide.

True ☐ False ☐

28. Give one word for the given statement.

The cell exiting G_1-phase to enter an inactive stage.________________

29. G_0-phase exists between mitotic phase and G_1 - phase.

True ☐ False ☐

30. The plant cells can show mitotic division in both ________________ cells.

31. Identify *A*, *B*, *C* and *D* in the figure of cell cycle.

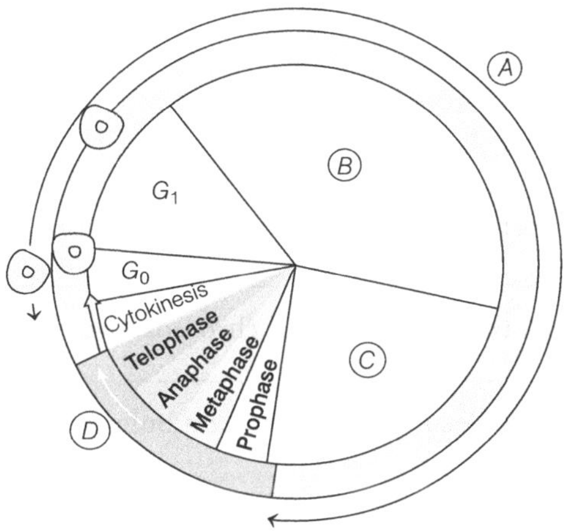

A. ________________

B. ________________

C. ________________

D. ________________

32. Name the plant species and stage at which mitosis is seen in haploid cells.

____

33. In animals, mitotic cell division is only seen in the ____________ somatic cells.

Topic 2 : Mitosis

34. Mitosis is also referred as ____________ division.

35. Write the stages name of karyokinesis in a sequential manner.

____

36. Which of the following is not a characteristic feature of mitosis in somatic cells?

Disappearance of nucleolus, Chromosome movement, Synapsis, Spindle fibres formation

____

37. In which stage of cell cycle, initiation of condensation of chromosome takes place?

____

38. The centriole moves to opposite poles of the cell in metaphase.
(*Rewrite the statement by changing the underlined word.*)

____

39. Consider the following figure of late prophase and answer the given questions.

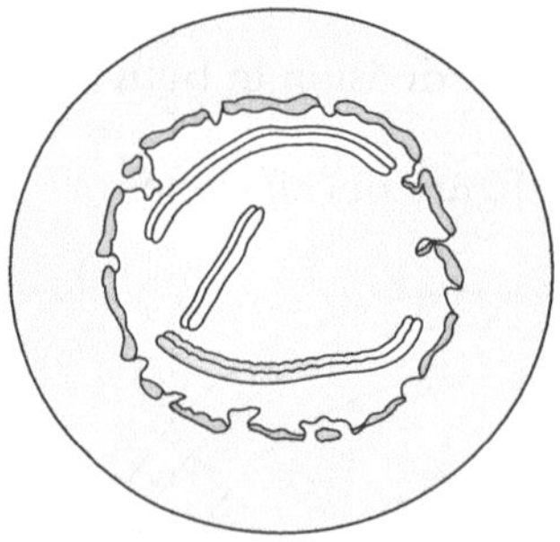

(i) Name the structure which are formed at this stage by the condensation of chromosomal material. ______________________________

(ii) Name the structure which attach two chromatids of chromosomes.

____

(iii) The centromeres at the opposite poles radiate out microtubules called.

____

40. How many of the cell organelles are not seen at the end of the prophase stage of mitotic cell division?

ER, Golgi body, Nucleolus, Lysosomes, Chloroplast

____

41. The complete disintegration of the ______________________ marks the initiation of second phase of mitosis.

42. The morphology of chromosomes can be studied at ____________ stage of cell division.

43. Metaphase chromosome is made up of two sister chromatids.

True ☐ False ☐

44. Write the technical term for the sentence given below.

Small disc-shaped structure at the surface of centromeres.______________________

45. The plane of alignment of the chromosomes at metaphase is referred to as

____

46. Observe the diagram given below and answer the questions that follows.

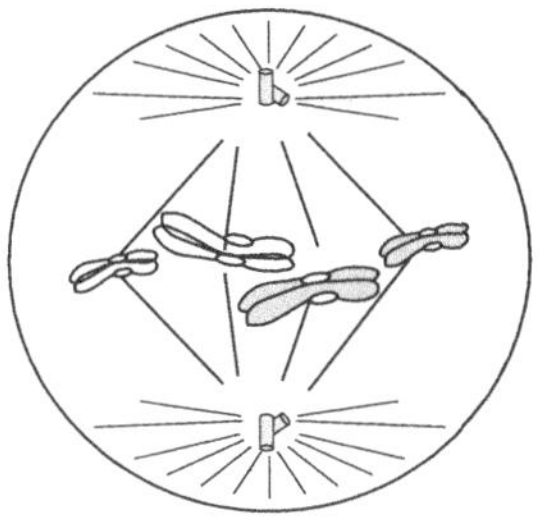

(i) Which phase of mitosis is dipicted in the given diagram? ______________

(ii) Spindle fibres attach to the ______________ of chromosome.

(iii) Chromosomes are moved to spindle equator and get aligned along metaphase plate through spindle fibre to both poles.

True ☐ False ☐

47. At which stage of mitosis, the two daughter chromatids separate from each other, migrate towards the opposite poles and are now referred to as chromosomes of the future daughter nuclei? ______________

48. During anaphase chromosomes move away from the ______________, the centromere of each chromosome is directed towards ______________ .

49. Observe the given diagram of mitotic stage, answer the following questions.

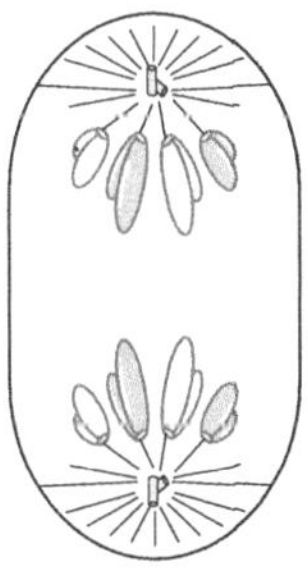

(i) Identify the stage. ______________

(ii) At this stage ______________ split and ______________ separate.

50. At the end of anaphase, each chromosome arranged at metaphase plate is split simultaneously.

True ☐ False ☐

51. Metaphase is the shortest phase of mitotic cell division.

True ☐ Falsc ☐

52. The mitotic anaphase differs from metaphase in possessing same number of chromosomes and half number of chromatids.

True ☐ False ☐

53. ______________ is the final stage of karyokinesis in which nuclear envelop develops around the chromosome clusters at each ______________ forming ______________ daughter nuclei

54. Look at the figure given below and answer the following questions.

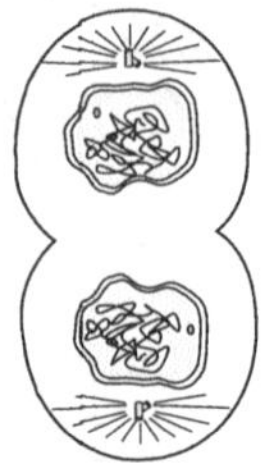

(i) Identify the stage of mitotic cell division.

____

(ii) Name the cell organelles which reform at this stage.

____

(iii) Chromosomes cluster at opposite spindle poles.

True ☐ False ☐

55. Match the following columns.

	Column I (Characteristics)		Column II (Phases of mitosis)
A.	Chromosomes become arranged in a horizontal plane at the equator	1.	Anaphase
B.	Daughter chromosomes move to opposite poles of the spindle	2.	Prophase
C.	Chromosomes become visible as fine like long threads	3.	Telophase
D.	Chromosomes lose their distinctiveness and gradually transform into chromatin network	4.	Metaphase

A.______ B.______ C.______ D.______

56. Give the technical term for the following sentence.
Segregation of duplicated chromosomes into two daughter nuclei.______________

57. Cytokinesis involves the segregation of nuclear content.

True ☐ False ☐

58. Cell furrow appears in the plasma membrane of animal cell during ______________

59. Cytoplasmic division occurs *via* cell plate formation in______________

60. When karyokinesis is not followed by cytokinesis, it results in the formation of

____

61. Study the diagram of amitosis carefully and answer the following questions.

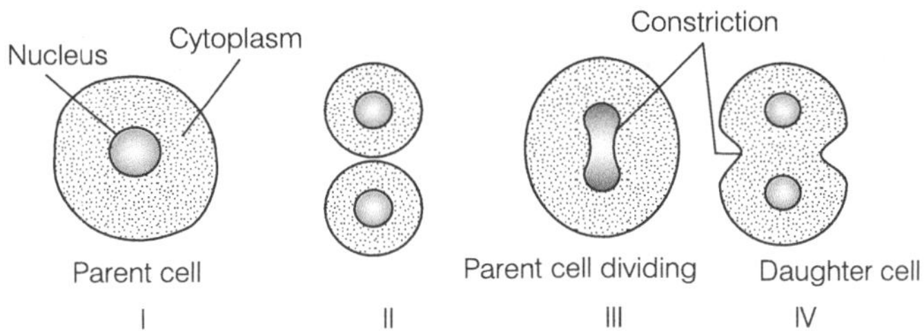

(i) Write the correct sequence of this event. ____________________

(ii) Give example of an organism which undergoes amitosis. ____________________

(iii) Nuclear envelope disintigrates.

True ☐ False ☐

62. Syncytium is the condition of multinucleate cells.

True ☐ False ☐

63. Formation of liquid endosperm in coconut is the result of not formation of ____________________ around daughter nuclei.

64. Mitosis usually results in the production of __________ daughter cells. Growth of ____________ organisms and __________.

65. Mitosis usually is restricted to the __________ cells.

66. In some social insects, mitosis can be seen in __________ cells.

67. Mitosis results into daughter cells with some different genetic constitution.

True ☐ False ☐

68. Give the technical term for the following statement.

The ratio between nuclear content and cytoplasm. ____________________

69. ____________ is maintained by process called cell division.

70. Some list of cells are given below, mark them as mitosis or meiosis on the basis of division exhibited by them.

(i) Upper layer of epidermal cells. ____________________

(ii) Germ cell. ____________________

(iii) Cells of lining of gut. ____________________

(iv) Blood cells. ____________________

71. The meristematic cells of plants continuously exhibit the process of

____

Topic 3 : Meiosis

72. The type of division that ensures the production of haploid phase in the life cycle of sexually reproducing organisms is ________

73. The kind of cell division in which chromosome number is reduced to half is called

____

74. Four haploid cells are formed at the end of meiosis-I only.

True ☐ False ☐

75. The precursor cells of meiosis are ________

76. Meiosis involves two sequential cycles of ________

77. The important events in meiosis are pairing of ________ and ________ between non-sister chromatids of homologous chromosomes.

78. During meiosis, DNA replication occurs thrice during interphase, M-phase and cytokinesis. (*Rewrite the statement by changing the underlined word.*)

____

79. Prophase-II of meiotic division is the longest and complex process as compared to similar stage in mitosis.

(*Rewrite the statement by changing the underlined word.*)

____

80. Choose the odd one out among the following.

Leptotene, Crossing over, Zygotene, Pachytene, Diplotene, Diakinesis

____

81. The chromosomes become gradually visible under the light microscope during ________ stage of meiosis.

82. Pairing of homologous chromosomes occurs during leptotene.

True ☐ False ☐

83. The compaction of chromosomes occurs during ________

84. Use the technical term for the following sentence.

The process of pairing of chromosomes. ________

85. The ________ complex is formed during zygotene stage of meiosis-I.

86. The tetrad or bivelant is formed during zygotene.

True ☐ False ☐

87. How many of the following events does not occur during zygotene?

Formation of synaptonemal complex, Pairing of chromosomes, Appearance of bivalents, Involvement of recombinase

____

88. The paired chromosomes are known as _______________

89. Pachytene is relatively longer than zygotene and leptotene.

True ☐ False ☐

90. Name the site where crossing over occurs between non-sister chromatids of the homologous chromosomes. _______________

91. Give the biological word for the following sentence.

Exchange of genetic material between the two homologous chromosomes.

____

92. The crossing over is a non-enzymatic complex process.

True ☐ False ☐

93. Complete the following analogy.

Zygotene : Bivalent : : Pachytene : _______________

94. Consider the following figure and answer the questions given below.

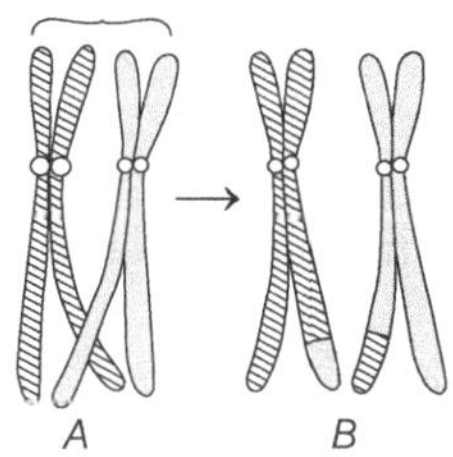

(i) Identify the event *A*._______________

(ii) At which stage event *A* occurs?_______________

(iii) Event *B* occurs in the substage _______________ of prophase-I of meiosis-I.

95. The begining of diplotene is recognised by the dissolution of the

____

96. Consider the following figure of cell cycle and name the structure marked as *A*.

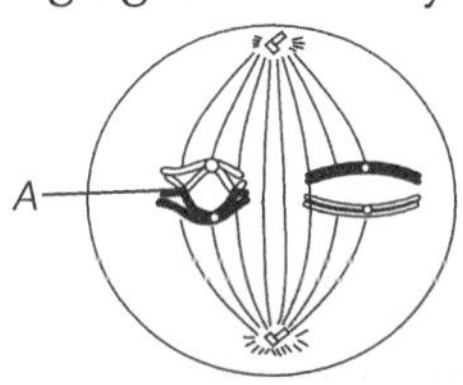

A. _______________

97. During meiosis-I, one of the daughter cells receive mixture of maternal and paternal chromosomes.

True ☐ False ☐

98. The crossed over homologous chromosomes get completely separated at the end of pachytene.

True ☐ False ☐

99. The X-shaped structure formed between the non-sister homologous chromosomes is called ______

100. The meiotic spindle condensation brings out separation of ______ chromosomes.

101. The diakinesis has intact nuclear envelope and nucleolus.

True ☐ False ☐

102. How many of the following events occurred during diakinesis.

Terminalisation of chiasmata, Dissolution of synaptonemal complex, Degeneration of nucleolus, Fully condensed chromosomes

______

103. Splitting of centromere and hence separation of chromatids occurs during

______

104. Name the stages of cell cycle which is represented by the diagram given below (*A* and *B*).

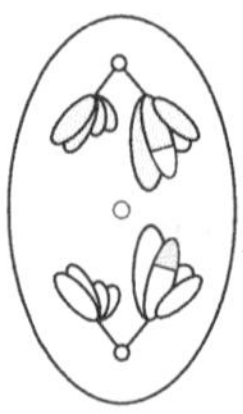

A

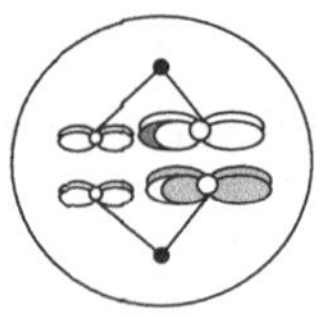

B

A. ______

B. ______

105. Identify the *A* and *B* of the given table considering the given figure.

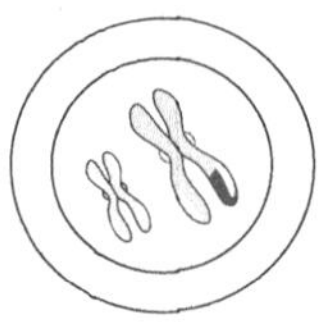

Number of homologous chromosomes	Number of chromatids	Phase of cell division
(*A*)	(*B*)	Telophase-I

A. ______ B. ______

106. Match the following columns.

	Column I		Column II
A.	Interkinesis	1.	Four haploid daughter cell
B.	Metaphase-I	2.	Nuclear membrane reappear
C.	Telophae-I	3.	Stage between two meiotic divisions
D.	Telophse-II	4.	Chromosome align at equatorial plate

A. ______ B. ______ C. ______ D. ______

107. After meiosis-I, the resultant daughter cells have __________ the amount of DNA in comparison to ___________ cell.

108. Match the following columns.

	Column I (Stages of meiosis)		Column II (Features)
A.	Prophase-I	1.	Bouquet stage
B.	Anaphase-II	2.	Centromeric division
D.	Metaphase-I	3.	Two kinetochores

A. ______________ B. _______________ C. _______________ D. _______________

109. Tetrad of cells are formed at the end of the __________________ of meiosis.

110. ______________ cell division occurs at the time of gamete formation in all sexually reproducing organism.

111. The genetic variability is maintained by the process of _______________________ and _______________ of meiotic cell divisions.

112. How many of the following is not the significance of meiosis?

Assortment of chromosomes, DNA replication, Crossing over, Mutation, Protein synthesis

____

113. Variations which are important for <u>mutation</u> are caused by meiosis.
(*Rewrite the statement by changing the underlined word.*) _________________________________

114. Meiosis results in reduction of chromosome number, maintaining consistency of chromosome number during sexual reproduction and production of genetic variability.

True ☐ False ☐

Answers

1. True
2. Single cell
3. (i) DNA replication, (ii)–Cell growth
4. Cell cycle
5. Cytoplasmic increase
6. Once
7. Genes
8. Nucleus
9. 24
10. False. Duration of cell cycle is different in different organisms.
11. 90 minutes
12. (i) Interphase
    (ii) M-phase/divisional phase
13. M-phase or mitosis phase
14. Interphase
15. 95%
16. (i) – Karyokinesis,
    (ii) – Cytokinesis
17. Interphase
18. M-phase is odd one out. Other three are stages of Interphase.
19. Interval
20. False. Enzymes needed for DNA replication are synthesised in G_1-phase.
21. A-3, B-4, C-1, D-2
22. 4C
23. False. Amount of DNA per cell becomes doubled during synthesis phase of interphase.
24. False. The number of chromosome does not increase during S-phase of cell cycle.
25. S-phase
26. G_2-phase
27. False. The heart cells enter the G_0-stage and do not exhibit division.
28. Quiescent stage (G_0-phase)

29. True
30. Diploid and haploid
31. *A*–Interphase, *B*–S-phase, *C*–G_2-phase, *D*–M-phase
32. Gametophyte of bryophyte, e.g. *Riccia.*
33. Diploid
34. Equational
35. Prophase, Metaphase, Anaphase, Telophase.
36. Synapsis occur during meiosis
37. Prophase
38. Prophase
39. (i) Chromosomes
(ii) Centromere
(iii) Asters
40. 3 (ER, Golgi body, nucleolus)
41. Nuclear membrane
42. Metaphase
43. True
44. Kinetochore
45. Metaphase plate
46. (i) Metaphase
(ii) Kinetochores
(iii) Ture
47. Anaphase
48. Equatorial plate, pole
49. (i) Anaphase, (ii) Centromeres, chromatids
50. False. At the onset of anaphase, each chromosome arranged at metaphase plate is split.
51. False. Anaphase is the shortest phase of mitotic cell division.
52. True
53. Telophase, pole, two
54. (i) Telophase, (ii) Nucleolus, Golgi complex and endoplasmic reticulum (iii) True.
55. A–4, B–1, C–2, D–3
56. Karyokinesis
57. False. Cytokinesis involves the segregation of cytoplasmic content.
58. Cytokinesis
59. Plant cell
60. Multinucleate cell
61. (i) I $\rightarrow$ III $\rightarrow$ IV $\rightarrow$ II
(ii) Meganucleus of *Paramecium*
(iii) False. Nuclear envelope remains intact.
62. True
63. Cell wall
64. Diploid, multicellular, cell repair
65. Diploid
66. Haploid
67. False. Mitosis results into daughter cells with identical genetic complement.
68. Nucleocytoplasmic ratio
69. Nucleocytoplasmic ratio
70. (i) Mitosis, (ii) Meiosis, (iii) Mitosis, (iv) Mitosis
71. Mitotic cell division
72. Meiosis
73. Meiotic cell division
74. False. Meiosis-I ends with the formation of two daughter cells.
75. Germ cell and diploid cells
76. Cell division and nuclear division
77. Homologous chromosome, recombination
78. Once during S-phase
79. Prophase-I
80. Crossing over is odd one out. Rest others are substages of prophase-I.
81. Leptotene
82. False. Pairing of homologous chromosomes occurs during zygotene.
83. Leptotene
84. Synapsis
85. Synaptonemal
86. True
87. 1 (Involvement of recombinase).
88. Homologous chromosomes
89. True
90. Recombination nodules
91. Crossing over
92. False. Crossing over is an enzymatic complex process.
93. Recombination nodule
94. (i) Crossing over, (ii) Pachytene, (iii) Diakinesis
95. Synaptonemal complex
96. *A*–Chiasmata
97. True
98. False. Crossed over homologous chromosomes get completely separated at the end of anaphase-I.
99. Chiasmata
100. Homologous
101. False. At the end of diakinesis, nuclear envelope and nucleolus disappear.
102. 3 (Terminalisation of chiasmata, degeneration of nucleolus, fully condensed chromosomes).
103. Anaphase-II
104. *A*–Anaphase-I, *B*–Metaphase-I
105. *A*–2, *B*–4
106. A-3, B-4, C-2, D-1
107. Twice, haploid
108. A-1, B-2, C-3
109. Telophase-II
110. Meiotic
111. Crossing over, recombination
112. 2 (DNA replication and protein synthesis).
113. Evolution
114. True

CHAPTER ELEVEN

TRANSPORT IN PLANTS

Topic 1 : Introduction and Means of Transport

1. The water taken up by the roots has to reach all parts of the plant except the tip of the growing stem.

   True ☐ False ☐

2. The ________________ of the leaves moved to all parts of the plant including the root tips embedded deep inside the soil.

3. Name the substances which are transported into flowering plants.

   ____

4. The substances moved by active transport and by ________________ supplemented by active transport for small distance.

5. Transport over longer distances proceeds through the vascular system is called ________________ (transportation/translocation).

6. The transportation of water and mineral is multidirectional in rooted plants.

   (*Rewrite the statement by changing the underlined word.*) ________________

7. Name any two substances which undergo multidirectional transport in plants.

   ____

8. ________________ synthesised in the photosynthetic leaves are transported all other parts of the plant including ________________.

9. The nutrient withdrawn from ________________ parts of the plants and moved to the ____________________ parts (senescent/growing).

10. Osmosis is the movement of molecules from regions of lower concentration to regions of higher concentration in a random fashion.
(*Rewrite the statement by changing the underlined words.*)

____

11. Diffusion is a ________ (slow/fast) process, it can account for only ________ (long/short) distance movement of molecules.

12. Consider the figure given below and answer the following questions.

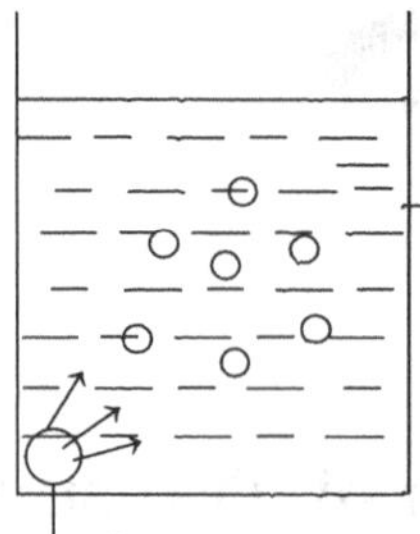

(i) Identify the process. ______________________________

(ii) The given process require expenditure of energy.

True ☐ False ☐

13. Name the only means for gaseous transport within the plant body. __________________

14. Rearrange the following on the basis of rate of diffusion of substances in decreasing order.

Gases, Solid, Liquid

____

15. The pressure exerted by particles of a substance, when they move from their higher concentration to lower is called as ____ .

16. What is the relationship between diffusion pressure, kinetic movement and number of diffusing particles? ____

17. Put the sign (< / >) in between the following equation.

Diffusion pressure of pure water ☐ Diffusion pressure of sugar solution.

18. What happened to the diffusion pressure of solution, when more solute is added?

____

19. The rate of diffusion is affected by the factors like the gradient of concentration, ____________________________, ____________________ and ______________.

20. What is the effect of following conditions on the rate of diffusion through membrane?

(i) Hydrophobic moiety ______________________________

(ii) Hydrophilic moiety ______________________________

21. Membrane protein provide site for substance containing hydrophilic moiety across the membrane, this process is called diffusion.
(*Rewrite the statement by changing the underlined word.*)

____

22. Facilitated diffusion transport the molecules from low to a high concentration.

True ☐ False ☐

23. In facilitated diffusion transport rate reached to ______________ when all of the ______________ transporters are being used.

24. Answer the following questions considering the given figure.

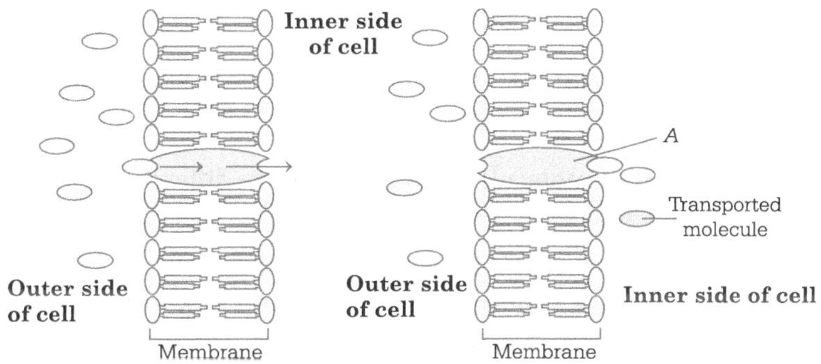

(i) Identify the labelled part *A* and the process of transportation.

____

(ii) How many of the given below molecules transport through the membrane *via* this process?

Glucose, Lipid, Na^+, Ca^{++}, Amino acid, Nucleic acid, Oxygen, H^+, Carbon dioxide

____

25. Facilitated diffusion is very ______________, it allows cell to select substances for uptake.

26. Facilitated diffusion is ______________ (sensitive/insensitive) to inhibitors which reacts with protein side chains.

27. The proteins form ______________ (tunnels/channels) in the membrane for molecules to pass through.

28. Some protein channels are always ______________ while others can be ______________, some are large, allowing a variety of molecules to cross.

29. Name the proteins that form large pores in the outer membranes of the plastids, mitochondria and some bacteria. ______________________

30. Which type of molecules pass through the porins?

____

31. In facilitated diffusion an intracellular molecule bound to the transport protein, these proteins then rotates to release the molecule inside the cell.
(*Rewrite the statement by changing the underlined word.*)

____

32. Name the porin which are usually specific for water permeability and exclude the passage of other solutes.______________________

33. Complete the following analogy with respect to the membrane protein.

Porin : Aquaporin : : ______________________ : α-helices protein

34. ______________________ are made up of __________ different types of aquaporins.

35. In ______________ (symport/uniport) two types of molecules move together in the ________ (same/different) direction.

36. In an uniport two types of molecules move together in different direction.

True ☐ False ☐

37. In __________________ transport system a molecule moves across a membrane independent of other molecules.

38. Match the following columns.

Column I (Types of transporter)	Column II (Figure of transporter)	Column III (Example of transporter)
(i) Uniport	(a) 	1. H^+ / K^+
(ii) Antiport	(b) 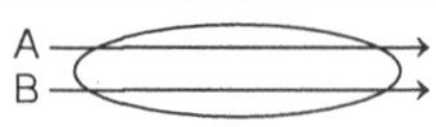	2. NH_4^+
(iii) Symport	(c) 	3. Na^+ / H^+

(i) ______________ (ii) ______________ (iii) ______________

39. The increase in temperature ____________ (decreases/increases) the rate of diffusion.

40. ____________ (Active/Passive) transport uses energy to transport and pump molecules against the concentration gradient.

41. Active transport is carried out by specific ________________

42. Pumps are proteins that use ________________ to carry substances across the ________________.

43. In uphill transport, the pump can transport substances from a high concentration to low concentration.

True ☐ False ☐

44. In which conditions, the rate of active transport reaches maximum?

____

45. Fill the given table (*A*-*G*) by applying Yes or No.

	Property	Simple diffusion	Facilitated diffusion	Active transport
(i)	Requires special membrane proteins	No	(*A*)	Yes
(ii)	Highly selective	No	(*B*)	(*C*)
(iii)	Transport saturates	No	Yes	(*D*)
(iv)	Uphill transport	(*E*)	(*F*)	Yes
(v)	Requires ATP energy	(*G*)	No	Yes

A. ________ B. ________ C. ________ D. ________

E. ________ F. ________ G. ________

Topic 2 : Plant-Water Relations

46. A watermelon has over 50% of water.
(*Rewrite the statement by changing the underlined word.*)

____

47. How many of the following soil water is available to the plants?

Capillary water, Gravitational water, Hydroscopic water

____

48. A mature corn plant absorbs almost ________ litres of water in a day.

49. Name the main limiting factor for plant growth and productivity. ________________

50. The greater the concentration of water in a system, the ________ is its kinetic energy or ________________.

51. Answer the following questions, considering the given experiment.

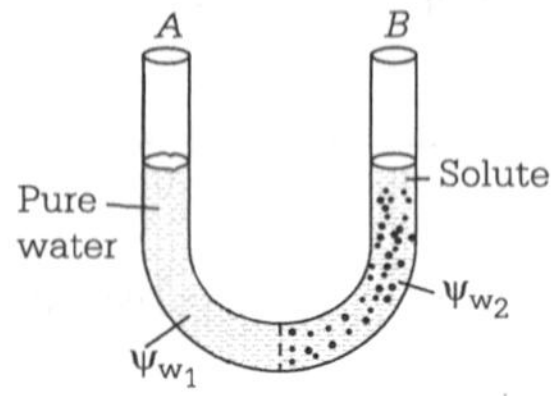

(i) In which direction water moves? ____________________

(ii) ψ_{w_2} has maximum value.

True ☐ False ☐

(iii) The water potential of pure water have the ____________________ value.

(iv) The value of ψ_{w_2} ________________ (increases/decreases) with addition of more solutes.

52. The water potential of pure water at standard temperature is <u>negative</u> .

(*Rewrite the statement by changing the underlined word.*)

____

53. All solutions have ________________ (lower/higher) water potential than pure water.

54. The magnitude of water potential of any solution lowers due to the dissolution of solute is called <u>matric</u> potential.

(*Rewrite the statement by changing the underlined word.*)

____

55. Solute potential is always positive.

True ☐ False ☐

56. Which of the following tubes possess more negative solute potential (ψ_s)?

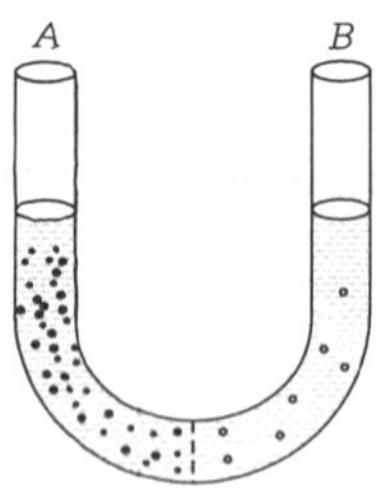

____

57. At which pressure water potential is equal to solute potential?

____

58. Solute potential in leaf cell sap is more ________________ during the day and less ________________ at night (positive/negative).

59. If a pressure greater than atmospheric pressure is applied to pure water or a solution, its <u>solute potential</u> increases.
(*Rewrite the statement by changing the underlined word.*)

____

60. Water enters in a plant cell due to ______________ causing a pressure built up against the cell wall, it makes the cell ______________________.

61. The pressure exerted by the protoplast against the cell wall is called matric potential.

True ☐ False ☐

62. Pressure potential is usually ______________________ (positive/negative).

63. Match the following columns.

	Column I		Column II
A.	Water potential	1.	It is negative in xylem
B.	Pressure potential	2.	In full turgid cell is zero
C.	Solute potential	3.	It is negative in phloem

A. ______________ B. ______________ C. ______________

64. ____________________ is an expression of adsorption of water by colloidal particles or surfaces in plant cells.

65. What is the relationship between ψ_w, ψ_s and ψ_p? ______________________

66. Consider the given data of two solutions and complete the analogy.

Solution A $\rightarrow \psi_s = 0$

Solution B $\rightarrow \psi_p = 0$, $\psi_s = -0.23$

$\psi_{w_A} : \psi_{w_B} :: 0 :$ __________

67. In which direction solutes moves, considering the given data of two solutions?

Solution $A \rightarrow \psi_w = 0$

Solution $B \rightarrow \psi_p = 0.23$ $\psi_s = -0.23$ ______________________

68. Identify the direction of water flow in the following figure.

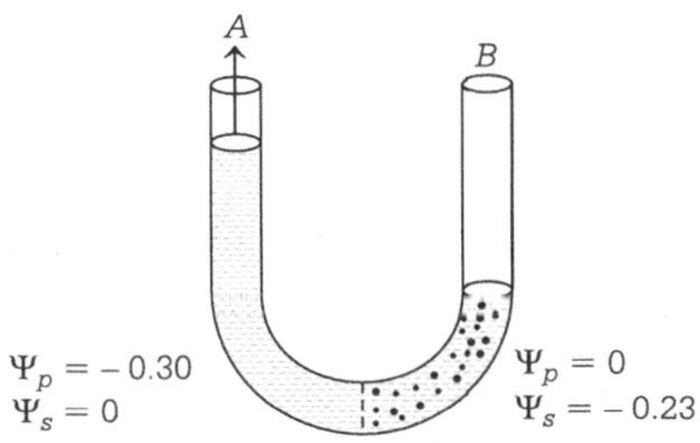

____

69. What happens if the turgor pressure becomes equal to the cell wall pressure?

____

70. What will be the direction of net movement of water if solution *A* enclosed in a semipermeable membrane, having solute potential – 30 bars and pressure potential 5 bars is submerged in a solution *B* with osmotic potential of – 10 bars and pressure potential is 0? ______________________________

71. Arrange the following solutions with respect to water potential in increasing order.

Pure water, 2% sucrose solution, 4% sucrose solution

____

72. The difference between the diffusion pressure of a solution and its pure solvent at a particular temperature and pressure is called ______________________________

73. Water always move from low DPD to high DPD.

True ☐ False ☐

74. What is the relation between DPD, Osmotic Pressure (OP) and Turgor Pressure (TP)?

____

75. Name the process of diffusion in which water moves across a differentially or selectively permeable membrane.______________________________

76. The net direction and rate of osmosis depends on both the ______________ and ______________ .

77. Water will move from its region of lower chemical potential to its region of higher chemical potential.

True ☐ False ☐

78. Answer the following questions, considering the given figure.

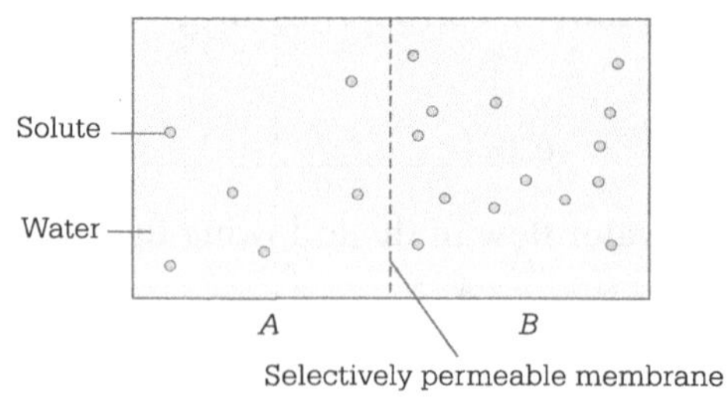

(i) Solution of which chamber has lower water potential? ______________________________

(ii) Solution of which chamber has a lower solute potential?______________________________

(iii) In which direction will osmosis occur?______________________________

(iv) At equilibrium which chamber will have lower water potential? ______________________________

79. Answer the following questions, considering the given experiment.

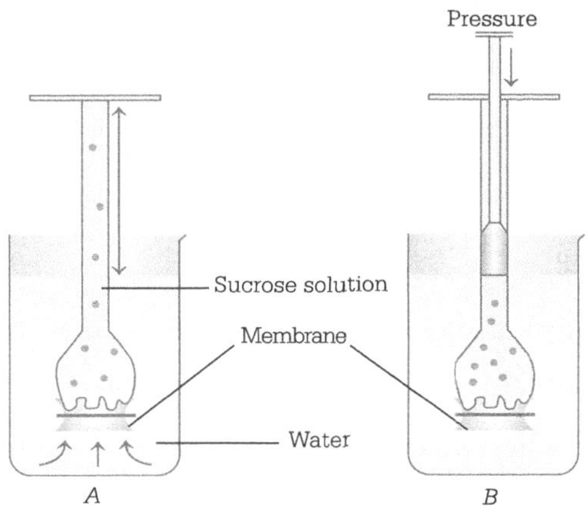

(i) The movement of water into the funnel will continue till the ____________ is reached.

(ii) In which funnel water potential is more? ____________________________

(iii) Due to the application of pressure in funnel *B*, the solute concentration decreases.

True ☐ False ☐

80. Water potential is the pressure required to completely stop the entry of water into an osmotically active solution across a semipermeable membrane.
(*Rewrite the statement by changing the underlined word.*)

____

81. Complete the following analogy with respect to osmosis.
Osmotic potential : Osmotic pressure : : Negative : ____________________

82. What is the relationship between osmotic pressure and concentration of solution?

____

83. The solute concentration of solution *A* is more than solution *B* which is separated by semipermeable membrane. In which solution greater pressure will be required to prevent water from diffusing in? ____________________________

84. ________________ is the osmotic entry of water into a cell when placed in pure water.

85. ______________________ is the withdrawal of water from a cell when placed in ______________________ .

86. Match the following columns.

	Column I		Column II
A.	Isotonic	1.	External solution is more concentrated
B.	Hypotonic	2.	External solution is more dilute than the cytoplasm
C.	Hypertonic	3.	External solution balances the osmotic pressure of the cytoplasm

A. ______________ B. ______________ C. ______________

87. Complete the following analogy with respect to cell shape when kept in different solution.

Hypertonic : Hypotonic : : ______________ : Swell.

88. Deplasmolysis occurs when water moves out of the cell and the cell membrane of a plant cell shrinks away from its cell wall.

(*Rewrite the statement by changing the underlined word.*)

____

89. Match the following columns.

	Column I (Plant cells)		Column II (Conditions)		Column III (Solutions)
(i)	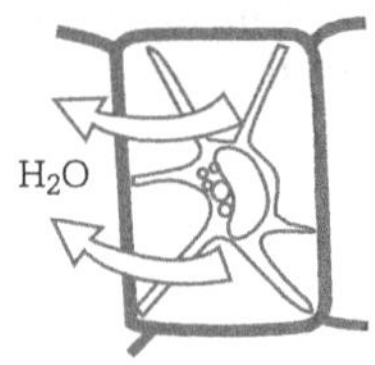	(a)	Turgid	(1)	Hypertonic
(ii)	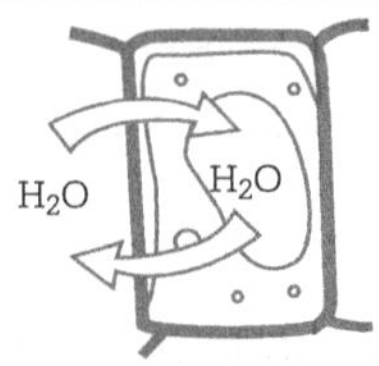	(b)	Plasmolysed	(2)	Isotonic
(iii)	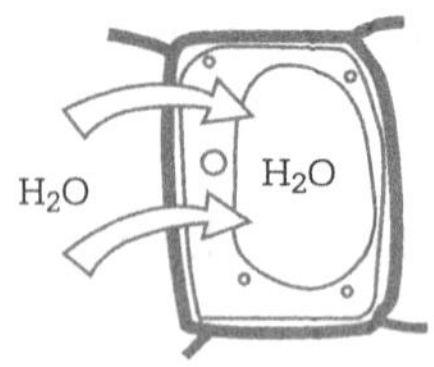	(c)	Flaccid	(3)	Hypotonic

(i) ______________ (ii) ______________ (iii) ______________

90. In ______________ (plasmolysed/flaccid) cell water moves out from the cytoplasm and then from vacuole.

91. The water when drawn out of the cell through ______________ into the extracellular fluid causes the protoplast to ______________ away from the walls.

92. When the cell is placed in ______________________ (isotonic/hypotonic) solution, there is no net flow of water towards the inside or outside the cells, it is said to be __________________ (turgid/flaccid).

93. When the cells are placed in <u>hypertonic</u> solution, water diffuses into the cell causing the cytoplasm to build up a pressure against the wall that is called <u>osmotic</u> pressure. (*Rewrite the statement by changing the underlined words.*)

____

94. The __________________ is responsible for enlargement and extension of cells.

95. Match the following columns.

	Column I (Stages of plasmolysis)		Column II (Features)
A	Limiting	1.	The protoplast withdraws itself from the corners
B.	Incipient	2.	Protoplast shrinks and withdraws from cell wall
C.	Evident	3.	The protoplast is in contact with cell wall

A. ____________ B. ____________ C. ____________

96. The cell of potato get bursted if it is kept in concentrated salt solution.

True ☐ False ☐

97. Pickles, meat and fish are preserved by salting. (*Give reason*)

____

98. When a plasmolysed cell placed in ____________________ solution, the protoplast again assumes its normal shape, this process is called ____________ .

99. Complete the following analogy with respect to cell water relations.

Plasmolysis : Deplasmolysis : : ______________ : Endosmosis

100. Name the special type of diffusion in which water is absorbed by solid colloidal substances causing them to increase in volume. ____________________________

101. Write the appropriate word for following statements.

(i) The solid particles which imbibe water. ____________________________

(ii) The liquid which is imbibed. ____________________________

102. Seeds, dry wood and root hair absorb water through ______________________

103. In prehistoric period man breaks rock and boulder by the process of <u>osmosis</u>. (*Rewrite the statement by changing the underlined word.*)

____

104. Due to the imbibition a pressure is developed is known as ______________ and heat is generated known as ____________.

105. Due to the ______________ seedlings would able to emerge out of the soil into the open.

106. A water potential gradient is not important for the imbibition.

True ☐ False ☐

107. There should be some forces of ____________ between imbibant and imbibed liquid or ____________ between imbibant and imbibed liquid.

Topic 3 : Long Distance Transport of Water

108. The movement of a molecule across a typical plant cell takes approximately ________.

109. In large and complex organism, substances have to be moved to ____________ distances because the site of ____________ or ____________ and ____________ are too far.

110. Name the process by which water, mineral and food are transported in plants.

____

111. ____________ is the movement of substances in bulk or in mass from one point to another due to ____________ between the two points.

112. Bulk flow can be achieved only by negative hydrostatic pressure gradient.

True ☐ False ☐

113. Complete the following analogy with respect to hydrostatic pressure.

(i) Negative : Positive : : ____________ : Straw

(ii) Negative : Positive : : Xylem : ____________

114. The bulk movement of substances through the conducting or vascular tissues of plants is called <u>transportation</u>.
(*Rewrite the statement by changing the underlined word.*)

____

115. How many of the following compounds translocated by xylem?

Organic solutes, Water, Mineral salts, Inorganic solutes, Organic nitrogen, Hormones

____

116. Root hairs are thin-walled slender extensions of root epidermal cells that greatly increases the surface area for absorption.

True ☐ False ☐

117. The ________________ (symplast/apoplast) is the system of adjacent cell walls that is continuous through the plant.

118. Name the pathway in which movement of water occurs exclusively through the intercellular spaces and the walls of the cells. ________________

119. Movement through the apoplast involve crossing the cell membrane.

True ☐ False ☐

120. Water evaporates into the ________________ (intercellular/intracellular) spaces or the atmosphere, tension develop in the continuous stream of water in the apoplastic movement.

121. Name the forces help in mass flow of water? ________________

122. The ________ (apoplastic/symplastic) system is the system of interconnected protoplasts.

123. Neighbouring cells are connected through cytoplasmic strands that extend through

____

124. Symplastic movement is aided by ________________ of individual cells.

125. Answer the following questions considering the given figure.

(i) Identify the labelled part *A*, *B*, *C*, *D*, *E* and *F*.

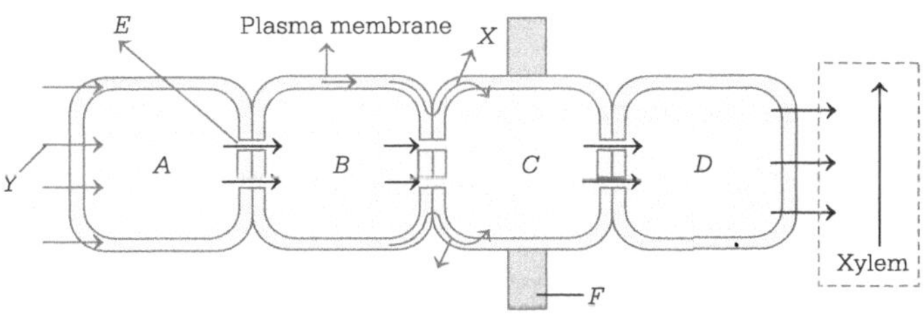

A. ________________
B. ________________
C. ________________
D. ________________
E. ________________
F. ________________

(ii) Which of the following pathway shows symplastic movement? ________________

(iii) In which pathway the movement of molecule is continuous throughout the plant except at the Casparian strips?

____

126. Give any two examples of cytoplasmic streaming.

____

127. Match the following columns.

Column I		Column II
A. Apoplast	1.	It is slower
B. Symplast	2.	It is faster
	3.	It consists of non-living parts of plant
	4.	It consists of living parts of plant

A. ________________ B. ________________

128. Two types of water movement pathway are given in Box *A* and Box *B*. Identify the type of pathway.

Box *A*	**Box *B***
Water move from one cell to another passing through the plasmodesmata.	Water move from plasma membrane and also cross the tonoplast.

A. ______________________ B. ______________________

129. Most of the water flow in the roots occur *via* the apoplast.

True ☐ False ☐

130. The cortical cell are __________ packed and hence offer no ____________ to water movement.

131. The endodermis is pervious to water because of a band of suberised matrix called the suberised strip.

(*Rewrite the statement by changing the underlined words.*)

____

132. Water molecules are unable to penetrate the endodermal layer, so they are directed to wall regions that are not __________, into the cells through the ____________.

133. The movement of water through the root layers is ultimately ______________ in the ______________.

134. In young roots, water enters directly into the ______________ or __________, these are non-living conduits and are the parts of the ______________.

135. Identify the labelled parts *A*, *B*, *C* and *D* from the given figure.

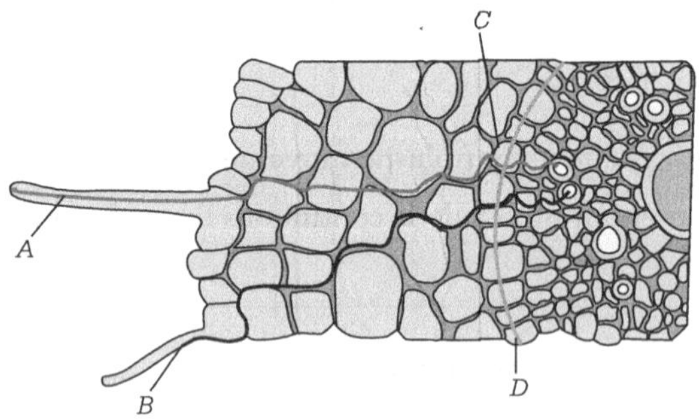

A. ______________________

B. ______________________

C. ______________________

D. ______________________

136. A mycorrhiza is a symbiotic association of a algae with the root system of lower plants. (*Rewrite the statement by changing the underlined words.*)

____

137. The mycorrhizal hyphae increase the surface area that absorb mineral ions and water.

True ☐ False ☐

138. *Pinus* have an ________________ association with the mycorrhizae. Their seeds cannot germinate and establish without it.

139. The vascular tissue of roots absorb water, due to this water potential of higher plants ____________________, hence the pressure inside the xylem also ______________ (increases/decreases).

140. The negative, pressure inside the xylem is called root pressure.

(*Rewrite the statement by changing the underlined word.*)

____

141. The ____________________ is accountable for pushing up water to small heights in the stem.

142. Why does the solution ooze out of the stem, when stem cut horizontally near the base.

____

143. Transpiration is the process in which water collects in the form of droplets around veins near the tip of grass blades and leaves of many herbaceous plants.

True ☐ False ☐

144. ________________ is the result of root pressure and low evaporation.

145. Root pressure play a major role in water movement in __________ trees (tall/short).

146. The root pressure maintain the continuous chains of ion molecules in the phloem.

(*Rewrite the statement by changing the underlined words.*)

____

147. The continuous chain of ____________________ molecules in the xylem break under the enormous tension created by ________________.

148. How many of the following seasons create maximum root pressure?

Winter, Summer, Rainy, Spring

____

149. Root pressure is favoured by ____________________ and ________________.

150. Root *A* placed in hypotonic solution and root *B* is placed in hypertonic solution. In which root, root pressure is maximum? ________________________________

151. What is the rate of water flow in the xylem? ____________________________

152. Cohesion-tension and transpiration pull theory is based upon ________________ developed in the xylem vessels and tracheids.

153. ____________________ (Adhesion/Cohesion) is the force due to which water molecules remain associated to the xylem surface.

154. Name the force which is responsible for the attraction between water molecules inside the xylem vessels. ____________________

155. Adhesion and cohesion provides ____________ and high ____________.

156. Adhesion and cohesion maintain the continuous water column in phloem.

True ☐ False ☐

157. The continuous water column in ____________ is pulled upward due to a force known as ____________.

158. Complete the following analogy with respect to ascent of sap theories.

Root pressure theory : Cohesion-tension theory : : ____________ : Dixon and Jolly

159. Give a collective term for the following theories.

Root pressure theory, Vital force theory, Capillary force theory, Pulsation theory

____

160. Complete the following flowchart, showing mechanism of transpiration pull.

Water is lost during transpiration → Turgor pressure of mesophyll cell ____________ and DPD ____________ → Mesophyll cell take water from ____________, hence their turgor pressure ____________ → Water is absorbed from ____________.

161. The transpiration pull theory suggests that the transport of water takes place at high pressure and low tension.

True ☐ False ☐

162. The air bubbles would break the continuity of water column, as these are formed under tension and low temperature.

(*Rewrite the statement by changing the underlined words.*)

____

163. Match the following columns.

	Column I (Theories)		Column II (Features)
A.	Root pressure	1.	The pulsatory activity of living cortical cell absorb water
B.	Transpirational pull	2.	The tracheids and vessels act as narrow capillary tube
C.	Vital force theory	3.	It is absent in gymnosperms
D.	Capillarity theory	4.	It helps in uptake to water up to a height of 130 m

A. ____________ B. ____________ C. ____________ D. ____________

Topic 4 : Transpiration

164. The evaporative loss of water in plants through stomata is known as ________________

165. The loss of water vapour in __________________, exchange of ________________ and in the leaf occurs through the stomata.

166. Match the following columns.

Column I (Types of transpiration)	Column II (Characteristics)
A. Cuticular	1. It contribute very little, but more than lenticular.
B. Bark	2. It contributes 3-10% of total transpiration.
C. Lenticular	3. It contributes approximately 95% of total transpiration.
D. Stomatal	4. It contributes only 0.1% of total transpiration.

A. ______________ B. ______________ C. ______________ D. ______________

167. The opening and closing of stomata is occur due to change in the turgidity of the <u>subsidiary cells</u>.

(*Rewrite the statement by changing the underlined words.*)

____

168. The inner wall of each guard cell, towards the pore or ______________, is thick and ______________ .

169. The opening of the stoma is aided by the orientation of the ________________ in the cell walls of the ___________________ .

170. Complete the following flowchart by putting *A* and *B* sentences.

Guard cells lose turgor → Elastic inner walls regain their shape → *A* → *B*

A. ________________________________ B. ________________________________

171. Answer the following questions considering the given figure.

(i) Identify the labelled parts *A*, *B* and *C* of stomata.

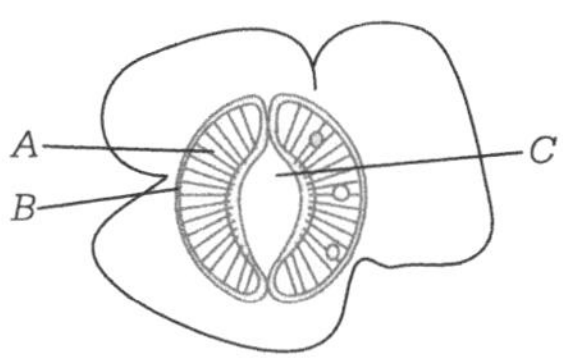

A. ___________________________

B. ___________________________

C. ___________________________

(ii) The microfibrils are ________ and arranged ________________ .

172. Some conditions of stomatal movement are given below. Identify in this situation stomata will be open or close.

(i) Osmotic pressure and diffusion pressure deficit increase in guard cell.

(ii) Turgor pressure of guard cell increase.

____

173. Complete the following flowchart of guard cell photosynthetic theory.

A
↓
Photosynthesis in guard cell
↓
B
↓
Increased osmotic pressure of guard cell
↓
C
↓
Guard cell become turgid
↓
D

A. ____________________ B. ____________________

C. ____________________ D. ____________________

174. Fill in the blanks with respect to guard cell photosynthetic production theory during night time.

(i) Sugar converted into ____________________ .

(ii) Osmotic pressure of guard cell ____________________.

(iii) Guard cell become ____________________ .

(iv) Stomata ____________________ .

175. The change in the ________ of the guard cell's cell sap is due to the presence or absence of ____________ .

176. Complete the following analogy with respect to pH of guard cell.

Day : Night : : Neutral : ____________

177. Low CO_2 concentration in the intercellular spaces induces ____________ of stomata.

178. Stomata will be closed when ____________ released to stop the K^+ exchange.

179. A flowchart is given below with respect to K^+ ion transport theory. Identify the steps V and VI.

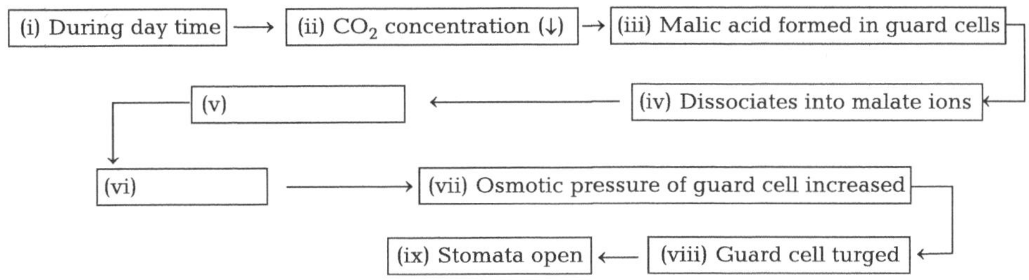

(v)________________ (vi) ________________

180. Complete the following analogy.

Stomata open :____________ : : Stomata close : ____________

181. The upper surface of dorsiventral leaf has a greater number of stomata.

True ☐ False ☐

182. How many of the following factors affect the rate of transpiration?
(i) Temperature, (ii) Humidity, (iii) Light, (iv) Plant height, (v) Wind speed, (vi) Distribution of stomata.

____

183. In ____________ (summers/winters) the vapour pressure gradient is high and hence transpiration is ____________ (high/low).

184. A low root/shoot ratio ____________ the rate of transpiration.

185. Transpiration helps in removed of excess water and better development of root system.

True ☐ False ☐

186. The substances that reduces the rate of transpiration are called ____________

187. Give a common collective term for the following chemicals.

Phenyl mercuric acetate, Abscisic acid, Dodecenyl succinic acid

____

188. The transpiration driven ascent of xylem sap depends mainly on the physical properties of water ____________, ____________ and ____________.

189. Match the following columns.

	Column I (Physical properties of water)		Column II (Characteristics)
A.	Cohesion	1.	Water molecules are attracted to each other in the liquid phase
B.	Adhesion	2.	Attraction between water molecules
C.	Surface tension	3.	Attraction of water molecules to polar surfaces

A. ____________ B. ____________ C. ____________

190. The system of xylem vessels from root to leaf vein can supply the water for respiration. (*Rewrite the statement by changing the underlined word.*)

____

191. Complete the following flowchart of mechanism of transpiration by identify A ans B.

(i) Ⓐconcentration of water vapour in atmosphere ← (ii) Ⓑfrom leaf cells ← (iii) Pulling of water molecules from xylem

A. _______________ B. _______________

192. Answer the following questions considering the given figure.

(i) Identify part A, B and C.

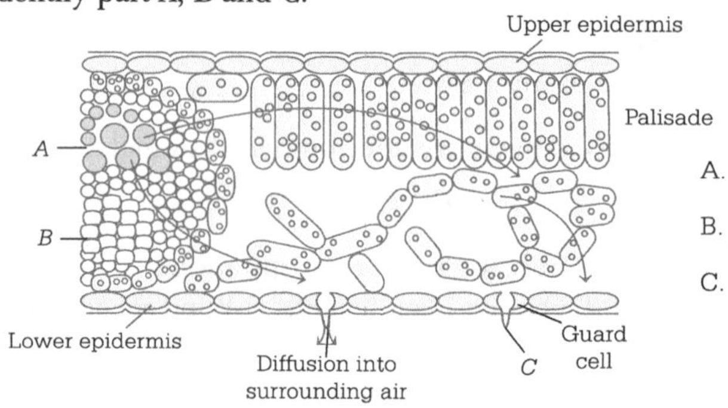

A. _______________

B. _______________

C. _______________

(ii) Evaporation from the leaf sets up a _______________ between outside air and the air spaces of the leaf.

193. Transpiration creates transpiration pull for photosynthesis.

(*Rewrite the statement by changing the underlined word.*)

____

194. Transpiration cools the leaf surfaces, maintains the __________ and __________ of the plant cells by keeping them __________ .

195. Photosynthesis is __________ by available water which can be swiftly __________ by transpiration.

196. Rainforests have high humidity. (*Give reason*)

____

197. The evolution of the __________ (C_3 / C_4) photosynthetic system is strategies for __________ (minimising/maximising) the water loss.

198. C_3 plant loses only half as much water as a C_4 plant for the same amount of CO_2 fixed.

True ☐ False ☐

199. The concentration of minerals in the soil is usually __________ than the concentration of minerals in the root.

Topic 5 : Uptake and Transport of Mineral Nutrients

200. Minerals are present in the soil as uncharged particles which cannot move across cell membranes.
(*Rewrite the statement by changing the underlined word.*)

____

201. Name the process by which most minerals enter the root into the cytoplasm of epidermal cells of root. ______________________________

202. What is responsible for the active uptake of ions in roots? ______________________________

203. Some ions also move into the epidermal cells ______________ (actively/passively).

204. What is present in the root hair cells which actively pump ions from the soil into the epidermal cells? ______________________________

205. ______________ of endodermal cells are control points, ______________, where a plant adjusts the quantity and types of solutes that reach the xylem.

206. Endodermis actively transport ions in one direction only. (*Give reason*)

____

207. Rearrange the following in increasing order with respect to osmotic pressure during osmotic absorption.

Root hair < Soil water < Cortical cell ______________________________

208. Absorption of water takes place even when osmotic pressure is ______________ than that of cell sap, this type of absorption is called non-osmotic absorption.

209. In ______________ absorption the movement of water from the soil into the plant is due to more negative water potential in the xylem.

210. What happens when turgor pressure decreases and diffusion pressure deficit increases in mesophyll cell? ______________________________

211. What is the role of the following regions in the mineral ions translocation.

Apical and lateral meristems, Young leaves, Developing flowers, Fruits and seeds, The storage organs

____

212. ______________ (Unloading/Uploading) of mineral ions occurs at the fine vein endings through diffusion and active uptake by chief sink cells.

213. Mineral ions are frequently remobilised from ______________ part to __________.

214. Name any one element which is not remobilised. ______________________________

215. Give the common term for the following molecules.

Amino acid, Phosphorus, Sulphur, Amines, Amide

____

216. Sucrose is transported by the vascular tissue from sink to a source.

True ☐ False ☐

217. Where does the food come from, when the buds of trees act as sink?

____

218. Complete the following analogy.

Xylem : __________ : : Unidirectional : __________

219. Food in the phloem sap can be transported in any ______________________________

220. Phloem sap is mainly __________ and ________________, but some other sugars, and __________ are also translocated through pholem.

221. Name the most accepted mechanism for the translocation of sugars from source to sink.

____

222. Glucose is prepared at the sink and it is converted to maltose.
(*Rewrite the statement by changing the underlined words.*)

____

223. Draw the path of sugar in plants from source to sink.

____

224. The process of loading of sucrose at the source produces a ______________ (hypotonic/hypertonic) condition in the phloem.

225. The sugar is loaded in phloem by passive transport.

True ☐ False ☐

226. Water in the adjacent xylem moves into the phloem by diffusion.
(*Rewrite the statement by changing the underlined word.*)

____

227. Rearrange the following parts in increasing order with respect to osmotic pressure.

Photosynthetic apparatus, Growing leaf, Companion cell, Sieve tube

____

228. Which transport is used to move the sucrose out of phloem sap into the cell which convert sugar into energy, starch or cellulose? ______________________________

229. When sugars are removed, the osmotic pressure ______________ and water ______________ out of the phloem.

230. Loading of the phloem sets up a ______________ that facilitates the mass movement in the phloem.

231. When the hydrostatic pressure in the sieve tube of phloem ______________ (increases/decreases), pressure flow begins and the sap moves through the phloem.

232. The loss of solute from phloem produces high water potential in the phloem this results in returning of into xylem.

True ☐ False ☐

233. Answer the following questions, considering the diagrammatic representation of mechanism of translocation.

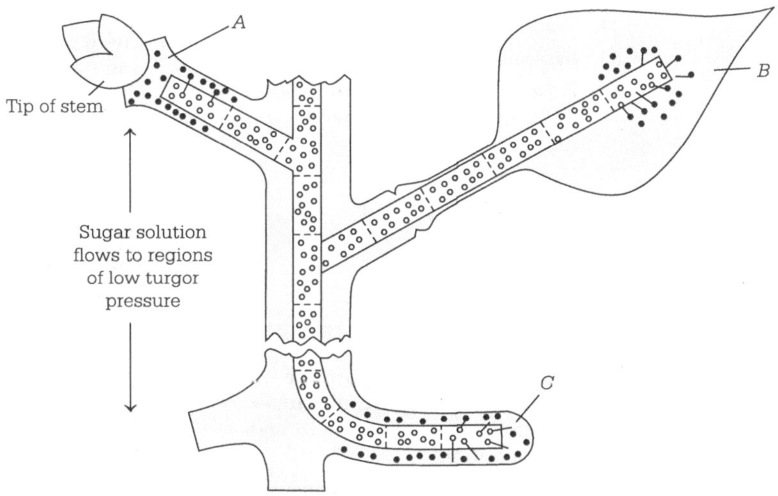

(i) At which of the following labelled parts sugar leave sieve tubes? ______________

(ii) At which of the labelled part turgor pressure is high? ______________

(iii) Name of the labelled part functions as source? ______________

234. Answer the following questions, considering the flowchart.

A ring of bark removed up to a depth of the phloem
↓ After few weeks
Stem becomes swollen above the ring

(i) What does the experiment show? ______________

(ii) This experiment is known as ______________

(iii) Why stem becomes swollen above the ring? ______________

Answers

1. False. The water taken by the roots has to reach all parts of the plant, up to the very tip of the growing stem.
2. Photosynthates
3. Water, mineral nutrients, organic nutrients, plant growth regulators and chemical signals.
4. Diffusion, cytoplasmic streaming
5. Translocation
6. Unidirectional
7. Organic compounds and mineral nutrients.
8. Organic compound, storage organs
9. Senescent, growing
10. Diffusion, higher, lower
11. Slow, short
12. (i) Diffusion, (ii) False. In this process no energy expenditure takes place.
13. Diffusion
14. Gases > Liquid > Solid
15. Diffusion pressure
16. Diffusion pressure is directly proportional to the kinetic movement and number of diffusing particles.
17. Diffusion pressure of pure water > Diffusion pressure of sugar solution.
18. Diffusion pressure decreases as more solute is added in a solution.
19. Permeability of the membrane, temperature, pressure
20. (i) Hydrophobic moiety soluble in lipids diffuse through the membrane faster, hence diffusion rate increases.

    (ii) Substances that have a hydrophilic moiety find it difficult to pass through the membrane, hence diffusion rate decreases.
21. Facilitated diffusion
22. False. Facilitated diffusion transport the molecules from high to a low concentration.
23. Maximum, protein
24. (i) *A*–Transport protein, Facilitated diffusion. (ii) 5 (Glucose, Na^+, Ca^{+2}, Amino acid, Nucleic acid).
25. Specific
26. Sensitive
27. Channels
28. Open, controlled
29. Porins
30. It allows the molecules up to the size of small proteins.
31. Extracellular
32. Aquaporins
33. β-barrel protein
34. Water channels, 8
35. Symport, same
36. False. In an antiport two types of molecules move together in different direction.
37. Uniport
38. (i) c-2, (ii) a-3, (iii) b-1
39. Increases
40. Active
41. Membrane protein
42. Energy, cell membrane
43. False. In uphill transport, the pump can trannsport substances from a low concentration to high concentration.
44. Active transport rate reaches a maximum when all the protein transporters are used.
45. *A*-Yes, *B*-Yes, *C*-Yes, *D*-Yes, *E*-No, *F*-No, *G*-No.
46. 92%
47. 1 (Capillary water)
48. Three
49. Water
50. Greater, Water potential
51. (i) From *A* to *B*, (ii) False. ψ_{W_1} has maximum value, (iii) Greatest, (iv) Decreases
52. Zero
53. Lower
54. Solute
55. False. Solute potential is always negative.
56. Tube *A* possess more negative solute potential.
57. Atmospheric pressure
58. Negative, negative
59. Water potential
60. Diffusion, turgid
61. False. The pressure exerted by the protoplast against the cell wall is called pressure potential (ψ_p).
62. Positive
63. A-2, B-1, C-3
64. Matric potential (ψ_m)
65. $\psi_w = \psi_s + \psi_p$
66. – 0.23
67. There will be no net flow of water as the ψ_w of solution *B* is also ($\psi_s + \psi_p = 0.23 - 0.23$) zero.
68. Water always move from high ψ_w to low ψ_w. It will move from *B* to *A* as the Ψ_w of *A* is $\Psi_p + \Psi_s = -0.30 + 0 = -0.30$ and Ψ_w of *B* is $0 - 0.23 = -0.23$.
69. The exchange of water does not takes place.
70. The movement of water is calculated as

    $\psi_w = \psi_s - \psi_p$

    for solution $A = -30 - 5 = -35$

    for $B = -10 - 0 = -10$

    The water will move from the area of high ψ_w to low ψ_w, hence it moves from *B* to *A*.
71. 4% sucrose solution < 2% sucrose solution < Pure water.
72. Diffusion Pressure Deficit (DPD)
73. True
74. DPD = OP – TP
75. Osmosis
76. Pressure gradient, concentration gradient
77. False. Water will move from its region of higher chemical potential to its region of lower chemical potential.
78. (i) Chamber *B*, (ii) Chamber *A*, (iii) From *A* to *B*, (iv) Both chambers shall have same water potential.

79. (i) Equilibrium, (ii) Funnel *A*, (iii) False. Due to the application of pressure in funnel *B*, the solute concentration increases.
80. Osmotic pressure
81. Positive
82. Osmotic pressure is directly proportional to the concentration of solution.
83. Solution A
84. Endosmosis
85. Exosmosis, hypertonic solution
86. A–3, B–2, C–1
87. Shrink
88. Plasmolysis
89. (i)-(b)-(1), (ii)-(c)-(2), (iii)-(a)-(3)
90. Plasmolysed
91. Diffusion, shrink
92. Isotonic, flaccid
93. Hypotonic, turgor
94. Turgor pressure
95. A-3, B-1, C-2
96. False. The cells of potato plasmolysed if it is kept in concentrated salt solution.
97. Salting kills the bacteria by inducing plasmolysis in their cells.
98. Hypotonic, deplasmolysis
99. Exosmosis
100. Imbibition
101. (i) Adsorbant, (ii) Imbibant
102. Imbibition
103. Imbibition
104. Imbibition pressure, heat of wetting
105. Imbibition pressure
106. False. Water potential gradient should occur between imbibant and adsorbant.
107. Attraction, affinity
108. 2.5 second
109. Long, production, absorption, storage
110. Mass or bulk flow
111. Mass flow, pressure difference
112. False. Bulk flow can be achieved either through a positive hydrostatic pressure gradient or a negative hydrostatic pressure gradient.
113. (i) Garden hose, (ii) Phloem
114. Translocation.
115. 4(Water, mineral salts, organic nitrogen, hormones).
116. True
117. Apoplast
118. Apoplastic pathway
119. False. Movement through the apoplast does not involve crossing the cell membrane.
120. Intercellular
121. Adhesive and cohesive forces
122. Symplastic
123. Plasmodesmata
124. Cytoplasmic streaming
125. (i) *A*–Epidermis, *B*–Cortex, *C*–Endodermis, *D*–Pericycle, *E*–Plasmodesmata, *F*–Casparian strip.
(ii) Y arrow shows symplastic movement.
(iii) Apoplastic pathway
126. Cells of the ***Hydrilla*** leaf and movement of chloroplast.
127. A–2, 3 and B–1, 4
128. Box *A*–Non-vacuolar symplast, Box *B*–Vacuolar symplast.
129. True
130. Loosely, resistance
131. Impervious, casparian
132. Suberised, membranes
133. Symplastic, endodermis
134. Xylem vessels, tracheids, apoplast
135. *A*–Symplastic path, *B*–Apoplastic path, *C*–Endodermis, *D*–Casparian strip
136. Fungus with the root system of higher plants.
137. True
138. Obligate
139. Increases, increases
140. Positive
141. Root pressure
142. The solution comes out due to the positive root pressure.
143. False. Guttation is the process in which water collects in the form of droplets around veins near the tip of grass blades and leaves of many herbaceous plant.
144. Guttation
145. Short
146. Water molecules in the xylem.
147. Water, transpiration
148. 2(Rainy and spring)
149. High absorption, low transpiration
150. Root *A*
151. 15 metres per hour
152. Physical forces
153. Adhesion
154. Cohesion
155. Pulling force, capillarity
156. False. Adhesion and cohesion maintain the continuous water column in xylem.
157. Xylem, transpiration pull
158. Priestley
159. All of these theories explain ascent of sap.
160. Decreases, increases, adjacent cell, decreases, xylem vessel
161. False. The transpiration pull theory suggests that the transport of water takes place at low pressure and high tension.
162. High temperature
163. A-3, B-4, C-1, D-2
164. Transpiration
165. Transpiration, oxygen, carbon dioxide
166. A-2, B-1, C-4, D-3
167. Guard cells
168. Stomatal aperture, elastic
169. Microfibrils, guard cells
170. *A*–Guard cells become flaccid, *B*–Stomata closes
171. (i) *A*–Microfibrils, *B*–Guard cell, *C*–Stomatal aperture,
(ii) Cellulosic, radially
172. Stomata will be open in this situation.
173. *A*–During day time, *B*–Produces sugars, *C*–Endosmosis into guard cells, *D*–Stomata open
174. (i) Starch, (ii) Decreases, (iii) Flaccid, (iv) Close
175. pH, CO_2
176. Acidic

177. Opening
178. Abscisic acid
179. (v) K^+ ions exchange from subsidiary cell, (vi) Increased pH of guard cell.
180. Day, night.
181. False. The lower surface of dorsiventral leaf has greater number of stomata.
182. 5 (i, ii, iii, v, vi)
183. Summers, high
184. Decreases
185. True
186. Antitranspirants
187. All of them are antitranspirants.
188. Cohesion, adhesion, surface tension
189. A-2, B-3, C-1
190. Photosynthesis.
191. *A*–Lower, *B*–Evaporation
192. (i) *A*–Xylem, *B*–Phloem, *C*–Stomatal pore
(ii) Pressure gradient
193. Absorption
194. Shape, structure, turgid
195. Limited, depleted
196. The humidity of rainforests is largely due to the cycling of water from roots to leaf to atmosphere and back to the soil.
197. C_4, minimising
198. False. C_4 plant loses only half as much water as a C_3 plant for the same amount of CO_2 fixed.
199. Lowers
200. Charged
201. Active absorption
202. Water potential gradient
203. Passively
204. Specific proteins
205. Transport proteins, control points
206. The root endodermis has the ability to actively transport ions in one direction only because of the layer of suberin.
207. Soil water < Root hair < Cortical cell
208. Higher
209. Passive
210. The mesophyll cells take water from adjoining cells.
211. All of these functions as chief sinks for the mineral elements.
212. Unloading
213. Senescent, younger leaves
214. Calcium
215. All of them are mobilised into organic form, phosphorus and sulphur also mobilised into organic form.
216. False. Sucrose is transported by the vascular tissue from source to a sink.
217. Sugar stored in roots mobilised to buds when it acts as sink.
218. Phloem, bidirectional
219. Direction
220. Water, sucrose, hormones, amino acids
221. Pressure flow hypothesis is the most accepted mechanism for the translocation of sugars from source to sink.
222. Source and it is converted to sucrose.
223. Leaf (Source) $\rightarrow$ Companion cell $\rightarrow$ Sieve tube $\rightarrow$ Growing tissue (Sink).
224. Hypertonic
225. False. The sugar is loaded in phloem by active transport.
226. Osmosis
227. Growing leaf < Sieve tube < Companion cell < Photosynthetic apparatus
228. Active transport
229. Decreases, moves
230. Water potential gradient
231. Increases
232. True
233. (i) A and C, (ii) B, (iii) B
234. (i) The experiment shows that phloem tissue is responsible for translocation of food.
(ii) Girdling.
(iii) The photosynthates does not transported to other parts below the ring, so there is accumulation of food just above the girdle region.

CHAPTER TWELVE

MINERAL NUTRITION

Topic 1 : Methods to Study the Mineral Requirements of Plants

1. ____________________ is a basic need of all living organisms.

2. Select the odd one out with respect to nutrient molecules.

Carbohydrates, Proteins, Fats, Nucleic acids

____

3. How many of the macromolecules given in the box required by living organisms for their growth and development?

Carbohydrates, Glycerol, Fats, Proteins, Water, Minerals

____

4. Plants can be grown to maturity in the absence of soil in a defined nutrient solution.

True ☐ False ☐

5. Julius von Sachs is associated with the discovery of ____________________ (hydroponics/tissue culture).

6. Soilless culture technique has been applied for how many of the plants/crops mentioned below?

Tomato, Mango, Seedless cucumber, Wheat, Lettuce

____

7. Plants grown using hydroponics cannot be cultivated anywhere else.

True ☐ False ☐

8. Name that technique of growing plants that does not require a growing medium.

____

9. The diagram given below represents a typical setup of _______________ (hydroponics/aeroponics).

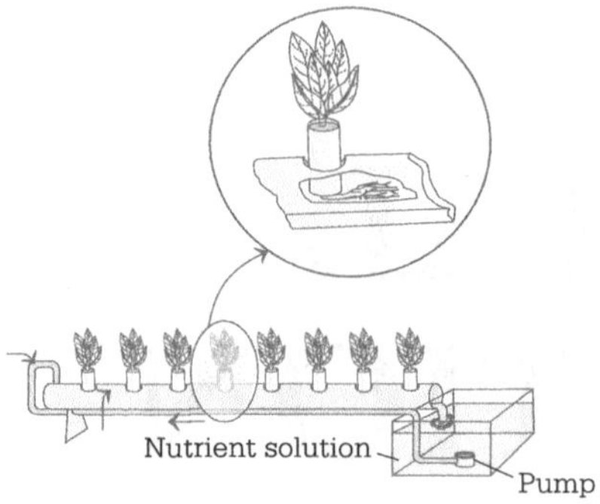

10. Observe the figure given below and answer the following questions.

(i) Identify the labels *A* to *D*.

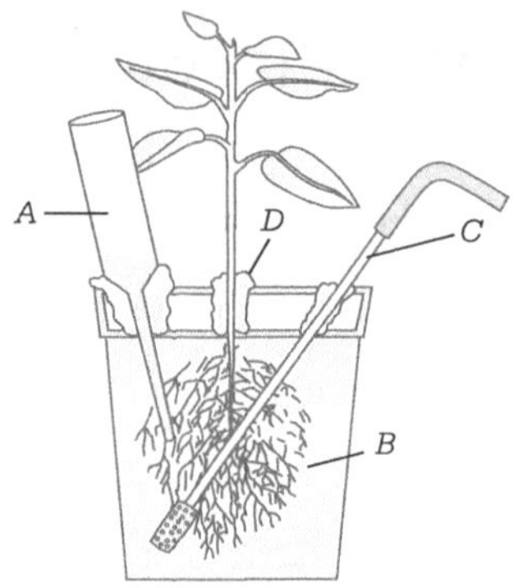

A. _______________

B. _______________

C. _______________

D. _______________

(ii) What is the function of component *C* in this setup?

____

11. In hydroponics, plant's _______________ (roots/shoots) are submerged in a solution of dissolved _______________ (minerals /oxygen).

12. The following flowchart represents the steps of the hydroponics technique.

Identify *A, B* and *C*.

Plants are grown in a *A* → A pump circulates *B* → Solution flows down the tube and returns to the *C*

____

13. What is the most likely consequence of poor aeration in hydroponics setup?

____

14. Hydroponics is useful in determining essential role of nutrients in plant metabolism.

True ☐ False ☐

Topic 2 : Essential Mineral Elements and Their Deficiency Symptoms

15. Minerals present in soil enters the plant through ______________________

16. Out of 105 elements, how many have been found in different plants? ___________

17. Strontium will be present in plants growing in close proximity to ______________________ (leather factories/nuclear test sites).

18. Give the biological term for the elements without which plants are unable to complete their life cycle.

    ____

19. C, H, O and N are called as structural elements. (*Give reason*)

    ____

20. An essential element must be ____________ (replaceable/required) in specific amount in plants.

21. Select the odd one out with respect to essential and non-essential elements.

    Selenium, Gold, Strontium, Nitrogen

    ____

22. The criteria for essentiality of an element are ______________, necessary for ____________ and plants do not complete their life cycle in ______________ of the element.

23. Essential elements are components of energy related compound in plants.

    True ☐ False ☐

24. An essential element have no influence on plant metabolism.
    (*Rewrite the statement by changing the underlined words.*)

    ____

25. Give appropriate headings for the difference tabulated below as *A* and *B*.

	A	B
Concentration in Plants	< 10 m mole kg^{-1}	> 10 m mole kg^{-1}
Functions	Synthesis of organic molecules and development of osmotic potential.	As enzyme cofactors, activators, inhibitors, etc.

    A. ______________ B. ______________

26. Pick the odd one out on the basis of nutrient requirement.

    Phosphorus, Calcium, Zinc, Sulphur

    ____

27. How many of the following are macronutrients present in plant tissues?

    C, H, Mg, Si, S, Co, P, Se, K, N, O, Ca

    ____

28. Essential micronutrients are ______________ (8/9) in total number.

29. List the microelements which are essential for a plant's normal growth and development.

____

30. Unlike macronutrients, the ______________ are required in trace amounts.

31. What happens when the supply of an essential element becomes limited?

____

32. Which of the following pairs of elements are considered as beneficial elements in higher plants?

Sodium and iron, Silicon and potassium, Cobalt and selenium

____

33. Complete the following analogy related to the element present in energy related chemical compounds.

Mg^{2+} : ______________ : : P : ATP

34. $\underline{Zn^{2+}}$ acts as an activator of many enzymes involved in photosynthetic carbon fixation. (*Rewrite the statement by changing the underlined word.*)

____

35. Complete the following analogy.

______________ : Alcohol dehydrogenase : : Mo : Nitrogenase

36. The ______________ regulates stomatal opening.

37. Essential elements regulate the ______________ of cell membrane.

38. Give any two functions performed by essential nutrients in plants.

(i) ____________________

(ii) ____________________

39. In plants, nitrogen is mainly absorbed as ______________ (NO_3^-/NO_2).

40. How many of the given biomolecules contain nitrogen as a major constituent?

Vitamins, Starch, Nucleic acids, Fats, Hormones, Proteins

____

41. An element obtained from both the mineral and non-mineral resources is ______________

42. Nitrogen is required mainly by metabolically active cells and <u>permanent tissue</u>. (*Rewrite the statement by changing the underlined words.*)

____

43. Phosphorus is absorbed by plants in the form of ______________ ($H_2PO_4^-$ / PO_4^{2-})

44. Plants do not absorb the organic form of phosphorus.

True ☐ False ☐

45. Phosphorus is also present in high energy biomolecules like ____________ .

46. Complete the following analogy.

Potassium : Maintains cellular turgidity : : ____________ : Synthesis of cell wall

47. ____________ is one of the major constituent of protoplasm.

48. In seeds, highest quantity of K^+ is present.

True ☐ False ☐

49. Which element tends to get accumulated in older leaves? ____________

50. Complete the table given below by filling the blanks marked (*A-F*).

Essential elements	**Regions where required**	**Functions**
Phosphorus	(*A*)	Energy transfer reactions
(*B*)	Meristematic tissues, bud, leaves and root tips	(*C*)
Nitrogen	All parts of plants	(*D*)
(*E*)	(*F*)	Formation of chlorophyll, carotenoids and nucleic acid

A. ____________ B. ____________

C. ____________ D. ____________

E. ____________ F. ____________

51. Phosphorus gets withdrawn from metabolically more active younger tissues.
(*Rewrite the statement by changing the underlined words.*)

____

52. Write the names of mineral elements associated with the functions mentioned below.

(i) Needed during the formation of mitotic spindle. ____________

(ii) Synthesis of pectin found in middle lamella of cell wall. ____________

(iii) Phosphorylation reactions. ____________

53. Ribulose bisphosphate carboxylase and phosphoenol pyruvate carboxylase, what is common to both of these enzymes? ____________

54. Out of the given options in bracket, select the correct forms of element in which these are absorbed from the soil.

(i) Calcium (Ca^{2+} / $Ca(OH)$) ____________

(ii) Sulphur (SO_4^{2-} / H_2SO_4) ____________

(iii) Magnesium ($MgCl_2$ / Mg^{2+}) ____________

55. Match the following columns.

	Column I (Elements)		Column II (Functions)
A.	Magnesium	1.	Synthesis of proteins
B.	Sulphur	2.	Organisation of mitotic spindle
C.	Calcium	3.	Maintenance of ribosome structure
D.	Potassium	4.	Major constituent of several vitamins

A. ______________ B. ______________ C. ______________ D. ______________

56. Pick the odd one out considering amino acids.

Cysteine, Methionine, Glycine, Taurine

____

57. Sulphur is a main constituent of how many molecules named below?

Ferredoxin, Carotene, Thiamine, ATP, Coenzyme-A, Cytochrome-*b*, Biotin

____

58. Due to the presence of ______________ (magnesium/sulphur), plants of family–Liliaceae have a characteristic odour.

59. Lipids, amino acids and proteins are ______________ of sulphur.

60. Manganese activates many enzymes involved in ______________, ______________ and ______________.

61. For which purpose, iron is required in plant body?

____

62. Complete the following analogy.

Iron : Fe^{3+} : : Manganese : ______________

63. How many of the following enzymes activates by manganese?

Glucokinase, Malic dehydrogenase, Oxalo dehydrogenase, Carboxylase, Nitrite reductase

____

64. $H_2O \xrightarrow{X} 2H^+ + 2e^- + 1/2\,O_2$, Name the reacion.

____

65. The above shown reaction (Q. 64) is facilitated by ______________ (Mn^{2+}/Mg^{2+}).

66. Complete the following analogy given below.

Zinc : Carboxylase : : Copper : ______________

67. Among the given classes of enzymes, zinc can activate how many of them?

Catalase, Dehydrogenases, PEPCase, Oxidases, Kinases, Plastocyanin

____

68. Complete the table given below.

Microelements	Functions	Required by
(A)	Involved in reactions of energy conversion.	Every plant part but abudantly in veins of leaves.
(B)	Root nodulation, calcium utilisation.	Leaves and seeds.
Molybdenum	(C)	All parts, but utilised mostly by roots.
(D)	(E)	Required by every part like leaf.

A. ______________ B. ______________

C. ______________ D. ______________

E. ______________

69. Boron is absorbed by plants as ________ or ________

70. Match the following columns.

	Column I (Major elements)		Column II (Functions)
A.	Boron	1.	Splitting of H_2O during photosynthesis
B.	Manganese	2.	Synthesis of auxin
C.	Molybdenum	3.	Component of nitrogenase
D.	Zinc	4	Pollen germination
E.	Iron	5.	Component of ferredoxin

A. ________ B. ________ C. ________

D. ________ E. ________

71. Availability of boron uptake increases with the pH.

True ☐ False ☐

72. Read the statements given below and select the element required for the given functions.

(i) Activation of enzymes involved in redox reactions. ______________

(ii) Helps in development of chlorophyll. ______________

(iii) Maintain the ribosome structure. ______________

(iv) Transportation of carbohydrates through the phloem. ______________

73. Element with variable valency can act as electron carriers. Give example of such elements.

____

74. Match the following columns.

	Column I (Role in plants)		Column II (Name of elements)
A.	Reduces nitrate to nitrite	1.	Boron
B.	Transfer electron from H_2O to PS-II	2.	Nickel
C.	Pollen germination	3.	Chlorine
D.	Component of urease	4.	Molybdenum

A. ______________ B. ______________

C. ______________ D. ______________

75. Similar to macronutrient K^+, the microelement ______________ also balance cationic-anionic ratio in plant cells.

76. Pick the odd one out based on its requirements in plants.

Fe^{2+}, BO_3^{3-}, SO_4^{2-}, Ni^{2+}

____

77. Give a biological terms for the concentration of an essential element below which plant growth gets retarded.

____

78. Deficiency symptoms of an element first appeared in young leaves. This indicates that the element is relatively ______________ (mobile/immobile).

79. Deficiency of elements manifests themselves as morphological changes. (*Give reason*)

____

80. Deficiency symptoms ______________(persist/disappear) when the deficient nutrient is supplied to the plant.

81. The deficiency symptoms of N, K and Mg first appears in senescent leaves. What does this indicate with respect to mobility of these elements? ______________

82. Elements like S and Ca^{2+} are immobile elements. What does this indicate?

____

83. Give a biological term for the sentence given below.
A deficiency disorder leading to death of tissue while being attached to the living plant.

____

84. Give a biological term for the sentence given below.
Externally visible pathological conditions produced due to the absence or diminished level of essential nutritive substances.

____

85. Necrosis is associated with deficiency of how many of the elements mentioned below?

Ca, Mg, N, Cu, K

____

86. Complete the following analogy given below.

Necrosis : Tissue death : : ____________________ : Yellowing of leaves

87. Yellowing of leaves is associated with loss of ____________ (xanthophylls/chlorophyll).

88. How many elements given in box inhibit cell division in plants due to their deficiency.

Nitrogen, Calcium, Potassium, Chlorine, Sulphur, Molybdenum

____

89. A plant exhibiting delayed flowering is most likely deficient in ____________________ (N, S, Mo/K, Mg and Cl^-).

90. Give a scientific term for premature fall of leaves and buds.____________________

91. Premature leaf fall is due to the deficiency of potassium.
(*Rewrite the statement by changing the underlined word.*)

____

92. The deficiency of an element will cause only a specific symptom.

True ☐ False ☐

93. Match the following columns.

	Column I		Column II
A.	Whiptail disease	1.	Zinc
B.	Heart rot disease	2.	Boron
C.	Reclamation disease	3.	Molybdenum
D.	Little leaf disease	4.	Copper

A. ______________ B. ______________

C. ______________ D.______________

94. A plant with following symptoms bronze leaves, wilting, chlorosis and necrosis is deficient in ________________

95. Complete the following analogy given below.

Die back disease : Copper : : Marsh spot disease : __________

96. A moderate increase in concentration of essential elements will lead to deficiency symptoms.
(*Rewrite the statement by changing the underlined words.*)

____

97. Appearance of brown spots surrounded by chlorotic veins indicate ________________ (boron/manganese) toxicity.

98. If manganese is present above its critical concentration, the plant will also manifest the deficiency symptoms of potassium.

True ☐ False ☐

99. Mineral toxicity occurs when a mineral ion in tissue reduces the ____________ (dry weight / wet weight), by approximately. _________ (10%/14%).

Topic 3 : Mechanism of Absorption of Elements

100. Minerals are absorbed as _______________ by the plants.

101. Complete the following analogy given with respect to absorption of mineral ions.

Passive transport : Apoplast : : Active transport : ____________________________________

102. The absorption of minerals due to difference in __________________________ without using ATP is passive absorption.

103. Passive absorption is slow, while active absorption is faster.

True ☐ False ☐

104. Name the energy dependent mineral uptake process. ____________________________

105. The movement of ion is _________________ (unidirectional/reversible) in active transport.

106. Give a scientific term for movement of ions from cell to cell.______________________

107. Ion channels and transmembrane proteins facilitate which type of movement of mineral ions?____

108. Efflux is the inward movement of ions into the cells.

True ☐ False ☐

109. Influx is the inward movement of ions into the cells.

True ☐ False ☐

110. Minerals absorbed by roots reach the leaves through _____________ (tracheary elements of xylem/phloem fibres).

111. Name the force responsible for ascending of xylem sap.__________________________

112. ___________________ process makes nutrients and minerals present in rocks available to plants.

113. The most common combination of fertilisers for soil enrichment is __________________ (N, P, K/ Ca, Mg, S).

Topic 4 : Metabolism of Nitrogen

114. Give a scientific term for cycle of events by which atmospheric N_2 gets converted to its various chemical forms. ____________

115. Atmosphere contains ____________ % of nitrogen.

116. Atmospheric nitrogen (N_2) is depicted as ____________ ($N \equiv N$ / $N = N$).

117. Given below are steps involved in nitrogen cycle in random sequence. Arrange these step in correct order.

Nitrification, Ammonification, Denitrification, Nitrogen-fixation

____

118. In nature, energy for conversion of N_2 to its oxides is provided by thunder and infrared radiations.
(*Rewrite the statement by changing the underlined words.*)

____

119. Pick the odd one out with respect to its chemical composition.

NO, NO_2, HNO_3, N_2O

____

120. List a few sources of atmospheric NO_x.

____

121. Study the flowchart of nitrogen cycle given below and identify *A-D*.

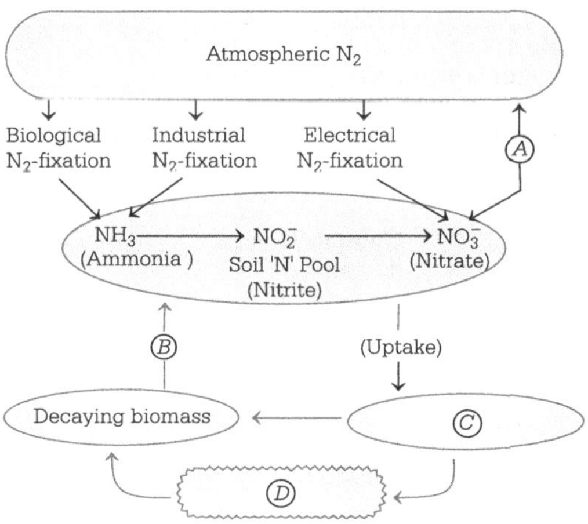

A. ____________

B. ____________

C. ____________

D. ____________

122. Study the reactions given below and answer the questions that follows.

(a) $2NH_3 + 3O_2 \longrightarrow (A) + 2H^+ + 2H_2O$

(b) $(A) + O_2 \longrightarrow (B)$

(i) Complete reactions by identifying *A* and *B*. ____________

(ii) Reaction (a) is facilitated by ____________ (electrical discharge/*Nitrosomonas*).

(iii) Reaction (b) is facilitated by bacteria ____________ these are ____________.

(iv) The above reactions represents the process of denitrification.

True ☐ False ☐

123. $\underset{\text{(In soil)}}{2NO_3} \longrightarrow \underset{\text{(Free nitrogen)}}{N_2}$

The above reaction indicates ____________, which is accomplishes by bacteria, i.e. ____________.

124. Biological nitrogen-fixation is performed by few prokaryotic organisms.

True ☐ False ☐

125. Give the collective term for nitrogen-fixing bacteria. ____________

126. The equation of biological N_2-fixation is given below.

$N_2 \xrightarrow{X} NH_4^+$, In this reaction *X* is ____________ (ammonia synthase/nitrogenase) enzyme.

127. Pick the odd one out with respect to nitrogen-fixation.

Azotobacter, Frankia, Clostridium, Rhodospirillum

____

128. Insectivorous plants fulfil their nitrogen requirements from ____________

129. Complete the table given below by identifying (*A-E*) with respect to types of nitrogen-fixers and their examples.

Column I (Nitrogen fixers)	**Column II** (Examples)
Chemotrophic N_2 fixers	(*A*)
(*B*)	*Anabaena* and *Nostoc*
Free-living anaerobic N_2 fixers	(*C*)
(*D*)	*Rhizobium*
Free-living aerobic N_2 fixers	(*E*)

A. ____________ B. ____________

C. ____________ D. ____________

E. ____________

130. Roots of *Alfalfa*, sweet clover, sweet pea, lenticels and garden pea form symbiotic association with which bacteria? ______

131. Symbiotic bacteria associate with host plant roots by forming ______

______

132. Root nodules of leguminous plants are red or pink in colour. (*Give reason*)

______

133. *Rhizobium* multiplies, colonises roots and gets attached to ______ and ______ cells.

134. Nodule formation is initiated in which part of the root?

______

135. Specialised nitrogen-fixing cells found in prokaryotes are archaeocytes.

True ☐ False ☐

136. Answer the questions related to the figures given below.

(i) In figure (d), identify the parts labelled as *A-C*.

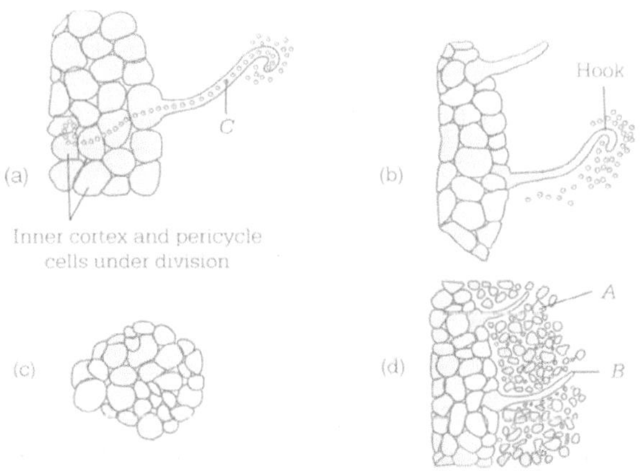

A. ______

B. ______

C. ______

(ii) Arrange the fig (a) to (d) in correct sequence of events in nodule development.

______

(iii) What does the part labelled as *D* indicate? ______

(iv) What is happening in figure (b)?

______

137. Name the first stable product of nitrogen-fixation. ______

138. Complete the following reaction by identifying *A, B* and *X*.

$$N_2 + 8e^- + 8H^+ + \textcircled{A} \xrightarrow{[X]} \textcircled{B} + H_2 + 16\,ADP + 16Pi$$

______

139. Leghaemoglobin present in root nodules regulates ______ (CO_2/O_2) supply to cells.

140. Leghaemoglobin protects nitrogenase enzyme. (*Give reason*)

____

141. Oxygen scavanger is ______________ (leghaemoglobin/nitrogenase).

142. Study the figure given below and answer the questions that follows.

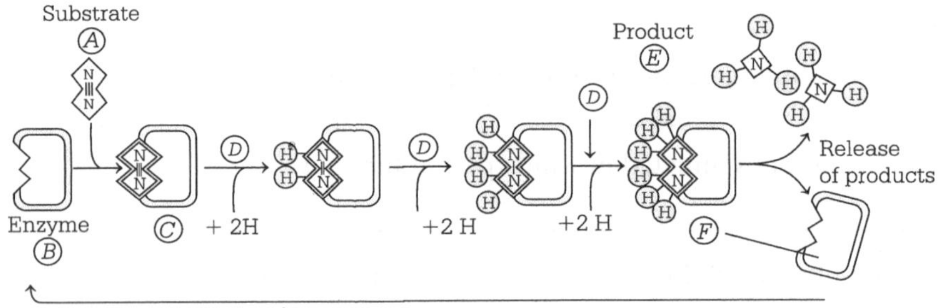

(i) Identify the following in above diagram.

- Substrate $A \rightarrow$ ______________
- Enzyme $B \rightarrow$ ______________
- Product $E \rightarrow$ ______________

(ii) D indicates a ______________ (reduction/oxidation) of substrate.

(iii) Write the event occurring at C.

____

(iv) At F, the enzyme is free to bind another substrate molecule.

True ☐ False ☐

143. When nitrogenase is not operational, host bacteria live as ______________ (aerobes/anaerobes).

144. A very high input of energy is required for each NH_3 produced from atmospheric nitrogen. Where does this requirement of energy is fulfilled?

____

145. NH_4^+ ions can get accumulated in plants as a source of nitrogen.

True ☐ False ☐

146. NH_4^+ ions are used in synthesis of ______________ in plants.

147. Study the reaction and answer the following questions.

α-ketoglutaric acid + NH_4 $\rightarrow$ NADPH $\xrightarrow{[X]}$ Glutamate + H_2O + NADP

(i) In the above reaction, name the process taking place. ______________

(ii) Identify the enzyme X. ______________

148. Match the following columns.

	Column I		Column II
A.	Nitrogen fixation	1.	Conversion of NH_4^+ into nitrite and nitrate
B.	Denitrification	2.	Conversion of nitrite or nitrate into atmospheric nitrogen
C.	Nitrification	3.	Conversion of atmospheric nitrogen into ammonia
D.	Ammonification	4.	Conversion of organic nitrogen into ammonia

A. ______________ B. ______________ C. ______________ D. ______________

149. Transfer of NH_2 of one amino acid to C=O of another amino acid is <u>nitrification</u>. (*Rewrite the statement by changing the underlined words.*)

____

150.

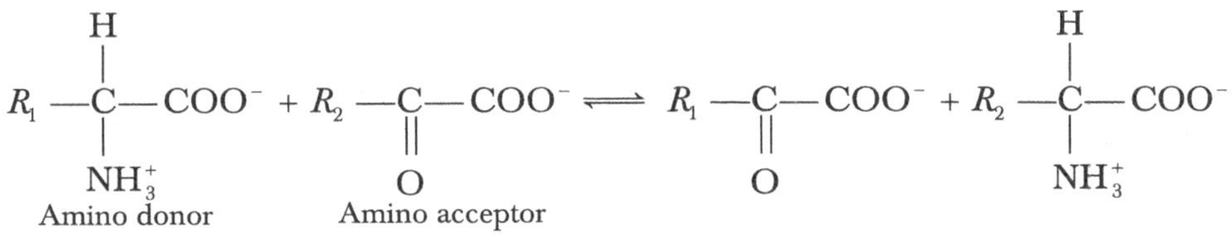

Name the enzyme which facilitates this reaction.

____

151. What happens when an amino acid combines with ammonia?

____

152. In amides, the __________ (OH/H^-) of amino group is replaced by __________ (NH_2^- / NH_4^+).

153. Amides contain less nitrogen than the amino acid.

True ☐ False ☐

154. Amides are <u>aspartic acid and glutamic acid</u>. (*Rewrite the statement by changing the underlined words.*)

____

155. Justify the statement, amides act as metabolic reservoirs.

____

156. Translocation of amides occurs through ______________ (phloem/xylem).

157. Ureides have high carbon to nitrogen ratio.

True ☐ False ☐

Answers

1. Nutrition
2. Nucleic acid is odd one out as it is a genetic material. Rest other molecules are nutrients.
3. 4(Water, minerals, carbohydrates, fats and proteins).
4. True
5. Hydroponics
6. 3 (Tomato, seedless cucumber and lettuce).
7. False. Plants grown using hydroponics can be cultivated anywhere.
8. Aeroponics
9. Hydroponics
10. (i) *A*–Funnel (for adding water and nutrients), *B*–Nutrient solution, *C*–Aerating tube, *D*–Cotton
    (ii) Component *C* provides oxygen for the normal growth and development of roots.
11. Roots, minerals
12. *A*-Tube, *B*-Nutrient medium, *C*-Reservoir
13. Poor aeration leads to decrease in oxygen levels, as a result cells will die and this can cause the entire plant to die as well.
14. True
15. Minerals present in soil enters the plant through roots.
16. More than 60
17. Nuclear test sites
18. Essential elements
19. C, H, O and N are called structural elements because they are the components of biomolecules of the cell.
20. Required
21. Nitrogen is odd one out, as it is essential element and others are non-essential elements.
22. Non-replaceable, metabolism absence
23. True
24. Direct influence
25. *A* – Macronutrient, *B* – Micronutrient
26. Zn is the odd one out because it is a micronutrient, while others are macronutrient.
27. 9 (C,H,O,N,P,K,S, Ca, Mg)
28. 8
29. Fe, Mn, Cu, Mo, Zn, B, Cl and Ni.
30. Micronutrients
31. The growth of the plants gets retarded.
32. Cobalt and selenium
33. Chlorophyll
34. Mg^{2+}
35. Zn^{2+}
36. Potassium (K)
37. Permeability
38. (i) It maintain the osmotic concentration of cell sap.
    (ii) It activates or inhibits the activity of enzyme.
39. NO_3^-
40. 4 (Proteins, nucleic acids, vitamins and hormones).
41. Nitrogen
42. Meristematic tissue
43. $H_2PO_4^-$
44. True
45. ATP
46. Calcium (Ca)
47. Potassium
48. False. In seeds, potassium is found in lesser quantity.
49. Calcium
50. *A* - Meristems or young tissues, *B* - Potassium, *C* - Maintains anion-cation balance, *D* - Support growth and metabolic activities, *E* - Magnesium, *F* - Growing areas of roots, stems and seeds.
51. Less active older tissues
52. (i) Calcium
    (ii) Calcium
    (iii) Phosphorus
53. Both of the enzymes are activated by magnesium.
54. (i) Ca^{2+}
    (ii) SO_4^{2-}
    (iii) Mg^{2+}
55. A-3, B-4, C-2, D-1
56. Glycine is the odd one out because unlike all others amino acid, it does not contain sulphur.
57. 5 (Ferredoxin, cytochrome-*b*, thiamine, coenzyme-A, biotin).
58. Sulphur (S)
59. Reservoir
60. Photosynthesis, respiration, nitrogen metabolism
61. Iron required for the synthesis of cytochrome and ferredoxin.
62. Mn^{2+}
63. 3 (Malic dehydrogenase, oxalo dehydrogenase, nitrite reductase).
64. Photolysis of water
65. Mn^{2+}
66. Cytochrome oxidase
67. 3 (Dehydrogenases, oxidases, kinases).
68. *A* - Iron, *B* - Boron, *C* - Nitrogen metabolism, *D* - Chlorine, *E* - Determining the cation-anion balance and solute concentration.
69. BO_3^{3-}, $B_4O_7^{2-}$
70. A–4, B–1, C–3, D–2, E–5
71. False. Uptake of boron decreases with increase in pH.
72. (i) Copper
    (ii) Iron
    (iii) Magnesium
    (iv) Boron
73. Fe ($Fe^{2+} \rightleftharpoons Fe^{3+}$) and Cu ($Cu \rightleftharpoons Cu^{2+}$)
74. A-4, B-3, C-1, D-2
75. Chlorine
76. SO_4^{2-} is odd one out, because it is a macronutrient, while others are micronutrients.
77. Critical concentration
78. Immobile
79. Each element has one or more specific structural or functional role in plants, in the absence of elements the plants show certain morphological changes.
80. Disappear

81. N, K and Mg are actively mobile elements.

82. These elements are a part of the structural component of the cell and hence are not easily released.

83. Necrosis

84. Deficiency symptoms

85. 4 (Ca, Mg, Cu and K).

86. Chlorosis

87. Chlorophyll

88. 4(Nitrogen, potassium, sulphur and molybdenum).

89. N, S and Mo

90. Abscission

91. Phosphorus

92. False. The deficiency of an element can cause multiple symptoms.

93. A-3, B-2, C-4, D-1

94. Chlorine

95. Manganese

96. Toxicity

97. Manganese

98. False. Excess of manganese induce deficiencies of iron, magnesium and calcium.

99. Dry weight, 10 %

100. Ions

101. Symplast

102. Electropotential gradient

103. False. Passive absorption is faster, while active absorption is slow.

104. Active transport

105. Unidirectional

106. Flux

107. Passive transport

108. False. Efflux is the outward movement of ions from cells.

109. True

110. Tracheary elements of xylem

111. Transpiration pull

112. Weathering

113. N, P and K

114. Nitrogen cycle

115. 78

116. $N \equiv N$

117. Nitrogen-fixation → Ammonification → Nitrification → Denitrification

118. Lightning and ultraviolet radiation

119. HNO_3 is the odd one out because it is nitric acid, while others are oxides of nitrogen.

120. Industrial combustions, forest fires, automobile exhausts and power-generating stations.

121. *A* - Denitrification, *B* - Ammonification, *C* - Plant biomass, D - Animal biomass

122. (i) $A - 2NO_2^-$, $B - 2NO_3^-$

(ii) *Nitrosomonas*

(iii) *Nitrobacter*, chemoautotroph

(iv) False. The reactions (a) and (b) represent nitrification.

123. Denitrification, *Pseudomonas.*

124. True

125. Diazotrophs

126. Nitrogenase

127. *Frankia* is the odd one out because it is symbiotic nitrogen fixer, while others are free-living nitrogen fixers. It forms nodule in non-leguminous plants, e.g. *Alnus* also

128. Insects

129. *A* - *Desulfovibrio*
B - Cyanobacteria
C - *Rhodospirillum/Clostridium*
D - Symbiotic bacteria
E - *Beijerinckia*

130. *Rhizobium*

131. Nodules

132. Root nodules of leguminous plants are red or pink in colour due to the presence of leghaemoglobin pigment.

133. Epidermal and root hair

134. Root cortex

135. False. Specialised nitrogen-fixing cells found in prokaryotes are heterocysts.

136. (i) *A* Soil particles, *B* - Root hair, *C* - Bacteria as infection thread

(ii) The correct sequence is - (d), (b), (a), (c).

(iii) *D* - Infection thread containing bacteria

(iv) Curling of root hair due to infection.

137. Ammonia

138. *A* - 16 ATP, *B*- $2NH_3$ and *X*- Mo- Fe nitrogenase

139. Oxygen (O_2)

140. Leghaemoglobin protect nitrogenase enzyme because it is sensitive to molecular oxygen.

141. Leghaemoglobin

142. (i) *A* - N_2, *B*- Nitrogenase, *E* - Ammonia

(ii) Reduction

(iii) At *C*, the substrate gets binded to the enzyme complex.

(iv) True

143. Aerobes

144. The required energy obtained form cellular respiration of the host cells.

145. False. Ammonium ions are toxic to cells, thus they do not accumulate in plants.

146. Amino acids

147. (i) Reductive amination

(ii) *X* - Glutamate dehydrogenase

148. A-3, B-2, C-1, D 4.

149. Transamination

150. Transaminase

151. It results in the formation of amides.

152. OH, NH_2^-

153. False. Amides contain more nitrogen than the amino acids.

154. Asparagine and glutamine

155. Amides get accumulated in the tissue of healthy plants when ammonia is assimilated and acts as reservoirs.

156. Xylem

157. False. Ureides have a high nitrogen to carbon ratio.

CHAPTER THIRTEEN

PHOTOSYNTHESIS IN HIGHER PLANTS

Topic 1 : Photosynthesis : Introduction and Early Experiments

1. Name the physio-chemical process by which plants use light energy to drive the synthesis of organic compound.

   ____

2. All living forms on earth depends on ______________ energy.

3. Photosynthesis is responsible for the release of ______________ into the atmosphere by green plants.

4. Photosynthesis transform ______________ energy into ______________ energy.

5. Consider the given experiment and identify in which part of the leaf photosynthesis takes place?

   Leaf partially covered with black paper → Exposed to sunlight → Test for starch formation

   ____

6. Consider the given experiment and answer the following questions.

   A leaf enclosed in test tube → Test tube contains (*A*) soaked cotton →

   Other half leaf exposed to (*B*) → Placed in sunlight → Test for the presence of starch

   (i) Identify *A* and *B*. ______________

   (ii) What is the role of *A*? ______________

   (iii) Which part of the leaf shows positive result for starch? ______________

7. Give one conclusion of the above given experiment in Q. 6.

   ____

8. Name the scientist who proved the essential role of air in the growth of green plants.

____

9. According to Priestley, plants restore the air whatever breathing animals and burning candles release.

True ☐ False ☐

10. The scientist, who discovered oxygen is _______________.

11. Jan Ingenhousz did an experiment with an aquatic plant and concluded that <u>all parts of plant release CO_2</u>.
(*Rewrite the statement by changing the underlined words.)*

____

12. Sunlight is essential for plant was concluded by _______________

13. Who provided evidence that in green parts of plant glucose is produced and stored as starch ? _______________

14. The first action spectrum of photosynthesis was observed by _______________.

15. Write the scientific name of the alga used by Engelmann. _______________

16. Name the scientist who used the following components for setting up an experiment related to photosynthesis.

Prism, *Cladophora*, Aerobic bacteria

____

17. Name the scientist who proved photosynthesis is a light dependent reaction.

____

18. In photosynthesis, oxygen from a an oxidisation compound reduces CO_2 to carbohydrates.

True ☐ False ☐

19. Write the overall equation for photosynthesis.

____

20. Write the empirical equation representing the process of photosynthesis.

____

21. _______________ is the hydrogen donor for purple and green sulphur bacteria.

22. Who demonstrated that O_2 comes from water instead from CO_2 during photosynthesis by using purple and green bacteria in their experiment?

____

23. Match the following columns.

	Column I (Scientist names)		Column II (Discoveries)
A.	Blackman	1.	Dark reaction
B.	Warburg	2.	Observed grana part of chloroplast
C.	Ruben and Kamen	3.	Study on *Chorella*
D.	Heitz	4.	O_2 comes from H_2O

A. ______________ B. ______________ C. ______________ D. ______________

Topic 2 : Site and Pigments of Photosynthesis

24. Chloroplasts align at the wall of ______________ cells to get optimum quantity of the incident light.

25. Find the odd one out with respect to chloroplast.

Fluid stroma, Stroma lamellae, Grana, Cisternae

____

26. Membranous system __________ (stroma/grana) of chloroplast responsible for the synthesis of sugar and which inturn forms starch.

27. The membrane system of chloroplast is responsible for ______________ and ______________ synthesis.

28. Name the reaction which are not directly light driven taking place in chloroplast.

____

29. Dark reaction are not dependent on ATP and NADPH.

True ☐ False ☐

30. Complete the following analogy.

Light reaction : ______________ : : Grana : Stroma : : ______________ : Starch

31. Consider the following figure and identify *A*, *B*, *C* and *D*.

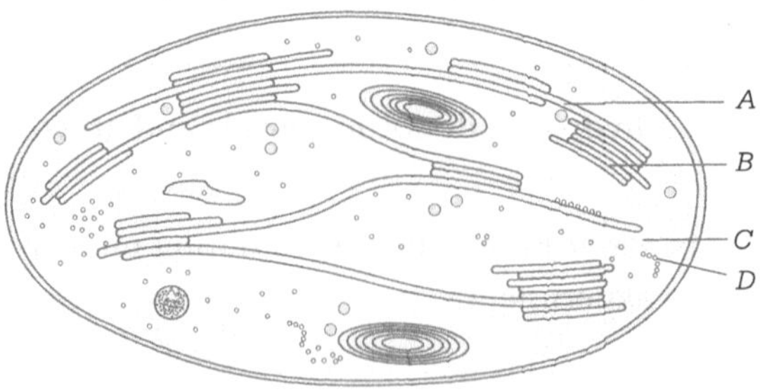

A. ______________

B. ______________

C. ______________

D. ______________

32. Write the technique that employed to separate leaf pigments of any green plant.

____

33. Match the following columns.

	Column I (Pigments)		Column II (Colour produce on chromatograph)
A.	Chlorophyll-*b*	1.	Yellow to yellow orange
B.	Carotenoids	2.	Yellow
C.	Chlorophyll-*a*	3.	Yellow-green
D.	Xanthophylls	4.	Blue-green

A. ______________ B. ______________ C. ______________ D. ______________

34. The blue and red regions of the spectrum shows ____________ (highest/lowest) rate of photosynthesis.

35. Choose the odd one out with respect to plant pigments.

Chlorophyll, Xanthophyll, Carotenoids, Phycoerythrin

____

36. Consider the following figure and identify the pigments *A*, *B* and *C*.

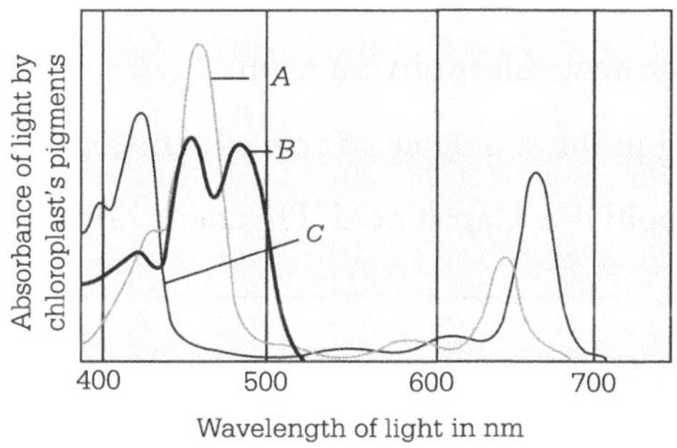

A. ______________ B. ______________ C. ______________

37. The absorption spectrum of chlorophyll-*a* is maximum.

True ☐ False ☐

38. ________________ is the chief plant pigment associated with photosynthesis.

39. Write the colour spectrum of the first action spectrum of photosynthesis.

____

40. A graph that plots the effect of different wavelenghts of light on the rate of photosynthesis is called ______________________

41. Absorption spectrum of chlorophyll-*a* and the action spectrum of photosynthesis is identical.

True ☐ False ☐

42. What is the wavelength of radiations in visible spectrum? ________________

43. Write the correct order of light given below on the basis of rate of photosynthesis in decreasing order.

Green, Red, Blue

____

44. Mostly photosynthesis takes place in the blue and red regions of VIBGYOR.

True ☐ False ☐

45. Write the empirical formula of chlorophyll-*a*.

____

46. Write the range of wavelength in which the rate of photosynthesis is maximum.

____

47. Give a collective term for the following pigments.

Chlorophyll-*b*, Xanthophylls, Carotenoids

____

48. ______________ pigments transfer the energy to chlorophyll-*a*.

49. Accessory pigments protect chlorophyll-*a* from ______________ .

50. Arrange the following in the sequence of energy transfer in photosynthesis.

Phycocyanin, Chlorophyll-*a*, Carotenoid, Phycoerythrin

____

Topic 3 : Light Reaction

51. Write the other name for light reaction.______________

52. Arrange the following in the order to its occurrence.

Water splitting, Light absorption, Oxygen release, Formation of ATP and NADPH

____

53. Write the full form of LHC.______________

54. Light Harvesting complexes (LHCs) are present within ______________ and ______________ these are made up of hundreds of pigment molecules bound to ______________ .

55. How many of the following constitute photosystems (PS)?

Reaction centre, Antennae, ATP synthase, H_2O

____

56. PS-I is located on the ______________ parts of grana.

57. PS-I has absorption peak at ______________ and PS-II has absorption peak at______________ .

58. Given figure depicts the Light Harvesting Complex (LHC), answer the following questions.

(i) Identify *A*, *B* and *C* in the figure given below.

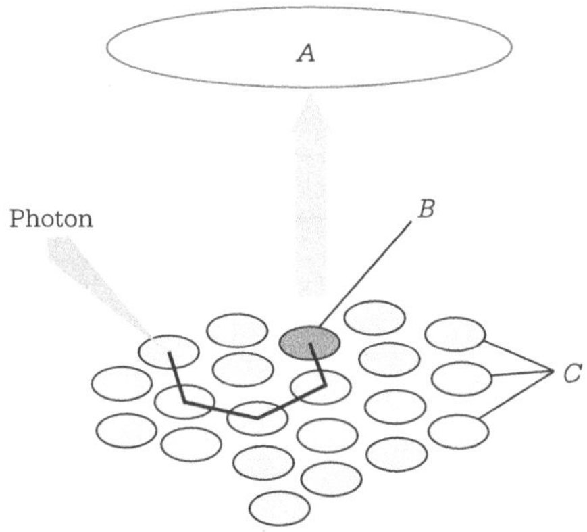

A. ______________________

B. ______________________

C. ______________________

(ii) Name the molecule which form part *B*. ______________________

59. Photophosphorylation is the formation of ATP in the presence of chemicals . (*Rewrite the statement by changing the underlined words.*)

____

Topic 4 : The Electron Transport

60. Electron transport system consists of ______________ proteins.

61. Movement of electron in the ETS is downhill in terms of an oxidation-reduction or redox potential scale.

True ☐ False ☐

62. When the chloroplast pigments absorb light electron become excited and move ______________ in the ETS.

63. The electrons in the reaction centre of ______________ are excited simultaneously with .

64. The primary acceptor of electron is located on the __________ (inner/outer) side of the membrane.

65. Electrons in the reaction centre of PS-I gets excited on receiving red light of __________ wavelength.

66. Arrange the following events of Z-scheme according to the transfer of electrons.

PS-II, Down the electron transport chain to PS-I, Uphill to the acceptor, Down hill to $NADP^+$, Transfer to another acceptor, Excitation of electrons

____

67. In Z-scheme of light reaction, the participating pigment systems are ______________ and ______________

68. The electrons transfer from PS-II to PS-I reduce______________________

69. Given below is the diagrammatic representation of the Z-scheme of light reaction of photosynthesis. Identify *A, B, C* and *D.*

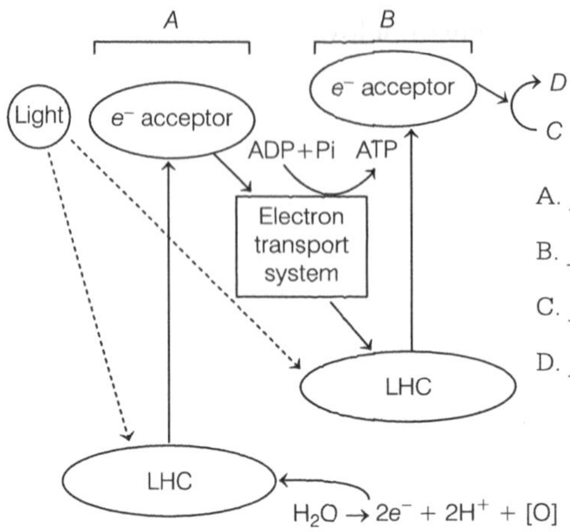

A. ______________

B. ______________

C. ______________

D. ______________

70. Write the chemical compound/molecule which supply electron continuously to PS-II.

____

71. During the light reaction, the one molecule of water splits into H^+, O_2 and electrons. (*Rewrite the statement by changing the underlined words.*)

____

72. If green plants are incubated with O^{18} labelled water, which molecule (photosynthesis product) will become radioactive? ______________

73. How many H^+ ions are formed from 12 water molecules during non-cyclic photophosphorylation?

12, 24, 36, 48

____

74. Select the appropriate words from the box given below to complete the statement.

Water, Oxygen, Carbon dioxide, Photosystem-I, Photosystem-II, Light harvesting centre

The electron removed from ______________ are provided by ______________ by splitting of ______________, which creates ______________.

75. The water splitting complex is associated with PS-II.

True ☐ False ☐

76. Water splitting complex is located on the ______________ (inner/outer) side of the membrane of ______________ (thylakoid/stroma).

77. The ETC of photosynthesis is bound to ______________ membrane.

78. ________________ membranes possess PS-I only, where as ________________ membranes possess both PS-I and PS-II. In cyclic photophosphorylation only ________ works.

79. Name the process of synthesis of ATP in cell. ________________

80. Name any two ATP synthesising organelles.

____

81. Write the process where two photosystems work in series from PS-II to PS-I.

____

82. When only PS-I is functional, electrons are circulated within the photosystem and the photooxidation occurs due to the cyclic flow of electrons.
(*Rewrite the statement by changing the underlined word.*)

____

83. Complete the following analogy.

Non-cyclic photophosphorylation : Cyclic photophosphorylation : : ________________ : ATP

84. Stroma lamellae lacks NADP reductase enzyme.

True ☐ False ☐

85. ________________ photophosphorylation can also occurs when wavelength beyond 680 nm is available for excitation.

86. Consider the following figure of cyclic phosphorylation, answer the given questions.
(i) Identify *A*, *B* and *C*.

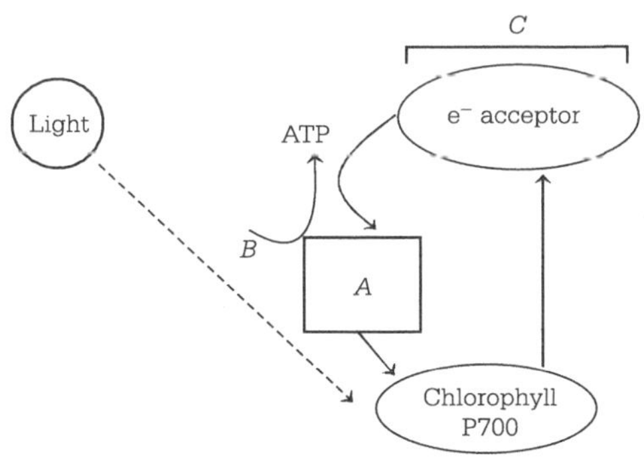

A. ________________
B. ________________
C. ________________

(ii) The excited electron pass onto $NADP^+$.

True ☐ False ☐

87. Name the hypothesis of ATP synthesis. ________________

88. ATP synthesis is linked to development of a proton gradient across the membrane.

True ☐ False ☐

89. During ATP synthesis, ____________________ accumulated towards the inside of the thylakoid membrane.

90. Protein accumulate in the intermembrane space of the chloroplast when electrons move through ETS during respiration.
(*Rewrite the statement by changing the underlined word.*)

____

91. Select the appropriate words from the box given below to complete the following paragraph

Outer, Lumen, Inner, Hydrogen, Thylakoid, ER, Potassium, Matrix

Splitting of water takes place on the ____________ side of the ______________ membrane by which proton or ___________ ions are produced that accumulate within the ___________ of the thylakoids.

92. The NADP reductase enzyme is located on the ____________ side of the membrane.

93. Identify *A, B, C, D* and *E* from the given figure of ATP synthesis through chemiosmosis.

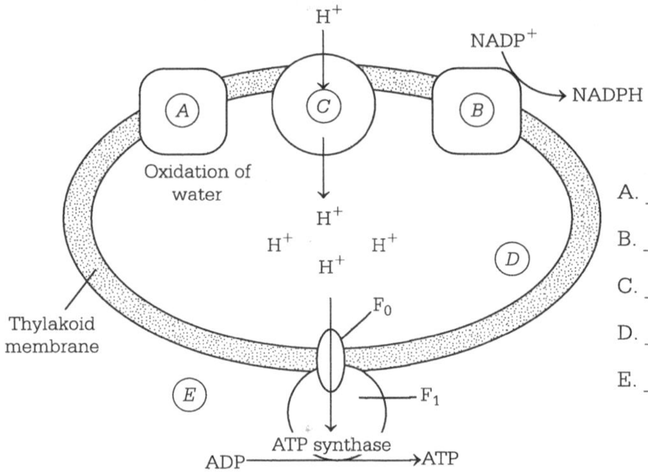

A. ______________
B. ______________
C. ______________
D. ______________
E. ______________

94. Complete the following analogy.
Proton increase : ________________ : : Proton decrease : Stroma

95. The primary acceptor of electron transfers its electron to an electron carrier and remove a proton from the stroma.

True ☐ False ☐

96. After passing on to the electron carrier on the inner side of the membrane, the proton is released into the ___________ (inner/outer) side or the ___________ (lumen/stroma) side of the membrane.

97. The pH of the lumen decreases, while proton gradiant is formed across thylakoid membrane.

True ☐ False ☐

98. Proton gradient is broken down due to ______________________ across the membrane to the stroma.

99. Name the transmembrane channel of the ATPase.______________________

100. Name the two parts of ATP synthase enzyme. ______________________

101. Write the location of CF_0 and CF_1 particles with respect to the thylakoid membrane. (Inside/Outside)

____

102. What is the role of CF_0?

____

103. The breakdown of the proton gradient provides ______________ to cause conformational change in the CF_1 particles of the ATPase.

104. Complete the following sequence.

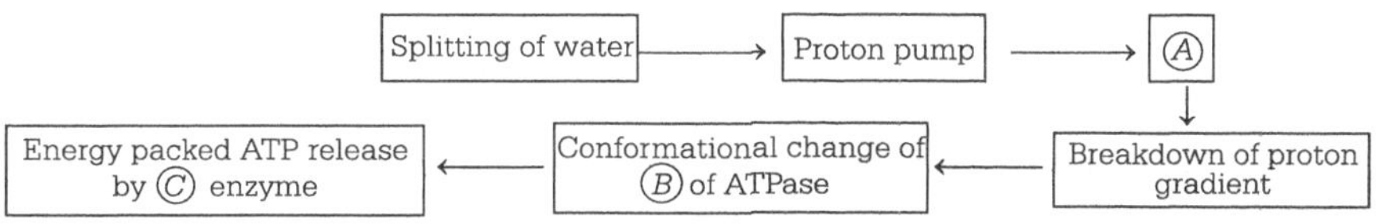

A. ____________ B. ____________ C. ____________

105. ATPase has channel that allows electron diffusion back across the membrane. (*Rewrite the statement by changing the underlined words.*)

____

106. ATP will be used immediately in the biosynthetic reaction for fixing ________ and synthesis of __________.

107. Write the major products of light reaction.

____

Topic 5 : Utilisation of ATP and NADPH and the Calvin Cycle

108. Name the phase where ATP and NADPH are used to drive the processes leading to the synthesise of sugar.

____

109. For one CO_2 molecule fixation, ________ (one/two/three) ATP and ________ (one/two/three) $NADPH_2$ are used in Calvin cycle.

110. Dark reaction is a misnomer name for biosynthetic phase.

True ☐ False ☐

111. Name the scientist who discovered 3 carbon organic acid as first CO_2 fixation product.

____

112. Answer the following questions with respect to mechanism of photosynthesis.

(i) Write the radioactive atom used by Calvin in his algal photosynthesis studies.

____

(ii) Write the full form of PGA. _______________

(iii) Write the chemical name of the first product identified by Calvin as first CO_2 fixation product. _______________

(iv) Write the two main pathways of CO_2 assimilation during photosynthesis.

_______________ .

113. Name the first CO_2 fixed product in C_4 plants. _______________

114. The pathway of CO_2 assimilation in which first product is a four carbon atom acid is called _______________ .

115. How many turns of Calvin cycle are needed to make one molecule of glucose?

____

116. Write the primary acceptor of CO_2 in Calvin cycle.

____

117. Calvin cycle does not occur in all photosynthetic plants.

True ☐ False ☐

118. Write the three main stages of Calvin cycle.

____

119. _______________ is the most crucial step of Calvin cycle.

120. In carboxylation, _______________ is utilised for carboxylation of RuBP.

121. Carboxylation results in the formation of three molecules of 3-PGA.

True ☐ False ☐

122. Name the step of Calvin cycle at which 3-PGA molecule converted into glyceraldehyde-3 -phosphate. _______________

123. Name the enzyme involved in carboxylation of RuBP. _______________

124. RuBisCO creats organic carbon from inorganic carbon dioxide in the air.

True ☐ False ☐

125. RuBisCO has oxygenation activity only.

True ☐ False ☐

126. Complete the overall equation of Calvin cycle by identifying *A* and *B*.

$6\,(A) + 6CO_2 + 18ATP + 12NADPH \rightarrow 6RuBP + (B) + 18ADP + 18P + 12NADP^+$

A. _______________ B. _______________

127. Reduction step in the Calvin cycle leads to the formation of _______________ .

128. Reduction involves utilisation of 2 molecules of ATP for ________________ and two of NADPH for __________ per CO_2 molecule fixed.

129. Fixation of six molecules of CO_2 are required for the formation of six molecules of glucose. (*Rewrite the statement by changing the underlined word.*)

____

130. Study the following figure and identify *A*, *B*, *C* and *D*.

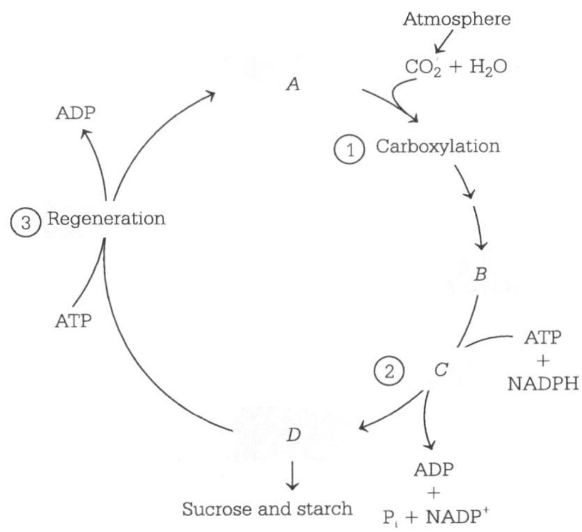

A. ________________

B. ________________

C. ________________

D. ________________

131. The regeneration step requires one ATP molecule for ____________ to form RuBP.

132. Complete the table with respect to Calvin cycle.

In	Out
6 CO_2	1 glucose
(*A*) ATP	(*B*) ADP
12NADPH	(*C*) NADP

A. ____________ B. ____________ C. ____________

Topic 6 : C_4 Pathway

133. C_4 plants also use C_3 pathway on the main biosynthetic pathway.

True ☐ False ☐

134. Name the important process that lacks in C_4 plants. ________________

135. Leaves of C_4 plant cannot tolerate high temperature.

True ☐ False ☐

136. C_4 leaves have greater productivity of energy.
(*Rewrite the statement by changing the underlined word.*)

____

137. Write the name of cells around vascular bundles of C_4 plants.

____

138. Write the technical name of the leaf anatomy of C_4 plants.

____

139. The bundle sheath cells of C_4 plants have large number of _______ thick walls to gas exchange and _______ intercellular space.

140. Write the other name for C_4 pathway.

____

141. Name the CO_2 acceptor in C_4 pathway. _______________

142. PEP is present in _______ cells.

143. Name the enzyme responsible for primary CO_2 fixation in C_4 pathway.

____

144. Match the following columns.

	Column I		Column II
A.	Pigment involved in transfer of electrons	1.	Cytochrome
B.	$NADPH_2$ is generated through	2.	CAM pathway
C.	CO_2 fixation occurs only night	3.	Photosystems

A. _______ B. _______ C. _______

145. Mesophyll cells have no _______ enzyme.

146. The oxaloacetic acid is present in the _______ cells.

147. OAA form malic acid and aspartic acid in the mesophyll cells.

True ☐ False ☐

148. Name the location where the C_4 acids are broken down to release CO_2.

____

149. The 3-carbon molecules are transported back to the mesophyll where it converts to aspartic acid.

(Rewrite the statement by changing the underlined words.)

____

150. The CO_2 released in the bundle sheath cells enter the Calvin cycle.

True ☐ False ☐

151. Name the enzyme present in bundle sheath cells. ______________________

152. Calvin cycle takes place in ______________ of C_4 plants.

153. Consider the following figure of Hatch and Slack pathway and identify *A*, *B*, *C* and *D*.

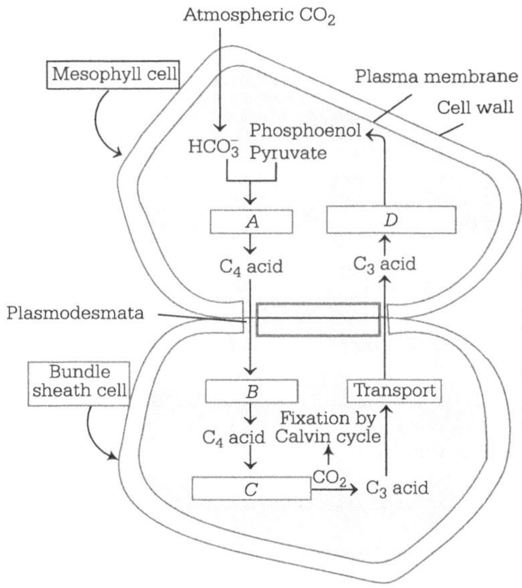

A. ______________________

B. ______________________

C. ______________________

D. ______________________

154. Complete the following analogy.

$C_3 : C_4 ::$ ______________________ : 4- carbon primary CO_2 acceptor.

155. How many of the following are C_3 plants?

Sugarcane, Potato, Oat, *Sorghum*, Maize, Wheat, Rice

____

156. How many of thc following plants show Kranz anatomy?

Sorghum, Maize, Sugarcane, Potato, Rice, Oat, Pearl millet

____

157. Find the odd one out.

Cleome, Papyrus, Orchardgrass, *Euphorbia*

____

Topic 7 : Photorespiration

158. Write the correct sequence of cell organelles used in photorespiration.

____

159. In C_4 plants photorespiration does not occur. (*Give reason*)

____

160. RuBisCO performs oxygenase activity at high CO_2 concentration.
(*Rewrite the statement by changing the underlined words.*)

____

161. Peroxisomes in photorespiration help in the synthesis of PGA.

True ☐ False ☐

162. Complete the following equation by identifying *A*.

$RuBP + CO_2 \xrightarrow{A} 2 \times 3\ PGA$ ______________________

163. Name the most abundant enzyme in the world.

____

164. The binding of CO_2 and O_2 to RuBisCO is uncompetitive mechanism.

True ☐ False ☐

165. Write the two products obtained by binding of RuBP with oxygen in C_3 plants.

____

166. Photorespiration in __________ plants results in the release of __________ with utilisation of __________ .

167. Photorespiration is a wasteful process in C_3 plants.

True ☐ False ☐

168. Consider the following figure and identify *A*, *B* and *C*.

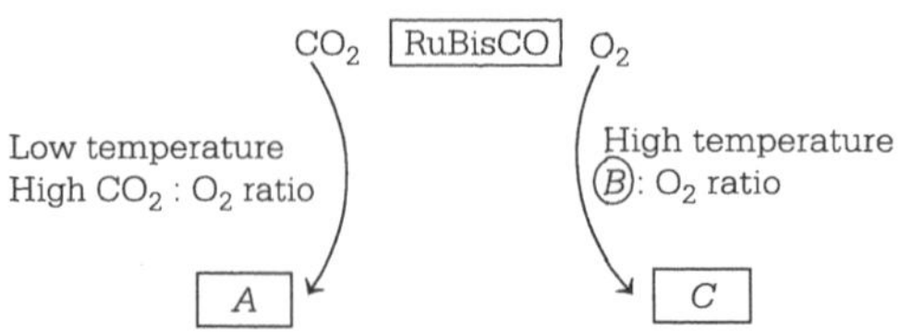

A. ______________________

B. ______________________

C. ______________________

169. Breaking down of C_4 acid in bundle sheath cell __________ intracellular CO_2 concentration.

170. The more CO_2 concentration ensures that the RuBisCO functions as a oxygenase, minimise carboxylase activity.
(Rewrite the statement by changing the underlined word.)

____

171. ______________ (High/Low) photorespiration would occur at low CO_2 concentration in C_3 plants.

172. Photorespiration is negligible in C_3 plants at low light intensities.

True ☐ False ☐

Topic 8 : Factors Affecting Photosynthesis

173. Find the odd one out with respect to factors affecting photosynthesis.

Sunlight availability, Mesophyll cells orientation, Number of chloroplasts, Orientation of leaves

____

174. Give a collective term for the following.

CO_2 concentration, Sunlight, Temperature, Water

____

175. The ______________ factors depend on the genetic predisposition and the growth of the plant.

176. Name the factor that come into effect when several factors affect any biochemical process.

____

177. Name the scientist who discovered the law of limiting factor.

____

178. The rate of the biochemical process will be determined by the factor that is nearest to its ______________ value.

179. Write the limiting factor among the following.

Green leaves, Optimal light, Optimum CO_2 concentration, Low temperature

____

180. The relationship between incident light and CO_2 fixation rate is ______________ (hyperbolic/parabolic/linear).

181. Light saturation occurs at 20% of full sunlight.

True ☐ False ☐

182. At higher light intensity, CO_2 fixation rate does not increase as the factors becomes altered.
(Rewrite the statement by changing the underlined word.)

____

183. Increase in incident light beyond saturation point causes breakdown of ________ and decrease in ________ .

184. Increase in CO_2 concentration upto ________ (0.05/0.04) per cent can cause an ________ (increase/decrease) in CO_2 fixation rates.

185. Availability of CO_2 level is limiting to the C_4 plants.

True ☐ False ☐

186. Name any two greenhouse crops.

____

187. C_3 plants respond to ________ (higher/lower) CO_2 concentration by showing ________ (increased/decreased) rates of photosynthesis leading to ________ (higher/lower) productivity.

188. The <u>light</u> reactions are more sensitive to temperature than <u>dark reaction</u>.
(*Rewrite the statement by changing the underlined words.*)

____

189. The C_4 plants respond to ________ (higher/lower) temperature and show ________ (higher/lower) rate of photosynthesis and C_3 have a much ________ (lower/higher) temperature optimum.

190. Water stress causes stomatal closer it reduces CO_2 availability.

True ☐ False ☐

191. Complete the following analogy.

Tropical plants : Temperate plants : : ________________ : Lower temperature optimum

192. The CO_2 fixation rate under high light conditions is ________ (high/low) for C_3 plants.

Answers

1. Photosynthesis
2. Sunlight
3. Oxygen
4. Light, chemical.
5. Photosynthesis occurred only in the green parts (uncovered part) of the leaves.
6. (i) *A*–KOH, *B*–Air
   (ii) KOH absorbs the carbon dioxide present in the test tube.
   (iii) The exposed part of leaf tested positive for starch.
7. The experiment showed that CO_2 is required for photosynthesis.
8. Joseph Priestley (1733-1804)
9. True
10. Priestley
11. Only green parts of plant release O_2.
12. Jan Ingenhousz
13. Julius von Sachs
14. TW Engelmann
15. *Cladophora*
16. Engelmann
17. Cornelius van Niel
18. False. In photosynthesis hydrogen from an oxidisation compound reduces CO_2 to carbohydrates.
19. $6\,CO_2 + 12H_2O \xrightarrow{\text{Light}} C_6H_{12}O_6 + 6H_2O + 6O_2$
20. $CO_2 + H_2O \xrightarrow{\text{Light}} [CH_2O] + O_2$
21. Hydrogen Sulphide (H_2S)
22. van Niel
23. A-1, B-3, C-4, D-1
24. Mesophyll
25. Cisternae is odd one out, as it is the part of mitochondria. Rest other options are the parts of chloroplast.
26. Stroma
27. ATP and NADPH.
28. Dark reactions
29. False. Dark reaction are dependent on ATP and NADPH.
30. Dark reaction, ATP/NADPH
31. *A* - Stroma lamellae, *B* - Grana, *C* - Stroma, *D* - Ribosomes.
32. Paper chromatography.
33. A - 3, B - 1, C - 4, D - 2
34. Highest
35. Phycoerythrin is odd one out because it is found in algae where as other pigments are found in higher plants.
36. *A* - Chlorophyll-*b*, *B* - Carotenoids, *C* - Chlorophyll-*a*.
37. False. Absorption spectrum of chlorophyll-*b* is maximum.
38. Chlorophyll-*a*
39. Blue and red regions of the spectrum.
40. An action spectrum
41. False. Absorption spectrum of chlorophyll-*a* and the action spectrum of photosynthesis is not identical.
42. 390-760 nm
43. Red > Blue > Green
44. True
45. $C_{55}H_{72}O_5N_4Mg$
46. 400 - 700 nm
47. Accessory pigments
48. Accessory
49. Photooxidation
50. Phycocyanin → Phycoerythrin → Carotenoid → Chlorophyll-*a*
51. Photochemical reaction
52. Light absorption → Water splitting → Oxygen release → Formation of NADPH and ATP.
53. Light Harvesting Complex.
54. PS-I, PS-II, protein
55. 2 (Reaction centre and antennae)
56. Non-appressed thylakoids and stroma thylakoids
57. 700 nm; 680 nm
58. (i) *A*–Primary acceptor, *B*–Reaction centre, *C*–Pigment molecules
    (ii) Chlorophyll-*a*
59. Presence of light
60. Cytochrome
61. True
62. Uphill
63. PS-I, PS-II
64. Outer
65. 700 nm
66. PS-II → Uphill to the acceptor → Down the electron transport chain to PS I → Excitation of electrons → Transfer to another acceptor → Downhill to $NADP^+$
67. PS-I and PS-II
68. $NADP^+$ to NADPH + H^+
69. *A*–Photosystem-II
    B–Photosystem-I
    C–$NADP^+$
    D–NADPH
70. H_2O
71. $2H^+$, [O] and 2 electrons.
72. O_2
73. H_2O gives $2H^+$; $12 \times 2 = 24\ [H^+]$
74. Photosystem-I, photosystem-II, water, oxygen.
75. True
76. Inner, thylakoid
77. Thylakoid
78. Stroma lamellae, grana lamella, PS-I
79. Phosphorylation
80. Mitochondrion and chloroplasts
81. Non-cyclic photophosphorylation.
82. Phosphorylation
83. ATP/NADPH
84. True
85. Cyclic
86. (i) *A*-ETS, *B*-ADP + Pi, *C*-PS-I
    (ii) False. The excited electron does not pass onto $NADP^+$, it is cycled back to PS-I.
87. Chemiosmotic hypothesis.
88. True
89. Proton
90. Mitochondria
91. Inner, thylakoid, hydrogen, lumen
92. Stroma
93. *A*–PS-II, *B*–PS-I, *C*–Cytochrome-b_6f, *D*–Lumen, *E*–Stroma

94. Lumen
95. False. The primary acceptor of electron transfers its electron to H carrier and remove a proton from the stroma.
96. Inner, lumen.
97. True
98. Movement of proton
99. CF_0
100. CF_0 and CF_1
101. CF_0 - Inside, CF_1 - Outside
102. It carries out facilitated diffusion of proton across the membrane.
103. Energy
104. *A*–Proton gradient, *B*–F_1 particles, *C*–ATPase
105. H^+ diffusion
106. Carbon dioxide (CO_2), sugars
107. ATP, $NADPH_2$ and O_2
108. Biosynthetic phase
109. Three, two
110. True
111. Melvin Calvin
112. (i) ^{14}C (Carbon-14).
     (ii) Phosphoglyceric acid
     (iii) 3-phosphoglyceric acid
     (iv) C_3 pathway and C_4 pathway
113. Oxaloacetic acid (OAA)
114. C_4 pathway
115. 6
116. Ribulose Bisphosphate (RuBP)
117. False. Calvin cycle occurs in all photosynthetic plants.
118. Carboxylation, reduction and regeneration.
119. Carboxylation
120. Carbon dioxide
121. False. Carboxylation results in the formation of two molecules of 3-PGA.
122. Reduction
123. RuBP carboxylase - oxygenase (RuBisCO)
124. True
125. False. RuBisCO has both carboxylation and oxygenation property.
126. *A*–RuBP, *B*–Glucose
127. Glucose
128. Phosphorylation, reduction
129. One molecule
130. *A*-Ribulose 1, 5 - bisphosphate, *B*-3-Phosphoglycerate, *C*-Reduction, *D*-Triose phosphate
131. Phosphorylation
132. *A*-18, *B*-18, *C*-12
133. True
134. Photorespiration
135. False. Leaves of C_4 plant can tolerate high temperature.
136. Biomass
137. Bundle sheath cells
138. Kranz anatomy
139. Chloroplast, impervious, no
140. Hatch and Slack pathway
141. Phosphoenolpyruvate (PEP)
142. Mesophyll
143. PEP Carboxylase
144. A - 1, B - 3, C - 2
145. RuBisCO
146. Mesophyll
147. True
148. Bundle sheath cells
149. Phosphoenol pyruvate.
150. True
151. RuBisCO enzyme
152. Bundle sheath cells
153. *A*-Fixation, *B*-Transport, *C*-Decarboxylation, *D*-Regeneration.
154. 3-carbon primary CO_2 acceptor
155. 4(Wheat, rice, oat, potato)
156. 4(Sugarcane, maize, *Sorghum*, pearl millet).
157. Orchardgrass is odd one out. Rest of the options are C_4 plants.
158. Chloroplast – Peroxisome – Mitochondria.
159. C_4 plants have mechanism that increases the concentration of CO_2 at the enzyme site.
160. Low CO_2 concentration.
161. False. Peroxisomes in photorespiration help in the oxidation of glycolate.
162. *A* - RuBisCO
163. RuBisCO
164. False. The binding of CO_2 and O_2 to RuBisCO is competitive mechanism.
165. Phosphoglycerate and phosphoglycolate
166. C_3, CO_2, ATP
167. True
168. *A*–Calvin cycle
     B–Low CO_2
     C–Photorespiration
169. Increase
170. Carboxylase, minimise oxygenase activity.
171. High
172. True
173. Sunlight availability is odd one out. Rest of the options are internal factors affecting photosynthesis.
174. External factors affecting photosynthesis.
175. Internal
176. Law of limiting factor
177. Blackman in 1905
178. Minimal
179. Low temperature
180. Linear
181. False. Light saturation occurs at 10% of full sunlight.
182. Limiting
183. Chlorophyll, photosynthesis
184. 0.05, increase
185. False. Availability of CO_2 level is limiting to the C_3 plants.
186. Tomato and bell pepper.
187. Higher, increased, higher
188. Dark, light reaction.
189. Higher, higher, lower.
190. True
191. Higher temperature optimum
192. Low

CHAPTER FOURTEEN

RESPIRATION IN PLANTS

Topic 1 : Do Plants Breathe?

1. How many of the given daily life activities need energy in all living organisms?

Absorption, Movement, Transport, Reproduction, Breakdown of food, Breathing

____

2. All the energy required for life processes is obtained by _______________ of some macromolecules that we call 'food'.

3. Green plants and cyanobacteria can prepare their own food by _______________ .

4. Give the scientific term for the process of trapping light energy and converting it into chemical energy that is stored in the bonds of carbohydrates like glucose, sucrose and starch in plants. _______________

5. Animals are _______________, they obtain food from plant directly, i.e. _______________ or indirectly, i.e. _______________ .

6. _______________ are dependent on dead and decaying matter for food like fungi.

7. Give the scientific term for the mechanism of breakdown of food materials within the cell to release energy and trapping of this energy for synthesis of ATP.

____

8. Respiration is a _______________ (catabolic/anabolic) and _______________ (endergonic/exergonic) process.

9. Complete the following analogy with respect to eukaryotes.

   Photosynthesis : Respiration : : Chloroplasts : ____________________

10. The breaking of C—C bonds of complex compounds through oxidation within the cells, leading to release of considerable amount of energy is called ____________________.

11. The compounds that are oxidised during the process of respiration are known as ______________________________.

12. How many of the following macromolecules can act as respiratory substrates to release energy in plants?

Carbohydrates, Fats, Proteins, Organic acids, Vitamins, Humus, Amino acids

   ____

13. Respiratory substrates release energy free into the cell in a single step.

   True ☐ False ☐

14. Name the energy currency of the cell.

   ____

15. Plant requires ______________________________ for respiration to occur and give out ________________________ during this process (oxygen/carbon dioxide).

16. For gaseous exchange, plants have ________________ and ________________ in their leaves and stems, respectively.

17. Plants exhibit respiration without respiratory organs. (*Give reasons*)

   ____

18. State whether following statements are true or false.

   (i) There is very little transport of gases from one plant part to another.

   True ☐ False ☐

   (ii) Roots, stems and leaves respire at rates far higher than animals.

   True ☐ False ☐

19. Complete the biochemical reaction of respiration.

   $C_6H_{12}O_6 + 6O_2 \longrightarrow$ ______________ + ______________ + ______________.

20. The gas required during combustion reaction is ________________.

21. Give two examples of each of the following.

   (i) Facultative anaerobes____

   (ii) Obligate anaerobes ____

Topic 2 : Glycolysis

22. In any case (aerobic or anaerobic), all living organisms retain the enzymatic machinery to partially oxidise glucose without the help of oxygen. This breakdown of glucose to pyruvic acid is called

____

23. During glycolysis, the one molecule of glucose (6C) is broken down into two molecules of

_______________ .

24. The term glycolysis has originated from two Greek words namely
Glycos, i.e. _______________ *Lysis*, i.e. _______________ .

25. The glycolysis is also known as EMP pathway. Here, EMP reffered as

____

26. Choose the odd one out from the given scientist names.

J Parnas, Calvin, Otto Meyerhof, Gustav Embden

____

27. In _______________ (aerobic/anaerobic) organisms, glycolysis is the only process in respiration.

28. Give the location for the following respiratory processes inside the cell.
(i) Glycolysis _______________
(ii) TCA cycle _______________
(iii) ETC _______________

29. Study the flowchart and identify *A* and *B*.

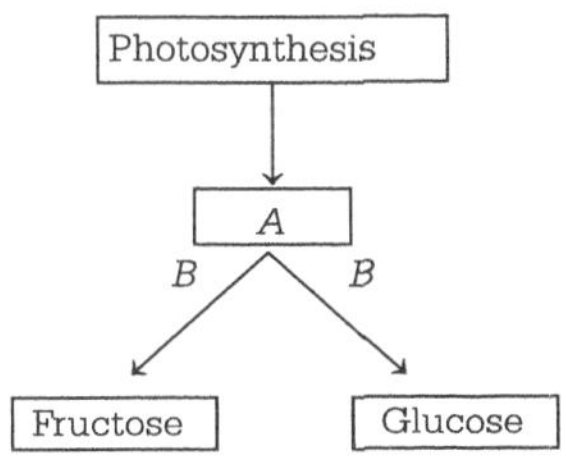

A. _______________
B. _______________

30. Give common terms for the following enzymes mentioned in the box given below.

Hexokinase, Aldolase, Isomerase, Phosphofructokinase

____

31. Each step of glycolysis uses different enzymes. (*Give reason*)

____

32. How many reactions are there in glycolysis?

____

33. The flowchart given below gives the steps involved in glycolysis. Fill in the missing molecules in place of *A, B* and *C.*

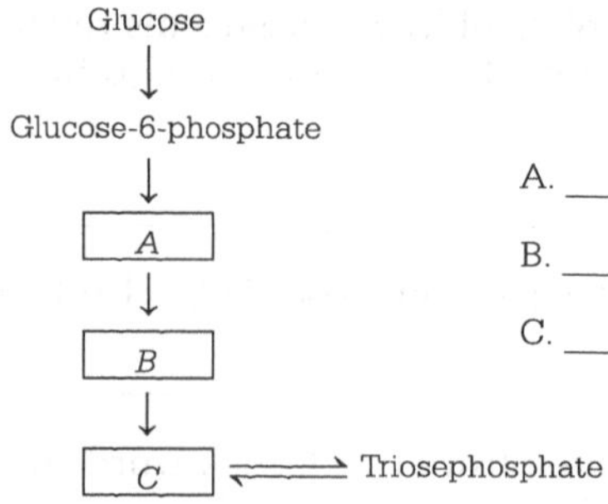

A. ____________________

B. ____________________

C. ____________________

34. Study the given steps of glycolysis and identify the enzymes *A, B* and *C* for carrying out these steps.

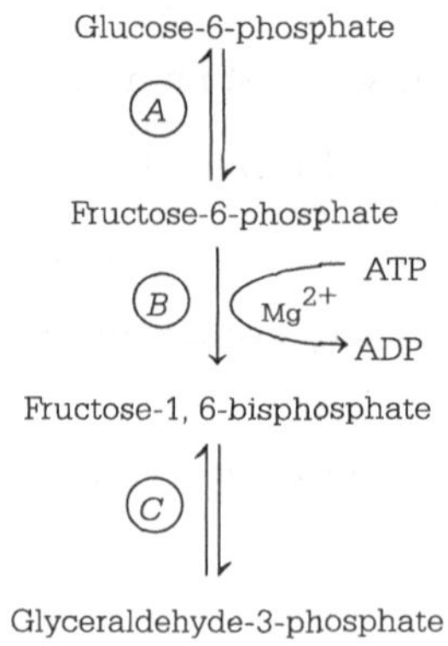

A. ____________________

B. ____________________

C. ____________________

35. The flowchart given below shows some steps in glycolysis. Identify the molecules *A* to *D.*

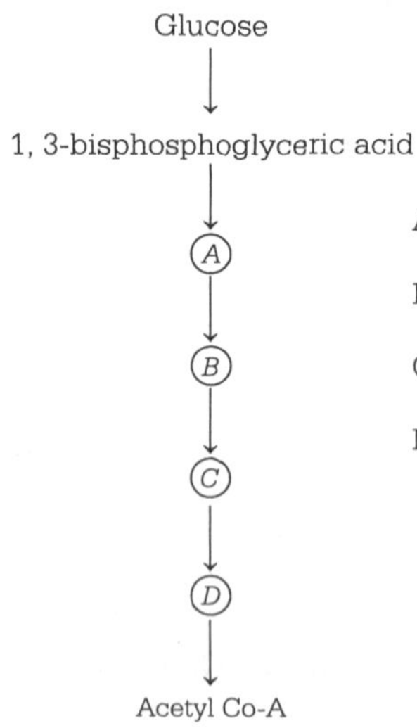

A. ____________________

B. ____________________

C. ____________________

D. ____________________

36. Write the reaction of glycolysis where a water molecule is removed.

____

37. Consider the following reaction and identify substrate *A*.

$$A \underset{\text{isomerase}}{\overset{\text{Triosephosphate}}{\rightleftharpoons}} \text{Glyceraldehyde-3-phosphate}$$

A. ______________________

38. Give the scientific term used to denote direct synthesis of ATP from ADP.

____

39. Complete the analogy with respect to ATP in glycolysis.

ATP used : 2 steps :: ATP formed : ____________ .

40. How many ATPs are formed during glycolysis?

____ .

41. NADH are ______________ (formed/consumed) during glycolysis.

42. What happens to the hydrogen atom and the electron that are removed during dehydrogenation and phosphorylation of 3 PGAL?

____

43. What will happen to glycolytic pathway if cell runs completely out of ATP?

____

44. How many of the following pairs are isomers?

PEP and pyruvate, 3PGA and 2PGA, PGAL and DHAP, Glucose and Fructose Glucose-6-phosphate and Fructose-6-phosphate

____

45. Give the full form of the following.

(i) ATP ______________________________________

(ii) NADH ____________________________________

(iii) PGAL ___________________________________

(iv) DHAP ___________________________________

(v) BPGA ____________________________________

(vi) PEP _____________________________________

46. At the end of glycolysis, ____________ is the net energy gain from one molecule of glucose *via* ______________, but there is also energy stored in the form of ____________ .

47. Given below are few steps of glycolysis. Fill in the boxes *A* to *D* by using ATP used and ATP formed.

(i) Glucose → Glucose-6-phosphate – [*A*]

(ii) BPGA → 3PGA – [*B*]

(iii) PEP → Pyruvic acid – [*C*]

(iv) Fructose-6-phosphate → Fructose-1, 6-bisphosphate – [*D*]

48. The end product reaction of glycolysis is given below.

$$\text{Glucose} \xrightarrow{\text{Glycolysis}} 2 \text{ Pyruvic acid} + 2\text{ATP} + 2\text{NADH}_2$$

True [ ] False [ ]

49. Match the following columns.

	Column I (Name of compounds)		Column II (Number of carbon atoms)
A.	DHAP	1.	6 carbon
B.	Glucose	2.	3 carbon
C.	Fructose-1, 6-bisphosphate		
D.	PEP		

A. ________ B. ________ C. ________ D. ________

50. In glycolysis, total ________ ATP molecules are produced and total gain of ATP molecules is ________.

51. There are three major ways in which different cells handle pyruvic acid produced by glycolysis. These are ________, ________ and ________.

52. How many molecules of NADH are produced when four molecules of phosphoglyceraldehyde are converted into four molecules of pyruvate?

____

53. ________ (OAA/Pyruvic acid/Acetyl Co-A) is the most important intermediate formed in all types of respiration.

54. The reaction forming 3-phosphoglyceric acid in glycolysis is ________ (oxidative phosphorylation/substrate level phosphorylation).

55. Name any three cells where glycolysis occurs.

____

Topic 3 : Fermentation

56. In many prokaryotes and unicellular eukaryotes, ______________________________ process takes place in anaerobic condition.

57. Complete the following flowchart showing fermentation.

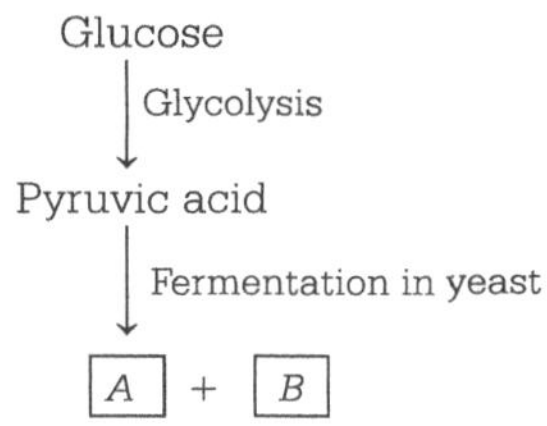

A. __________________________

B. __________________________

58. Study the reaction and answer the following questions.

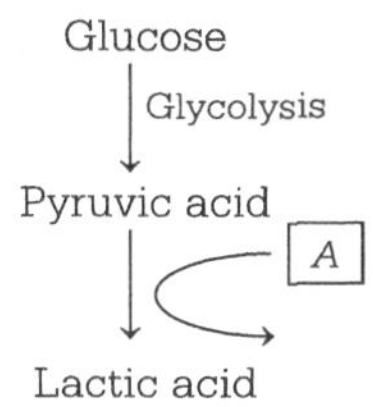

(i) What is the energy rich compound denoted as '*A*'? ________________________

(ii) In which organism this reaction occurs? ______________________________

(iii) Which animal cell also shows this reaction? ___________________________

(iv) Write the enzyme responsible for this process. __________________________

59. In alcoholic fermentation (by yeast), the reduction of pyruvic acid needs two enzymes, i.e. ________________________ and _______________________ .

60. In term of efficiency at converting the energy of glucose in ATP, anaerobic conditions are much less efficient than aerobic.

True ☐ False ☐

61. Write numerical value to fill in the blanks.

(i) In both lactic acid and alcoholic fermentation not much energy is released, less than ________________ per cent of the energy in glucose is released out and not all of it is trapped as ATP.

(ii) Yeast poison themselves to death when the concentration of alcohol reaches about ________________ per cent.

(iii) $C_6H_{12}O_6 + 6O_2 \longrightarrow 6CO_2 + 6H_2O +$ __________________ kcal

(iv) During the conversion of pyruvic acid to acetyl Co-A ___________________ molecule of NADH are produced.

62. Fermentation process is used in ______________ industry, ____________ of tea and tabacco and production of ______________.

63. Choose the odd one out with respect to the part that can respire in the absence of oxygen.

Stem, Leaves, Root, Seed

____

Topic 4 : Aerobic Respiration and Tricarboxylic Acid Cycle

64. Give the scientific term for following statement.

The process that leads to a complete oxidation of organic substances in the presence of oxygen and releases CO_2, H_2O and energy needed for cellular metabolism.

____

65. For aerobic respiration to take place within the '*X*', the product of glycolysis, i.e. pyruvic acid is transported from the cytoplasm into the '*X*'.

(Name the cell organelle '*X*')

____

66. Name the two crucial events in aerobic respiration.

____

67. Pyruvate is formed by the glycolytic ______________ (anabolism/catabolism) of carbohydrates in the cytosol, after it enters mitochondrial ______________ (stroma/matrix) undergoes oxidative ______________ (carboxylation/ decarboxylation) by a complex set of reaction.

68. Complete the following reaction.

$$\text{Pyruvate} + \boxed{A} + \boxed{B} \xrightarrow[\boxed{C}]{Mg^{2+}} \text{Acetyl Co-A} + \boxed{D} + \boxed{E}$$

A. ____________ B. ____________ C. ____________ D. ____________

E. ____________

69. Write the full form of the following.

(i) TCA ______________________________

(ii) ETS ______________________________

(iii) OAA ______________________________

70. Name the scientist who elucidated TCA cycle.

____

71. Write another name of TCA cycle.

____

72. In prokaryotes TCA cycle occurs in______________________________

73. In two turns of Krebs cycle, numbers of molecules of CO_2, NADH, $FADH_2$ and ATP produced are ________________, ________________, ________________ and ________________, respectively.

74. Match the following columns.

	Column I		Column II
A.	4 carbon compound	1.	Oxaloacetic acid
B.	3 carbon compound	2.	Pyruvate
C.	5 carbon compound	3.	α-ketoglutaric acid
D.	6 carbon compound	4.	Glucose-6-phosphate

A. ______________ B. ______________ C. ______________ D. ______________

75. The TCA cycle starts with the condensation of acetyl group with oxaloacetic acid (OAA) to water to yield ______________________.

76. Complete the sequence of events in Krebs cycle in the following flowchart.

Acetyl-Co-A
↓
Citric acid
↓
[A]
↓
[B]
↓
Fumaric acid
↓
[C]
↓
[D]

A. ______________________________

B. ______________________________

C. ______________________________

D. ______________________________

77. Write the link reaction between aerobic and anaerobic respiration.

____

78. The first 5-C dicarboxylic acid in Krebs cycle which is used in nitrogen metabolism is <u>citric acid</u>.

(*Rewrite the statement by changing the underlined word.*)

____

79. How many enzymes are used in Krebs cycle? Name them.

____

80. Write the missing molecule *X* in the chemical reaction given below.

Citric acid ⇌ [X] → α-ketoglutarate

X. ____

81. Consider the following reaction and identify *X* and *Y*.

$$\text{Succinyl Co-A} \xrightarrow[\text{GDP} \;\curvearrowright\; \boxed{Y}]{\boxed{\text{Enzyme } X}} \text{Succinic acid}$$

X.____________________ Y.____________________

82. The first member of TCA cycle is ____________________ .

83. Complete the following reaction given below.

$$\text{Pyruvic acid} + \boxed{A} + \boxed{B} + \boxed{C} + \text{ADP} + \text{Pi} \xrightarrow[\text{matrix}]{\text{Mitochondrial}} \boxed{D} + \boxed{E} + 4\text{H} + \text{FADH}_2 + \text{ATP}$$

A. ____________ B. ____________ C. ____________

D.____________ E. ____________

84. Name the molecules from which the given molecules are formed in Krebs cycle.

(i) OAA ____________________

(ii) Malic acid ____________________

(iii) Fumaric acid ____________________

85. Given below is flowchart of Krebs cycle. Name *X*, *Y* and *Z*.

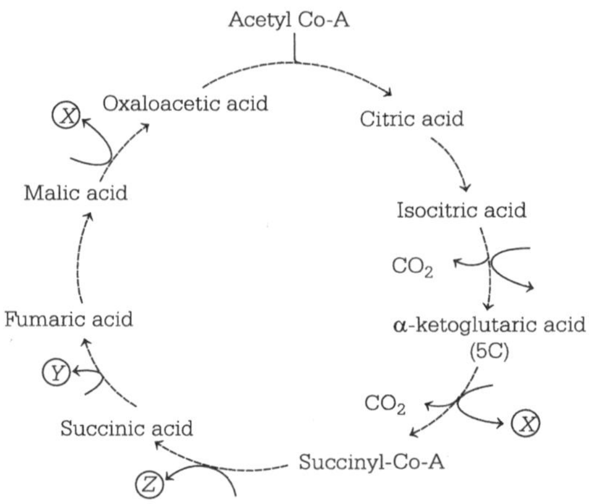

X. ____________________

Y. ____________________

Z. ____________________

86. Krebs cycle is also called ____________________ as it is a common pathway for carbohydrates, fats and proteins/amino acids.

87. Substrate level phosphorylation occurs during which step of Krebs cycle?

____

88. Write the types of reaction occur in the following reactions.

(i) Succinyl CoA → OAA ____________________

(ii) Citrate → Isocitrate ____________________

(iii) Oxalosuccinate → α-ketoglutaric acid ____________________

89. Write the name of the process mentioned.

(i) The complete oxidation of pyruvate by the stepwise removal of all the hydrogen atoms, leaving three molecules of CO_2. ______

(ii) The passing on of the electrons removed as part of the hydrogen atoms to molecular oxygen with simultaneous synthesis of ATP. ______

Topic 5 : Electron Transport System (ETS) and Oxidative Phosphorylation

90. During passage of electron over ETC, pH of matrix increases.

True ☐ False ☐

91. Cytochromes are found in ______ (cristae/matrix) of mitochondria.

92. Match the following columns.

	Column I (Complexes)		Column II (Respective names)
A.	Complex-I	1.	ATP synthase
B.	Complex-II	2.	Cytochrome-*c* oxidase
C.	Complex-III	3.	Cytochrome-bc_1
D.	Complex-IV	4.	Succinate dehydrogenase
E.	Complex-V	5.	NADH dehydrogenase

A. ______ B. ______ C. ______ D. ______
E. ______

93. Cytochrome-*c* oxidase complex contains cytochromes ______, ______ and ______ copper atoms.

94. Cytochrome ______ (*a*/*b*/*c*) acts as a mobile carrier for transfer of electrons between complex-III and IV.

95. Give the location of ubiquinone.

______

96. Electrons from NADH produced in the mitochondrial matrix during TCA cycle are oxidised by complex-I, and electrons are then transferred to <u>cytochrome-*c*</u>.

(*Rewrite the statement by changing the underlined word.*)

______

97. During ETS, the number of ATP molecules synthesised depends on the nature of

______.

98. When the electron pass from one carrier to another *via* complex-I to IV in ETS, they are coupled to complex-V to produce ______.

99. Identify *A, B, C, D, E* and *F* in the given figure of electron transport chain.

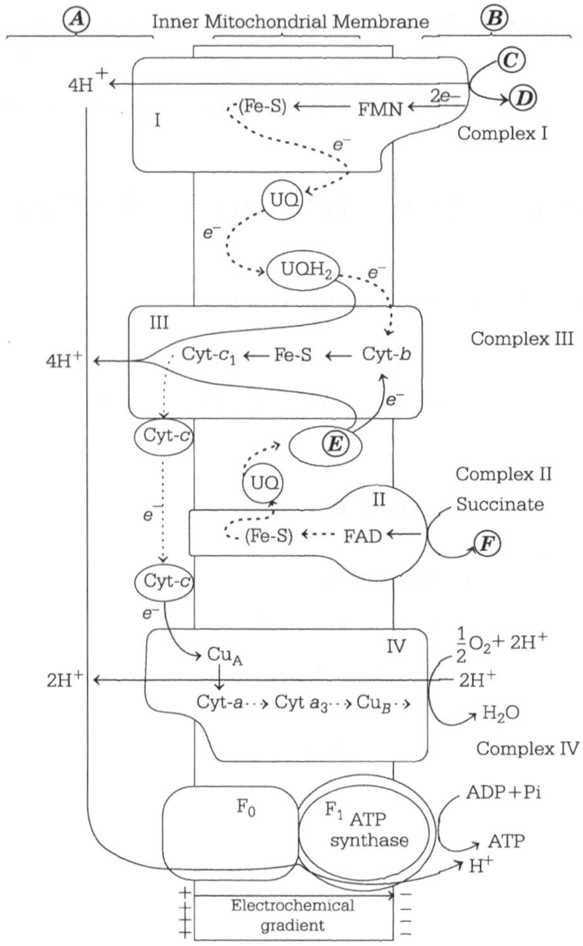

A. ____________________

B. ____________________

C. ____________________

D. ____________________

E. ____________________

F. ____________________

100. Complete the following analogy with respect to ATP molecules produced.

NADH : ____________ ATP :: $FADH_2$: ____________ ATP

101. What is the role of oxygen in ETS? (*Give answer in one line*)

____

102. For production of proton gradient, photosynthesis requires ____________ while, respiration requires ____________.

103. Give the scientific term for the process of formation of ATP by transfer of phosphate group from a substrate to ADP.

____

104. Chemiosmotic theory of ATP synthesis in chloroplasts and mitochondria is based on ____________ (membrane potential/proton gradient).

105. Oxygen is reduced to water in ____________ process.

106. Coenzyme NAD^+ and FAD are connected with respiratory reactions as they act as ________________.

107. The two major components of ATP synthase are __________ and __________.

108. In ATP synthase,

(i) Peripheral membrane protein complex is ________________

(ii) Integral membrane protein complex is ________________

109. Give the functions of the following.

(i) F_0 particle ________________

(ii) F_1 head piece ________________

110. Given below is the diagrammatic presentation of ATP synthesis in mitochondria. Identify *A* to *C*.

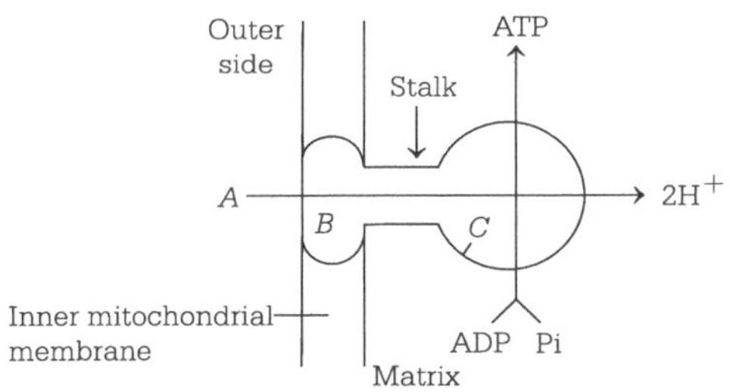

A. ________________

B. ________________

C. ________________

111. The passage of protons through the channel formed by F_1 is coupled to the catalytic site of the F_0 component.

True ☐ False ☐

112. Give one difference between photophosphorylation and oxidative phosphorylation with respect to oxygen.

____

Topic 6 : Respiratory Balance Sheet and Amphibolic Pathway

113. Fill in the blanks about the assumptions made to calculate net gain of ATP.

(i) There is a sequential, orderly pathway functioning, with one substrate forming the next and with __________, __________ and __________ pathway following one after another.

(ii) The NADH synthesised in glycolysis transferred into the mitochondria and undergoes __________.

(iii) __________ of the intermediates compounds in the respiratory pathway are utilised to synthesise any other compound.

(iv) Only __________ is being respired, no other alternative substrates are entering in the pathway at any of the intermediary stages.

114. The net gain of ATP molecules during aerobic respiration from one molecule of glucose is __________.

115. Match the following columns.

Column I (Stages of aerobic respiration of a glucose molecule)	Column II (ATP produced through ETS)
A. Glycolysis	1. 6
B. Formation of acetyl Co-A	2. 3
C. Krebs cycle	3. 22

A.________________ B.________________ C.________________

116. Give common term for the molecules mentioned in the box given below.

Cyanides, Antimycin-A, Carbon monoxide

____

117. As per chemiosmotic coupling hypothesis in mitochondria, protons accumulate in outer membrane space.
(*Rewrite the statement by changing the underlined word.*)

____

118. How many ATP molecules will be generated in a plant system during complete oxidation of 40 molecules of glucose?

____

119. Out of the three respiratory substrates (carbohydrates, fat and protein), which one is the most favoured substrate?

____

120. Respiratory pathway is catabolic. (*Rewrite the statement by changing the underlined word.*)

____

121. The figure given below describes inter-relationship among metabolic pathways showing respiration mediated breakdown of different organic molecules to CO_2 and H_2O. Identify *A-D.*

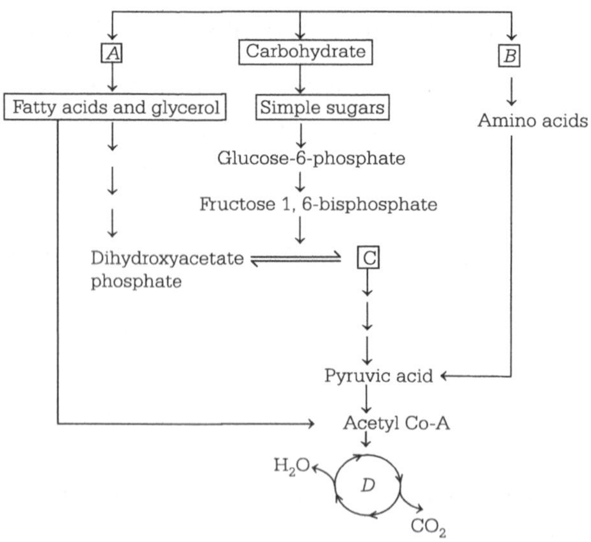

A. ______________ B. ______________ C. ______________ D.______________

122. What is the next molecule formed in the respiratory pathway for the given substrates?

(i) Glycerol ____________________

(ii) Proteins ____________________

(iii) Fatty acids ____________________

(iv) Fats ____________________

Topic 7 : Respiratory Quotient

123. Complete the following formula.

Respiratory Quotient $= \frac{X}{Y}$

X. ____________________ Y . ____________________

124. If the volume of CO_2 liberated during respiration is more than the volume of O_2 used then the respiratory substrate will be ____________________ .

125. Name an instrument used to measure rate of respiration and respiratory quotient.

____

126. Write the value of respiratory quotient of the following substrates.

(i) Tripalmitin ____________________

(ii) Organic acids ____________________

(iii) Fats ____________________

(iv) Proteins ____________________

(v) Carbohydrates ____________________

(vi) Oxalic acid ____________________

127. When respiratory quotient can be

(i) Zero ____________________

(ii) Infinity ____________________

Answers

1. 5 (All of them need energy except breakdown of food because it releases energy.)
2. Oxidation
3. Photosynthesis
4. Photosynthesis
5. Heterotrophs, herbivores, carnivores
6. Saprophytes
7. Cellular respiration
8. Catabolic, exergonic
9. Mitochondria
10. Respiration
11. Respiratory substrate
12. 4 (Carbohydrates, fats, proteins, organic acids).
13. False. Respiratory substrates release energy (in the form of ATP) in a series of slow stepwise reactions controlled by enzymes.
14. ATP molecules
15. Oxygen, carbon dioxide
16. Stomata, lenticels
17. Each plant part takes care of its own gaseous needs. Plants do not present great demands for gaseous exchange. Almost all living cells in a plant have their surface expose to air.
18. (i) True
    (ii) False. Roots, stems and leaves respire at rates far lower than animals.
19. $6CO_2 + 6H_2O$ + Energy
20. O_2 (Oxygen)
21. (i) *Staphylococci, Escherichia coli*
    (ii) *Bacteroides fragilis, Clostridium tetani*
22. Glycolysis
23. Pyruvic acid (3C)
24. Sugar, splitting
25. EMP–Embden, Meyerhof and Parnas.
26. Calvin is odd one out. All of the given scientists proposed scheme of glycolysis except Calvin, he proposed Calvin cycle of photosynthesis.
27. Anaerobic
28. (i) Cytosol
    (ii) Mitochondrial matrix
    (iii) Mitochondrial inner membrane
29. *A*–Sucrose, *B*–Invertase
30. Enzymes of glycolysis
31. Each step involves a different chemical reactions.
32. 10
33. *A*-Fructose-6-phosphate, *B*-Fructose-1, 6-bisphosphate, *C*-Glyceraldehyde-3- phosphate.
34. *A*-Phosphohexose isomerase, *B*-Phosphofructokinase, *C*-Aldolase
35. *A*–3-phosphoglyceric acid, *B*–2-phosphoglyceric acid, *C*–Phosphoenol pyruvate, *D*–Pyruvic acid
36. 2-phosphoglycerate→ 2 Phosphoenol Pyruvate (PEP) + H_2O
37. *A*-Dihydroxyacetone-3-phosphate
38. Substrate level phosphorylation
39. 2 steps
40. 4 molecules of ATP
41. Formed
42. 3PGAL loses H^+ to NAD to form $NADH_2$
43. Glycolysis will stop as its first and third step need ATP.
44. 4 pairs are isomers (PEP and pyruvate are not isomers).
45. (i) Adenosine Triphosphate
    (ii) Nicotinamide Adenine Dinucleotide
    (iii) 3-Phosphoglyceraldehyde
    (iv) Dihydroxyacetone Phosphate
    (v) 1, 3-Bisphosphoglycerate
    (vi) Phosphoenol Pyruvate
46. 2ATPs, substrate level phosphorylation, NADH + H^+
47. *A*–ATP utilised, *B*–ATP formed, *C*–ATP formed, *D*–ATP utilised
48. True
49. A-2, B-1, C-1, D-2
50. 10, 8
51. Alcoholic fermentation, lactic acid fermentation and aerobic respiration.
52. 4
53. Pyruvic acid
54. Substrate level phosphorylation
55. Muscle cells, nerve cells and RBCs
56. Fermentation
57. *A*–Ethanol, *B*–CO_2
58. (i) *A*– NADH + H^+
    (ii) Some bacteria
    (iii) Muscle cells during exercise
    (iv) Lactate dehydrogenase
59. Pyruvic acid decarboxylase and alcohol dehydrogenase.
60. True
61. (i) 7
    (ii) 13
    (iii) 686
    (iv) 1
62. Brewing, curing, vinegar
63. Seed is odd one out as it is respired in the absence of oxygen.
64. Aerobic respiration
65. Mitochondria
66. TCA (Tricarboxylic Acid) and ETS (Electron Transport System)
67. Catabolism, matrix, decarboxylation
68. *A*–Co-A, *B*–NAD^+, *C*–Pyruvate dehydrogenase, *D*–CO_2, *E*–NADH + H^+
69. (i) Tricarboxylic Acid Cycle
    (ii) Electron Transport System
    (iii) Oxaloacetic Acid
70. Hans Krebs
71. Krebs cycle or Citric acid cycle
72. Cytoplasm
73. $4CO_2$, 6NADH, $2FADH_2$, 2ATP
74. A-1, B-2, C-3, D-4
75. Citric acid

76. *A*–α–ketoglutaric acid, *B*–Succinic acid, *C*–Malic acid, *D*–OAA
77. Pyruvic acid → Acetyl Co-A + CO_2 + $NADH_2$
78. α-ketoglutaric acid
79. 8 (Citrate synthase, aconitase, isocitrate dehydrogenase, α-ketoglutarate dehydrogenase, succinyl synthetase, succinate dehydrogenase, fumarase, malate dehydrogenase)
80. *X*–Isocitrate
81. *X*–Succinate dehydrogenase, *Y*–GTP
82. Oxaloacetic acid
83. *A*–4NAD^+, *B*–FAD^+, *C*–2H_2O, *D*–3CO_2, *E*–4NADH
84. (i) Malic acid
    (ii) Fumaric acid
    (iii) Succinic acid
85. *X*–NADH + H^+, *Y*–$FADH_2$, *Z*–GTP
86. Metabolic sink
87. Conversion of succinyl Co-A to succinic acid
88. (i) Oxidation
    (ii) Isomerism
    (iii) Decarboxylation
89. (i) Tricarboxylic acid cycle
    (ii) Electron transport chain
90. True
91. Cristac
92. A-5, B-4, C-3, D-2, E-1
93. a, a_3 and 2
94. Cytochrome-*c*
95. Within the inner membrane of mitochondria.
96. Ubiquinone
97. Electron donor
98. ATP
99. *A*-Inter-membrane space, *B*-Matrix, *C*-NADH+ H^+, *D*-NAD^+, *E*-UQH_2, *F*-Fumarate
100. 3 molecules, 2 molecules
101. Oxygen derives the whole process by removing hydrogen from the system. It is the final hydrogen acceptor resulting in the formation of H_2O.
102. Light energy, energy of oxidation or reduction
103. Oxidative phosphorylation
104. Proton gradient
105. Electron transport chain
106. Hydrogen carrier
107. F_1, F_0
108. (i) F_1, (ii) F_0
109. (i) It forms the channel through which protons cross the inner membrane.
     (ii) It provides site for the synthesis of ATP from ADP and Pi.
110. *A*-2H^+, *B*-F_0, *C*-F_1.
111. False. The passage of protons through the channel formed by F_0 is coupled to the catalytic site of the F_1 component.
112. In photophosphorylation, oxygen is evolved, while in oxidative phosphorylation, oxygen is used.
113. (i) Glycolysis, TCA cycle and ETS
     (ii) Oxidative phosphorylation
     (iii) None
     (iv) Glucose
114. 38
115. A-2, B-1, C-3.
116. Respiratory poisons or inhibitors
117. Intermembrane space.
118. 1520 (38 ATP by 1 glucose $40 \times 38 = 1520$ ATP).
119. Glucose (carbohydrate)
120. Amphibolic.
121. *A*-Fats, *B*-Proteins, *C*-Glyceraldehyde-3-phosphate, *D*-Krebs cycle
122. (i) PGAL
     (ii) Amino acids
     (iii) Acetyl Co-A
     (iv) Glycerol and fatty acid
123. $RQ = \frac{\text{Volume of } CO_2 \text{ evolved } (X)}{\text{Volume of } O_2 \text{ consumed } (Y)}$
124. Organic acid
125. Respirometer
126. (i) 0.7
     (ii) More than one
     (iii) Less than one
     (iv) Less than one
     (v) 1
     (vi) 4
127. (i) In succulent plants
     (ii) During anaerobic respiration

CHAPTER FIFTEEN

PLANT GROWTH AND DEVELOPMENT

Topic 1 : Growth

1. During plant growth, ______________________ phase precedes flowering.
2. Development is the sum total of two processes, named ____________________
3. Refer to the given below diagrams and answer the following questions.
   (i) In the given diagrams, name the structures labelled as *A-D*.

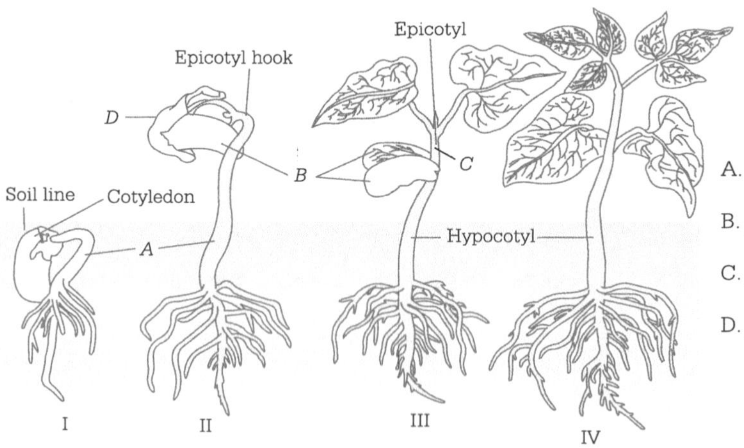

A. ____________________

B. ____________________

C. ____________________

D. ____________________

   (ii) The given diagrams represent germination and seedling development in <u>maize</u>.
   (*Rewrite the statement by changing the underlined word.*)

   ____

   (iii) Which stage occurs after the stage IV? (Senescence/Death) ____________________

4. The factors which govern and control developmental processes in plants are either intrinsic or extrinsic.

   True ☐ False ☐

5. Read the following characteristics and identify the biological term.
   I. Irreversible permanent increase in size of an organ or its parts or even of an individual cell.
   II. Most fundamental and conspicuous characteristic of a living being.

   ____

6. Growth is accompanied by ______________ processes that occur at the expense of ______________ .

7. An example of growth is ______________
   (leaf expansion /swelling of wood piece in water).

8. Name the tissue which enables plants to retain the capacity of unlimited growth throughout their life. ______________

9. How many of the listed characters are displayed by the cells of meristems?

Capacity to divide continuously, Self-perpetuation, Present throughout plant body

   ____

10. Name the type of growth wherein new cells are always being added to the plant body by the activity of meristem. ______________

11. If the meristem ceases to divide, ______________ occurs.

12. Refer to the given diagram and answer the following questions.
    (i) Name the parts labelled as *A*, *B* and *C*.

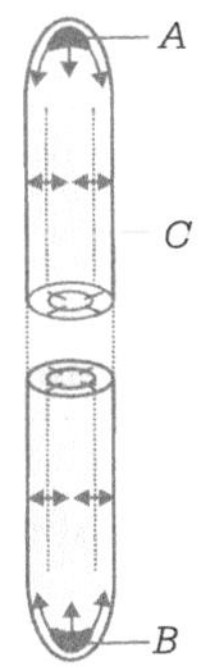

    A ______________
    B. ______________
    C. ______________

    (ii) Among the parts *A*, *B* and *C*, choose the odd one out with respect to the type of growth contributed by them. ______________
    (iii) Along with part *C*, name other two structures which appear later in life of dicotyledonous plants and gymnosperms. ______________

13. Complete the following analogy given below with respect to different meristems and their function.

    Apical meristem : Elongation of plant : : ______________ : Increase in girth of associated organs.

14. At cellular level, the growth is principally a consequence of increase in the amount of ______________________.

15. How many of the listed parameters are used to measure the growth in plants?

Increase in fresh weight, Increase in length, Dry weight, Increase in cell number, Increase in protoplasm directly

____

16. Match the following columns.

Column I (Various structures of plants)	Column II (Parameters of growth)
A. Maize root apical meristem	1. Increase in size of the cell upto 3,50,000 times
B. Cell in a watermelon	2. Increase in the length
C. Dorsiventral leaf	3. Increase in cell number from one to 17500 per hour
D. Pollen tube	4. Increase in surface area

A.__________ B.__________ C.__________ D.__________

17. Name the three phases of the period of growth.

____

18. Consider the following diagram of root tip and answer the questions that follows.

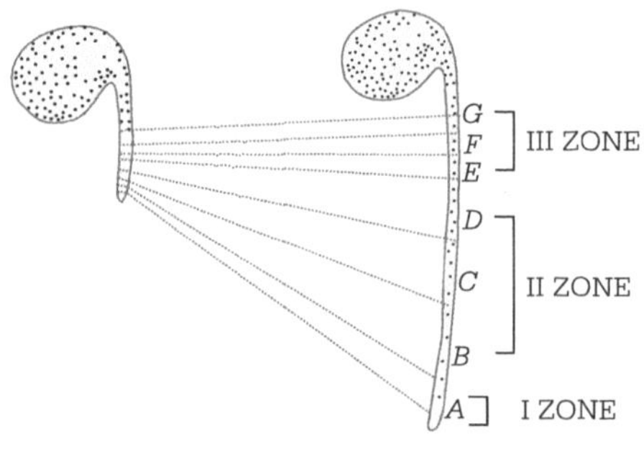

(i) The given diagram shows the parallel line technique for the detection of zones of ______________________.

(ii) Match the following columns considering the above figure.

Column I	Column II
A. I	1. Phase of elongation
B. II	2. Phase of maturation
C. III	3. Meristematic phase of growth

A.__________ B.__________ C.__________

(iii) Choose the odd one out with respect to the cell characteristics of zone I of the above figure.

Constantly dividing, Rich in protoplasm, Large nuclei, Thickened cell walls, Abundant Plamodesmata

____

(iv) How many of the listed characteristics are likely to be found in the cells of zone II?

Increased vacuolation, Cell enlargement, Cell attain maxmimum size, New cell wall deposition

____

(v) Mention any two characteristics of cells found in zone III.

(a) _______________

(b) _______________

19. Study the diagram given below and identify the *A* and *B* pattern of growth.

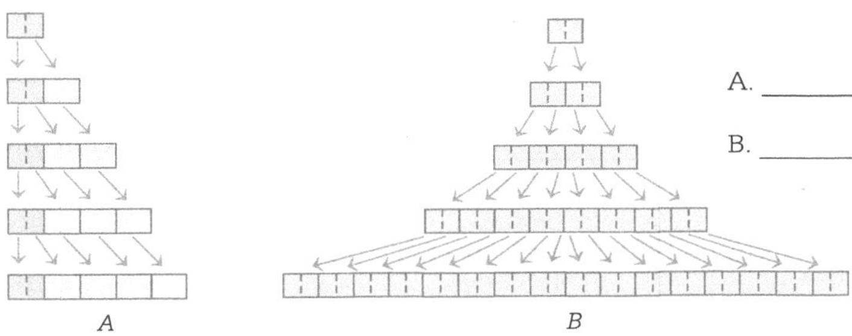

A. _______________

B. _______________

20. In <u>geometric growth</u>, following mitotic cell division in actively dividing cells only one daughter cell continues to divide while other differentiates and matures.
(*Rewrite the statement by changing the underlined word.*)

____

21. The given diagram shows the stages during embryo development. What does *A* and *B* represent in it?

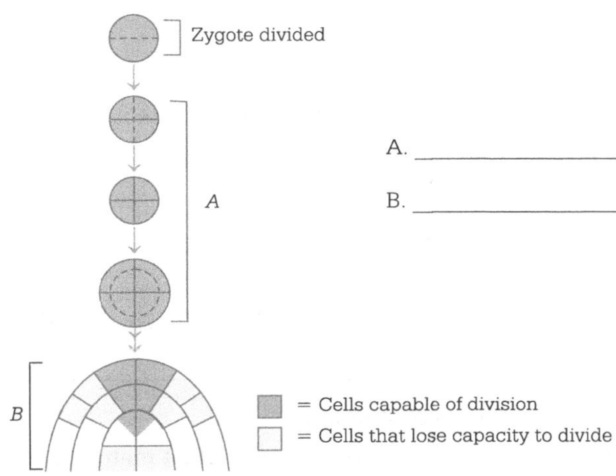

A. _______________

B. _______________

22. Refer to the graph and answer the questions that follows.

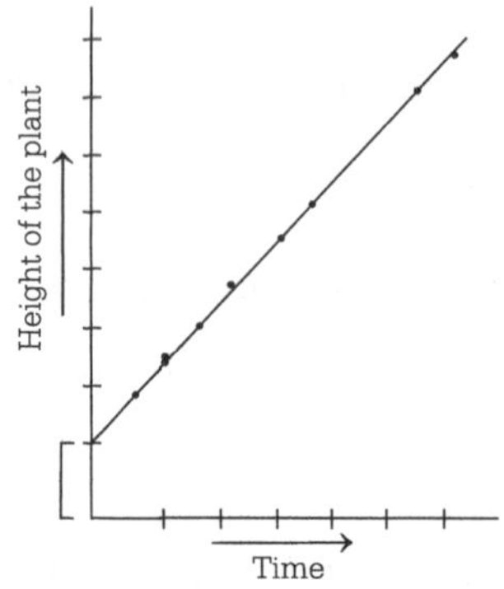

(i) Name the type of growth rate shown in the graph. ____________________

(ii) The given graph can be obtained in case of ____________________
(embryo differentiation/root elongation)

23. Give the mathematical expression for the graph in previous question.

____

24. In case of geometric growth rate, ____________________ (no cell/both the daughter cells) divide following the mitotic cell division.

25. What type of graph is likely to be obtained in case of geometric growth rate?

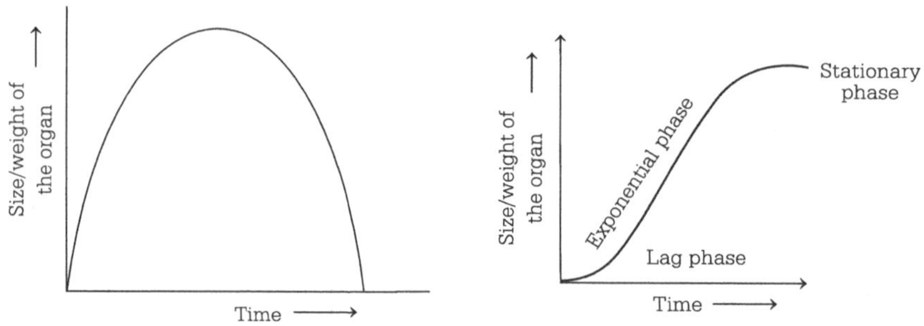

I. Parabola II. Sigmoid

____

26. In ____________________ growth rate, stationary phase is reached due to the limited nutrient supply. This growth rate is characteristics of living organisms growing in ________________ environment.

27. A tree showing seasonal activities and cells in culture exhibit ____________________ (sigmoid/linear) growth curve.

28. Give the mathematical expression for exponential/geometric growth rate.

____

29. Give technical term for the following sentence.
The measure of the ability of the plant to produce new plant material.

____

30. The absolute and relative growth rates give the quantitative comparison between the growth of living system.

True ☐ False ☐

31. Complete the following analogy given below with respect to growth patterns.

Absolute growth rate : Total growth per unit time. : Relative growth rate :

____

32. Refer to the given diagram and answer the questions that follows.

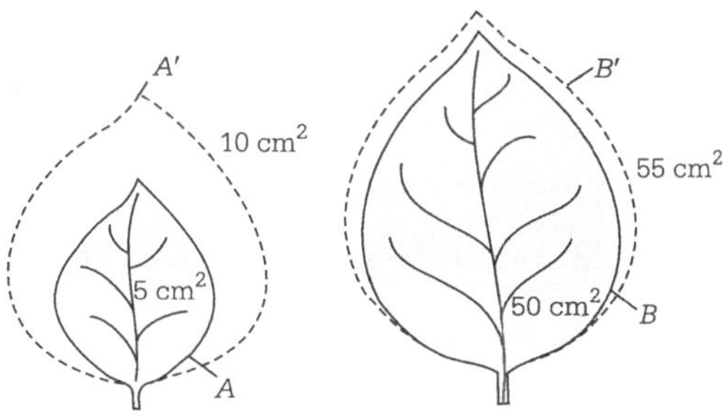

(i) The growth of two leaves A and B is A' and B' in the given time is an example of ______________(absolute growth/relative growth) in area.

(ii) Which leaf show much higher relative growth rate? (A or B)

____

33. Name the device that is used to measure growth rate.

____

34. Name any three essential elements for growth.

____

35. Match the following columns.

Column I (Elements)	Column II (Functions)
A. Water	1. Synthesise protoplasm and acts as a sources of energy
B. Oxygen	2. Cell enlargement and provides medium for enzymatic activities
C. Nutrients	3. Release metabolic energy through cellular respiration

A. __________ B. __________ C. __________

36. Any deviation from the optimum temperature range could be detrimental for plant survival

Yes ☐ No ☐

Topic 2 : Differentiation, Dedifferentiation and Redifferentiation

37. Give biological term for the following sentence.

    Process of maturation of cells derived from root apical and shoot-apical meristems and cambium differentiate and mature to perform specific functions.

    ____

38. Maturation of cells derived from apical meristems and cambium is associated with structural changes in cell walls and ______________ .

39. The tracheary elements carry water to long distances in plants. To perform this function, their cells would possess following anatomical features during differentiation except one.

Abundant protoplasm, Strong and elastic cell walls, Lignocellulosic secondary walls

    ____

40. Complete the following analogy given below related to differentiated cells.

    Differentiated cells regain the capacity of division : Dedifferentiation
    : : ______________________________ : Redifferentiation

41. Formation of interfascicular cambium and cork cambium from fully differentiated parenchyma cells is a example of <u>redifferentiation</u>.
    (*Rewrite the statement by changing the underlined word.*)

    ____

42. Complete the following analogy given below with respect to formation of vascular tissue.

    Differentiation : Formation of tracheary elements : : ______________ : Specialisation of vascular cambium into secondary xylem and phloem.

43. A tumour is an example of ______________ .

44. The parenchyma cells that are made to divide under controlled laboratory conditions during plant tissue culture represents ______________________ (dedifferentiation/redifferentiation).

45. Give biological term for the sentence given below.

    The ability of a single cell to divide and produce all of the differentiated cells.

    ____

46. The two types of open growth in plants are

    ____

47. The differentiation in plants is open because cells/tissues arising out of the __________ have __________ at maturity.

48. The final structure at maturity of a cells is determined by the location of cell within.

True ☐ False ☐

49. Complete the following analogy given below with respect to cell position after differentiation.

Cell away from root apical meristem : Root cap cells : : Cells at the periphery :

__________ .

Topic 3 : Development

50. Give biological term for the sentence given below.

All changes that an organism goes through during its life cycle.

____

51. In plants, development phase is the period between seed germination and __________ (flowering/senescence)

52. The given diagram shows sequence of developmental processes in a plant cell, identify *A-D*.

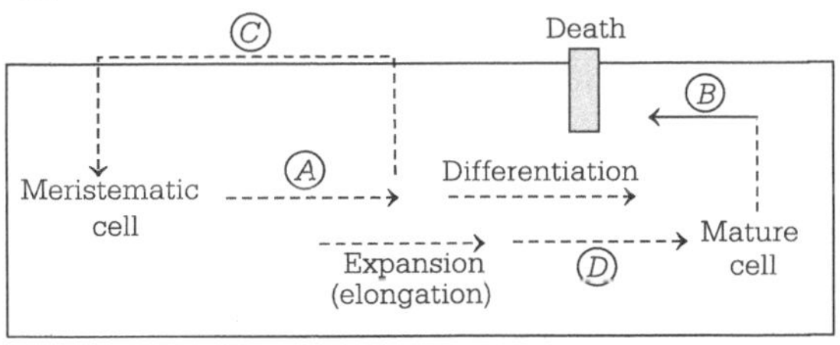

A. __________

B. __________

C. __________

D. __________

53. The ability of plants to follow different pathways in response to environment or phases of life to form different structures is called <u>heterozygosity</u>.

(*Rewrite the statement by changing the underlined word.*)

____

54. Choose the odd one out with respect to heterophylly.

Cotton, Buttercup, Coriander, Larkspur

____

55. Match the following columns.

	Column I	Column II
A.		1. Adult larkspur
B.		2. Buttercup in water habitat
C.		3. Juvenile larkspur
D.		4. Terrestrial buttercup

A. ________________ B. ________________ C. ________________ D. ________________

56. Complete the flowchart given below.

Development

- Intrinsic factors
  - *A* → e.g. genetic
  - *B* → e.g. plant growth regulators
- Extrinsic factors
  - *C* → e.g. light, temperature, oxygen, etc.

A. ________________ B. ________________ C. ________________

Topic 4 : Plant Growth Regulators

57. PGRs are small, simple molecules of diverse ______________. These are also called ______________ .

58. Match the following columns.

Column I	Column II
A. Indole compounds	1. Kinetin
B. Adenine derivatives	2. ABA
C. Carotenoids derivatives	3. C_2H_4
D. Terpenes	4. GA
E. Gases	5. IAA

A. ______________ B. ______________ C. ______________

D.______________ E.______________

59. Choose the odd one out with respect to phytohormones.

Kinetin , Purine, N^6-furfurylamino purine, Ethylene

____

60. The phytohormones which promote cell division, enlargement, tropic growth, flowering, fruiting and seed formation are called ______________ . The PGRs which promote dormancy and abscission in response to wounds and stresses are called

____

61. How many of the listed phytohormones are plant growth promoters?

Auxins, Gibberellins, Abscisic acid, Cytokinins

____

62. Discovery of phytohormones began with the observation of ______________ when they observed that the coleoptiles of ______________ respond to ______________ illumination.

63. Refer to the given diagram and answer the questions that follows :

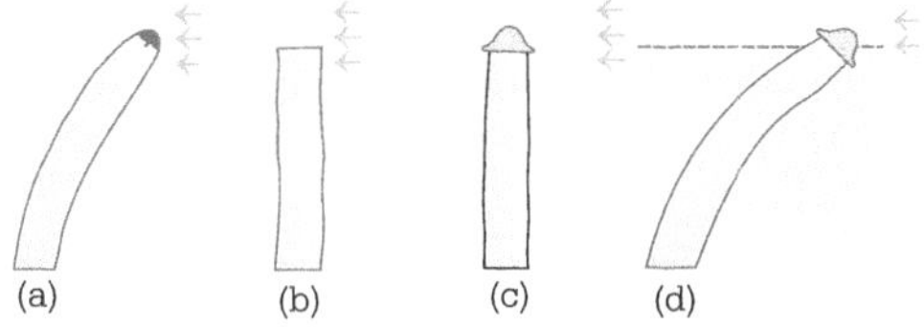

(i) What does the experiment demonstrate?

____

(ii) What does the arrow indicate in the diagram?

____

64. Match the following columns.

	Column I		Column II
A.	FW Went	1.	Reported symptoms of bakane disease in rice seedling.
B.	E. Kurosawa	2.	Confirmed release of volatile substances from ripened oranges.
C.	Skoog and Miller	3.	Crystallised cytokinesis promoting active substance, kinetin.
D.	Cousins	4.	Isolated auxin from tips of coleoptiles of oat seedlings.

A.________ B. ________ C. ________ D. ________

65. In F Skoog experiment, the tobacco stem of callus would proliferate, if nutrient medium is supplemented with yeast extract along with auxins.

Yes ☐ No ☐

66. Auxins was first isolated from the oat seedling.

True ☐ False ☐

67. Auxins are ________ and ________ compounds having certain growth regulating properties. These are generally produced by ________ of the stems and roots.

68. Expand the following.

(i) IBA ________

(ii) NAA ________

(iii) 2,4-D ________

(iv) IAA ________

(v) GA ________

(vi) ABA ________

69. Complete the following analogy given below.

Natural auxins : IAA and IBA : : Synthetic auxins : ________ .

70. Name the hormone which initiates rooting in stem cutting, flowering in pineapples and promote abscission of older leaves and fruits.

____

71. Name the phenomenon in which the growing apical bud inhibits the growth of lateral or axillary buds.

____

72. In tea plantation and hedge making, apical dominance is inhibited by the removal of ________.

73. What does the diagram given below depict?

(a) Plant with apical bud (b) Plant with apical bud removed

____

74. Auxin induces ____________ in tomatoes, 2, 4-D is widely used as ____________ as it kills ______________ and does not affect mature monocotyledonous plants.

75. Auxins control xylem differentiation and help in cell division.

True ☐ False ☐

76. <u>GA_1</u> is the first gibberellin to be discovered and intensively studied form.

(*Rewrite the statement by changing the underlined word.*)

____

77. Complete the following analogy given below with respect to function of plant hormone.

Grapes : GA increases length of axis and length of stalk : : Apples :

____

78. Name the fungal pathogen which caused foolish seedling disease in rice seedlings.

____

79. Fruits can be left on trees longer so as to extend the shelf life because GA delays ______________ (senescence/parthenocarpy).

80. What is the function of GAs in brewing industry?

____

81. Gibberellin helps to increase the yield of sugarcane by 20 tonnes per acre by increasing the ______________ .

82. Spraying juvenile conifers with GAs hastens the ______________ (seed production/maturity period).

83. Give biological term for the following sentence.

Process of internode elongation just prior to flowering.______________

84. Choose the odd one out with respect to functions of gibberellin.

Beet, Cabbage, Sugarcane, Lettuce

____

85. Name the hormone which acts as antagonistically to ABA.

____

86. Cytokinins have specific effect on ________________ and were discovered as chemical called ________________ from the autoclaved herring ________________ .

87. F Skoog observed that the internodal segments of tobacco stem proliferate in the presence of auxin and ________________ (coconut milk /ethylene).

88. Complete the following analogy given below with respect to chemical nature of cytokinin.

Synthetic cytokinin : Kinetin : : Natural cytokinin : ________________ .

89. Name any two sources of zeatin.

____

90. Choose the odd one out with respect to the site of cytokinin production.

Root apices, Mature leaves, Young fruits, Developing shoot buds

____

91. Cytokinins help to produce new leaves, ________________ in leaves, lateral shoot growth and shoot formation. It helps to overcome the ________________ and promote ____________ .

92. The gaseous PGR, ethylene is synthesised by tissues undergoing senescence and ripening fruits.

Yes ☐ No ☐

93. Name the hormone which promotes horizontal growth of seedling, fruit ripening and senescence and abscission of leaves and flowers.

____

94. Ethylene inhibits respiratory climactic.

True ☐ False ☐

95. Hormones __________ and __________ are found in high concentrations in a mature slightly overripped fruits.

96. Identify whether the ethylene promote (P) or inhibit (I) the listed processes.

(i) Seed and bud dormancy (P/I). ________________

(ii) Germination in peanut seeds (P/I). ________________

(iii) Internode/petiole elongation in deep water rice plant (P/I). ________________

(iv) Root growth and root hair formation (P/I). ________________

97. Apples are generally wrapped in wax paper. (*Give reason*)

____

98. Mention the function of ethylene in pineapples.

____

99. Name the most widely used compound as source of ethylene in agriculture.

____

100. Choose the odd one out with respect to the ethephon functioning.

Hasten ripening in tomato and apple, Inhibit abscisssion in flowers and fruits, Promote female flowers in cucumber

____

101. Read the following statements and identify the PGR.

(i) General plant growth and metabolism inhibitor.

(ii) Acts as an antagonist to GAs.

(iii) Known as stress hormone.

____

102. ABA inhibits seed germination and closure of stomata.

True ☐ False ☐

103. ABA helps the seed to withstand desiccation by inducing ______________ .

104. Give the common term for the following hormones.

Inhibitor-B, Abscission II, Dormin

____

105. Match the following columns.

	Column I (Phytohormones)		Column II (Plant parts where they are synthesised)
A.	IAA	1.	Tissue undergoing senescence
B.	Cytokinins	2.	Shoot apices
C.	Ethylene	3.	Root apices

A ____________ B. ____________ C. ____________

106. How many of the following listed events involve more than one PGRs?

(i) Dormancy in seed or buds

(ii) Abscission

(iii) Senescence

(iv) Apical dominance

____

Topic 5 : Photoperiodism

107. Give biological term for 'the effect of duration of light exposure on the growth and development of plants and flowering'.

____

108. Match the following columns.

	Column I (Photoperiodic plants)		Column II (Characteristics)
A.	Long day plants (LDPs)	1.	No correlation between ligth exposure duration and flowering induction
B.	Short day plants (SDPs)	2.	Exposure of light period exceeds critical duration
C.	Day neutral plants (DNPs)	3.	Exposure of light period is less than critical duration

A. ________ B. ________ C. ________

109. Refer to the given diagram and answer the following questions.

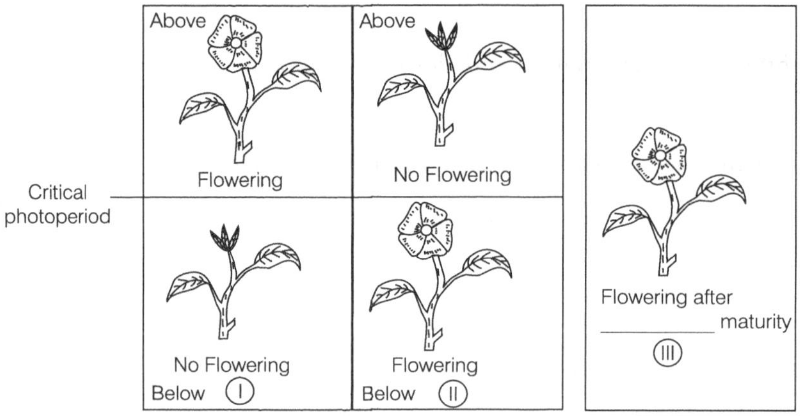

(i) Identify the types of plants shown in I, II and III. ________

(ii) When does the flowering in plant III initiated? ________

110. Choose the odd one out with respect to SDP or LDP.

Dahlia, Rice, Sugarcane, Tobacco, Potato, Wheat, Strawberry, *Xanthium*

____

111. How many of the listed plants are long day plants?

Spinach, Sugarbeet, Oat, Lettuce, Radish, *Chrysanthemum*, Sunflower, Cucumber

____

112. Flowering in certain plants is dependent on the duration of both light and dark exposures.

True ☐ False ☐

113. Name the site of light/dark duration perception in plants. ________

114. Complete the following analogy with respect to phytohormone.

Pfr : Absorb far red light : : Pr : ________.

115. __________ hormone migrates from __________ to __________________ for inducing flowering when plants are exposed to necessary inductive photoperiod.

Topic 6 : Vernalisation and Seed Dormancy

116. Name the phenomemon in which the flowering is either quantitatively or qualitatively dependent on exposure to low temperature.

____

117. Vernalisation prevents precocious __________________________ late in growing season and it shortens the ___________________ phase. It refers to promotion of flowering by a period of _____________________.

118. Give any three examples of plants having winter and spring varieties.

____

119. Winter varieties fail to flower, if they are planted in <u>autumn</u>.
(*Rewrite the statement by changing the underlined word.*)

____

120. Match the following columns.

Column I (Seasons)	**Column II** (Events in winter varieties of plants)
A. Autumns	1. Small seedlings are formed
B. Winter	2. Harvest
C. Spring	3. Planted
D. Summer	4. Seeds begin to grow

A. _______________ B. _______________ C. _______________ D. _______________

121. The monocarpic plants that flower and die in the second season are called perennials.
True ☐ False ☐

122. Name the category to which following plants belong.

Sugarbeet, Cabbage, Carrot

____

123. What would happen if the growing biennial plant is subjected to a cold treatment?

____

124. Name the phenomenon in which the certain seeds fail to germinate even when external conditions are favourable.

____

125. Seed dormancy is controlled by ______________________________ (intrinsic factors/extrinsic factors).

126. Dormancy of seed is associated with ______________________________ embryo, _______________________ and _____________________ .

127. Name the plant which lack any dormancy period.

____

128. Mention any two methods to overcome the seed coat barrier in dormant seeds.

(i) ______________________________

(ii) ______________________________

129. Effect of inhibitory substances can be removed by subjecting the dormant seeds to ______________ (high temperature/chilling conditions).

130. Gibberellic acid and nitrates help to suppress the effect of inhibitory substances in dormant seeds.

True ☐ False ☐

Answers

1. Vegetative
2. Growth and differentiation
3. (i) *A*–Hypocotyl
   B–Cotyledons
   C–Epicotyl
   D–Seed coat.
   (ii) Bean (dicot).
   (iii) Death
4. False. The factors which govern and control developmental processes in plants are both intrinsic and extrinsic.
5. Growth
6. Metabolic, energy
7. Leaf expansion
8. Meristematic tissue (meristems)
9. 2 (Capacity to divide continuously, self perpetuations). Meristems are not found throughout the plant body, but at certain location only.
10. Open form of growth
11. Differentiation
12. (i) *A*-Shoot apical meristem
    B-Root apical meristem
    C-Vascular cambium
    (ii) C is odd one as it contributes to secondary growth whereas *A* and *B* contribute to primary growth in plants.
    (iii) Lateral meristem and cork cambium.
13. Lateral meristem
14. Protoplasm.
15. 4 (All of them except increase in protoplasm directly). It is difficult to measure the increase in protoplasm directly. Therefore, some quantity is measured which is more or less proportional to it.
16. A-3, B-1, C-4, D-2
17. Meristematic phase, elongation phase and maturation phase.
18. (i) Elongation
    (ii) A-3, B-1, C-2
    (iii) Thickened cell wall is odd term because cells in zone I possess primary (thin) cell walls.
    (iv) 3 (Increased vaccuolation, cell enlargement, new cell wall deposition)
    (v) (a) Maximal size of cell and wall thickening,
    (b) Protoplasmic modifications
19. *A*–Arithmetic growth
    B–Geometric growth
20. Arithmetic growth
21. *A*–Geometric phase,
    B–Arithmetic phase
22. (i) Arithmetic growth
    (ii) Root elongation
23. $L_t = L_o + rt$, where, L_t = Length at time 't', L_0 = Length at time zero, r = growth rate/elongation per unit time.
24. Both the daughter cells
25. II, sigmoid-shaped
26. Geometric, natural
27. Sigmoid
28. $W_1 = W_0 e^{rt}$, where, W_1 = Final size (Weight, Height, Number, etc.),
    W_0 = Initial size, r = Growth rate,
    t = Time of growth,
    e = Base of natural logrithms.
29. Efficiency index
30. True
31. Growth per unit initial parameter at a given time.
32. (i) Absolute growth.
    (ii) *A*. Both leaves gained 5 cm area, but leaf *A* gained 100% of its original area while leaf *B* gained only 20% of its area.
33. Auxanometer or time lapse photography.
34. Water, oxygen, macro and micronutrient
35. A-2, B-3, C-1.
36. Yes
37. Differentiation
38. Protoplasm
39. Abundant protoplasm would not be found in differentiated tracheary cells whereas these cells would lose their protoplasm.
40. Dedifferentiated cells lose the capacity of division.
41. Dedifferentiation
42. Redifferentiation
43. Differentiation
44. Dedifferentiation
45. Totipotency
46. Determinate and indeterminate.
47. Same meristem, different structure
48. True

49. Epidermis
50. Development
51. Senescence
52. *A*-Plasmatic growth, *B*-Senescence, *C*-Cell division, *D*-Maturation
53. Plasticity
54. Buttercup is odd one out as it shows heterophylly due to environment whereas other three show heterophylly in different stages of life cycle.
55. A-3, B-2, C-4, D-1
56. *A*-Intracellular, *B*-Intercellular, *C*-Environment
57. Chemical composition, Phytohormones.
58. A-5, B-1 , C-2, D-4 , E-3
59. Ethylene is odd one out as it is a gaseous hormone while other three are adenine derivatives.
60. Plant growth promoters, Plant growth inhibitors
61. 3 (Auxins, gibberellins, cytokinins). Abscisic acid is plant growth inhibitor.
62. Charles Darwin and Francis Darwin, canary grass, unilateral
63. (i) The experiment demonstrates that tip of coleoptile is the source of auxin.
    (ii) The arrows indicate the direction of light.
64. A-4, B-1, C-3, D-2
65. Yes
66. False. Auxins was first isolated from human urine.
67. Natural, synthetic, growing apices
68. (i) Indole Butyric Acid
    (ii) Naphthalene Acetic Acid
    (iii) 2, 4-Dichlorophenoxyacetic acid
    (iv) Indole-3-Acetic Acid
    (v) Gibberellic Acid
    (vi) Abscisic acid
69. NAA and 2, 4-D
70. Auxins
71. Apical dominance
72. Shoot tips
73. Apical dominance in plants growth of lateral buds into branches after decapitation in figure (b)
74. Parthenocarpy, herbicide, dicot weeds
75. True
76. GA_3.
77. GA elongates and improves the fruit's shape.
78. *Gibberella fujikuroi*
79. Senescence
80. Speeding up the malting process.
81. Length of the stems.
82. Maturity period
83. Bolting
84. Sugarcane is odd one out as it is not rosette plant. GAs promote stem elongation in it whereas other three plants are rosette which show bolting by application of GAs.
85. Gibberellic acid
86. Cytokinesis, kinetin, Sperm DNA
87. Coconut milk
88. Zeatin
89. Coconut milk and corn kernels.
90. Mature leaves is odd one out as cytokinins are produced in the region of rapid cell division.
91. Chloroplast, adventitious apical dominance, nutrient mobilisation
92. Yes
93. Ethylene
94. False. Ethylene promotes respiratory climactic or rise in rate of respiration.
95. ABA, ethylene
96. (i) Inhibit (ii) Promote (iii) Promote (iv) Promote
97. It is because wax prevents entry of O_2 to apple, this checks the cellular respiration and overripening of fruit by preventing ethylene fomation.
98. Initiate flowering and synchronizing fruit set.
99. Ethephon
100. Inhibit abscission in flowers and fruits is odd one out as ethephon accelerates this process.
101. Abscisic acid
102. False. ABA inhibits seed germination but promotes the closure of stomata.
103. Dormancy
104. Abscisic acid (Inhibitor hormone)
105. A-2, B-3, C-1
106. 2(Abscission and senescence are promoted by both ethylene and ABA).
107. Photoperiodism
108. A-2, B-3, C-1.
109. (i) I-Long day plant, II-Short day plant, III-Day neutral plant
     (ii) After vegetative maturity.
110. Wheat is odd one out because it is long day plant while others are short day plants.
111. 5 (Spinach, sugarbeet, oat, lettuce, radish). *Chrysanthemum* is short day plant, while sunflower and cucumber are day neutral plants.
112. True
113. Leaves
114. Absorb red light
115. Florigen, leaves, shoot apices
116. Vernalisation
117. Reproductive development, vegetative, low temperature
118. Wheat, barley, rye
119. Spring
120. A-3, B-1, C-4, D-2
121. False. The monocarpic plants that flower and die in the second season are called biennials.
122. Biennial plants
123. Photoperiodic flowering response is stimulated.
124. Seed dormancy
125. Intrinsic factors
126. Immature, abscisic acid, phenolic acid.
127. *Rhizophora* (mangrove)
128. (i) Vigorous shaking,
     (ii) Mechanical abrasions using knives or sandpapers.
129. Chilling conditions
130. True

CHAPTER SIXTEEN

DIGESTION AND ABSORPTION

Topic 1 : Digestive System

1. Name any five major components of our food? ____________________

2. Food provides enzymes and organic materials for growth and repair of heart.

   True ☐ False ☐

3. The water plays an important role in anabolic processes and also prevents dehydration of the body.
   (*Rewrite the statement by changing the underlined word.*)

   ____

4. ____________________ (Biomicromolecules/Biomacromolecules) in food cannot be utilised by our body in their original form.

5. ____________________ are broken down and converted into simple substances in the digestive system.

6. Name the process of conversion of complex food substances to simple absorbable forms.

   ____

7. Digestion is carried out by our digestive system by ____________________ and ____________________ methods.

8. What does a human digestive system consists of?

   ____

9. Identify the labelled parts as *A-F* in the following figure of digestive system.

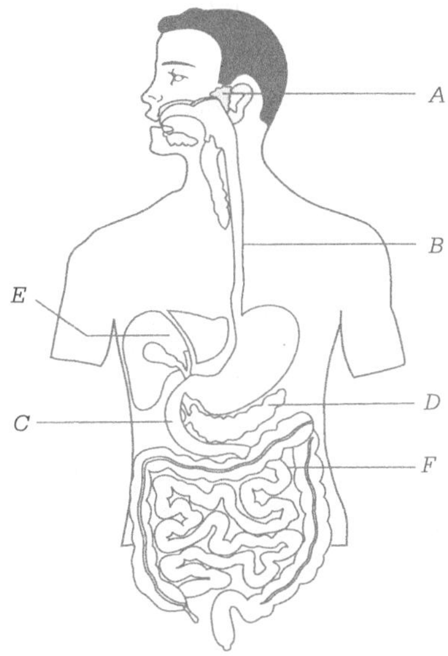

A. ______________________

B. ______________________

C. ______________________

D. ______________________

E. ______________________

F. ______________________

10. The anterior and posterior opening of alimentary canal in humans is ______________ and ______________, respectively.

11. In humans alimentary canal, the mouth leads to the trachea.
(*Rewrite the statement by changing the underlined word.*)

____

12. Correct the following sequence of organs of alimentary canal.

Mouth, Buccal cavity, Stomach, Oropharynx, Oesophagus, Large intestine, Small intestine, Anus

____

13. Match the following columns.

	Column I (Parts of alimentary canal)		Column II (Features)
A.	Foregut	1.	It contains pharynx, oesophagus, stomach and intestine upto colon region.
B.	Midgut	2.	It contains colon, rectum and anus.
C.	Hindgut	3.	It includes mouth and buccal cavity.

A. ______________ B. ______________ C. ______________

14. Name two structures found in oral cavity.

____

15. The soft palate or ______________ contains a curved fold of muscle hanging down called as ______________.

16. Each tooth is embedded in a socket of jaw bone, this type of attachment is called ______________(thecodont/bunodont).

17. Majority of mammals including human being form two sets of teeth during their life, a set of temporary milk or ______________ teeth that get replaced by a set of permanent teeth.

18. The type of dentition in which two sets of teeth appear in the lifetime of mammals is known as heterodont.
(*Rewrite the statement by changing the underlined word.*)

____

19. How many teeth are found in

(i) Human adult ______________

(ii) Human child ______________

20. Permanent teeth in humans are of four different types, i.e.

____

21. The presence of different teeth in humans is known as ______________ dentition.

22. The arrangement of teeth in each of the upper and lower jaw in the order M, Pm, C, I is represented by a dental formula.

True ☐ False ☐

23. Complete the following analogy.

Human(adult) : : $\frac{2123}{2123}$: Human(child) : ______________.

24. Consider the following image of the arrangement of different types of teeth, answer the following questions.

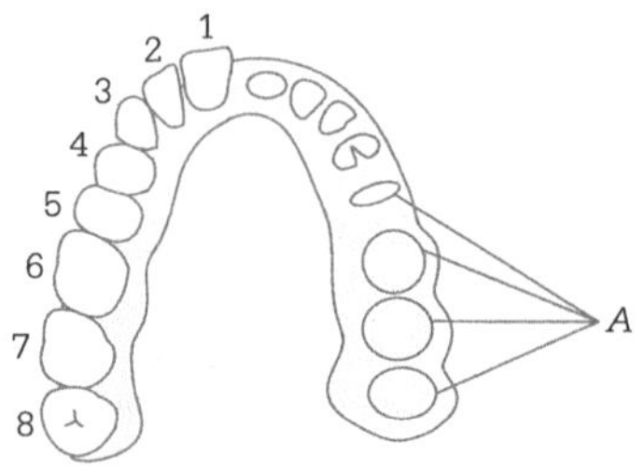

(i) Identify molars and premolars in the given diagram? __________

(ii) Complete the following analogy.

Number of incisors in each half : Two : : Number of canines in each half : __________

(iii) Which teeth possess three roots? __________

(iv) Identify the labelled part '*A*'. __________

25. Find the odd one out related to dentition.

Frog, Snake, Lizard, Man

____

26. Complete the following table.

Types of teeth		Features		Examples	
(i)	(A)	(a)	Tooth embedded in the jaw bone socket	1.	Crocodile
(ii)	Acrodont	(b)	Tooth is fixed in jaw bone without socket	2.	(B)
(iii)	(C)	(c)	Tooth is fused with inner side of jaw ridge	3.	Lizard

A. ________________ B. ________________ C. ________________

27. (i) The hard chewing surface of the teeth is made up of ________________ which helps in the ________________ of food.

(ii) Identify the labelled parts as *A–D* in the given section of a tooth.

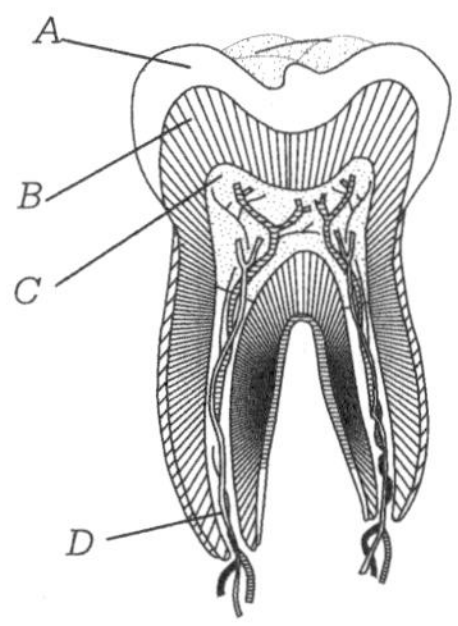

A. ________________

B. ________________

C. ________________

D. ________________

28. The pharynx is a freely movable muscular organ in human alimentary canal. (*Rewrite the statement by changing the underlined word.*)

____

29. Name the structure which attaches the tongue to the floor of the oral cavity.

____

30. The ____________ (palate/papillae) is the small projection on the upper surface of the tongue, some of which bear ________________ (taste buds/glands).

31. Which of the following labelled parts of taste zone (*A–D*) on human tongue provide bitter, sour, salt and sweet taste, respectively?

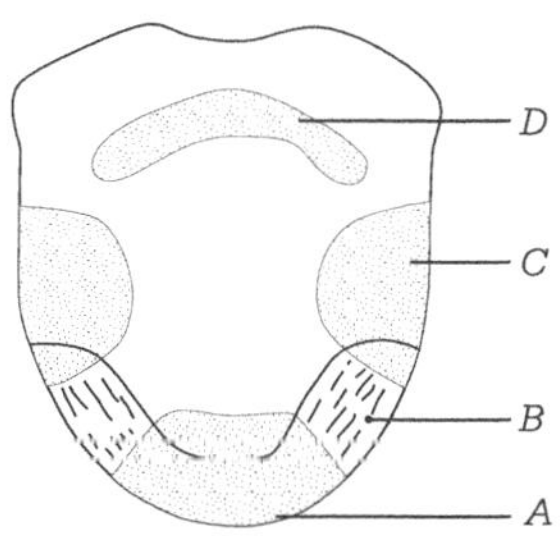

A. ________________

B. ________________

C. ________________

D. ________________

32. The tube of alimentary canal which serves as a common passage for food and air is called ________________.

33. How many of the following structures open into pharynx?

Stomach, Oesophagus, Pancreas, Trachea

____

34. A cartilaginous flap called tonsil, prevents the entry of food into the pharynx during swallowing.
(*Rewrite the statement by changing the underlined words.*)

____

35. The opening of the windpipe is known as ______________ .

36. The ______________ is a thin, long tube which extends posteriorly and passes through the neck, thorax and diaphragm.

37. What is the shape of stomach in humans? ______________

38. A muscular gastro-oesophageal sphincter regulates the opening of oesophagus into duodenum.

True ☐ False ☐

39. Where is the J-shaped structure of alimentary canal located in the human body?

____

40. Stomach has four major parts, i.e. ______________

41. Consider the following figure of human stomach and answer the following questions.

(i) Identify the labelled parts as *A*, *B*, *C* and *D*.

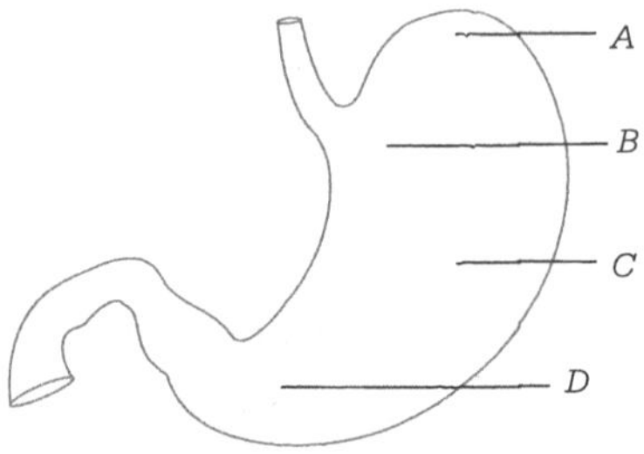

A. ______________

B. ______________

C. ______________

D. ______________

(ii) In which labelled part oesophagus opens? ______________

(iii) Which of the labelled part represents main central region? ______________

42. The ______________ (fundic/pyloric) portion of human stomach opens into the first part of ______________ (small/large) intestine.

43. Match the following columns.

	Column I (Regions of small intestine)		Column II (Features)
A.	Duodenum	1.	C-shaped
B.	Ileum	2.	Long and coiled
C.	Jejunum	3.	Highly coiled

A. ______________ B. ______________ C. ______________

44. Rearrange the following portions of small intestine from anterior to posterior.

Ileum, Duodenum, Jejunum

____

45. The opening of the stomach into the duodenum is guarded by the _______________ sphincter.

46. Name the region of small intestine which opens into large intestine.

____

47. Name the part of alimentary canal which possesses caecum, colon and rectum.

____

48. _______________ (Colon/Caecum) is a small blind sac which hosts some symbiotic microorganisms.

49. A vestigial structure called villi arises from caecum. It is a narrow finger-like tubular projection.

True ☐ False ☐

50. The caecum opens into the _______________ .

51. Mark the following parts of large intestine in the given figure of colon.

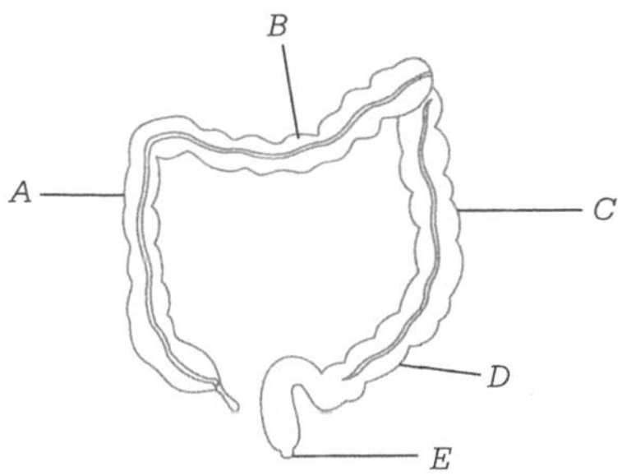

(i) Ascending colon _______________

(ii) Transverse colon _______________

(iii) Descending colon _______________

(iv) Sigmoid colon _______________

52. The _______________ part of colon opens into the rectum which further opens out through the _______________.

53. The opening of anus is controlled by two sphincter, i.e. internal sphincter and external sphincter.

True ☐ False ☐

54. Name four different layers of human alimentary canal.

____

55. ________________ is the outermost layer of alimentary canal and is made up of a thin ________________ with some connective tissues.

56. Name the layer of alimentary canal which is made up of smooth muscles and arranged into an inner circular and an outer longitudinal. ________________

57. Other than circular and longitudinal layers, some regions contain ________________ muscle layer.

58. Which of the following is odd one out with respect to serosa layer of alimentary canal?

Stomach, Oesophagus, Jejunum, Ileum

____

59. The ________________ (longitudinal/circular) layer of muscularis is more thick.

60. Complete the following analogy with respect to the oblique muscle fibre.

Stomach : Present : : ________________ : Absent

61. Consider the following diagrammatic representation of transverse section of gut and answer the following questions.

(i) Identify the labelled parts *A*, *B*, *C* and *D*.

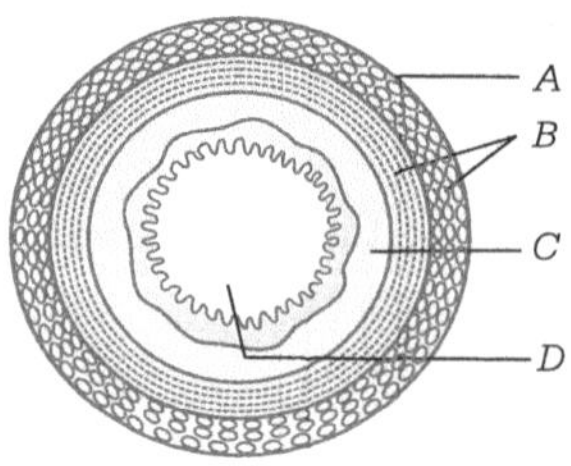

A. ________________

B. ________________

C. ________________

D. ________________

(ii) Name the layer which is formed of loose connective tissue and contains nerves, blood and lymph vessels. ________________

62. The sub-mucosa layer of duodenum contains mucus secreting gland, i.e.________________

63. ____________ (Mucosa/Sub-mucosa) layer forms irregular folds called rugae in the ________________ (stomach/intestine) and small finger-like folding called villi in the ________________ (large intestine/small intestine).

64. Arrange the following walls of alimentary canal from inner side to outer side.

Sub-mucosa, Serosa, Circular muscularis, Mucosa, Longitudinal muscularis

____

65. The cells lining the villi produce numerous microscopic projections called ________________, giving a brush border appearance and increase surface area enormously.

66. Villi are supplied with a network of capillaries and a large lymph vessel called the ____________________(crypts/lacteal).

67. Identify labelled parts *A*, *B* and *C* from the following diagram of a section of small intestine.

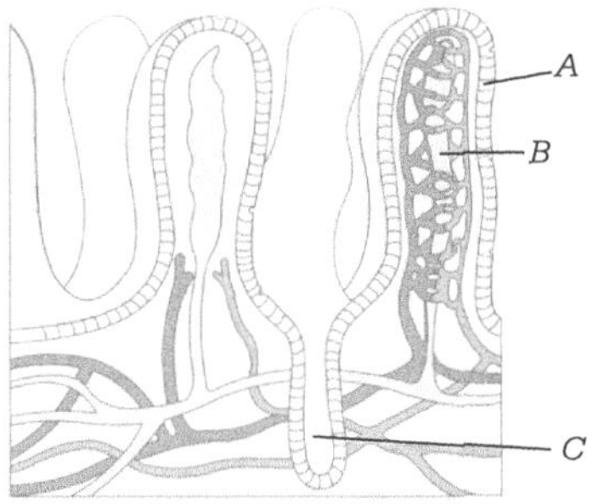

A. ____________________

B. ____________________

C. ____________________

68. Which structure is present inside the part labelled as *C* in the above given diagram of small intestine?

____

69. (i) ______________ forms glands in stomach and crypts in between the bases of ____________________ in intestine.

(ii) Intestinal mucosal epithelium has ______________________________ which secretes mucus and helps in lubrication.

70. Name three glands associated with the alimentary canal of humans.

____

71. How many of the following glands secrete saliva?

Oxyntic gland, Parotid gland, Paneth cells, Sub-maxillary gland, Brunner's gland, Sub-lingual gland

____

72. Mention the location of parotid, sub-maxillary and sub-lingual glands in buccal cavity.

____

73. Arrange the following glands according to their size from smallest to largest.

Sub-mandibular gland, Parotid gland, Sublingual gland

____

74. Oxyntic glands secrete serous fluid-like substances along with the enzyme salivary amylase or ptyalin.
(*Rewrite the statement by changing the underlined word.*)

____

75. Salivary glands situated just inside the buccal cavity secrete salivary juice into the pharynx.

True ☐ False ☐

76. Liver is the ____________ gland of the body weighing about _________ to _______ kg in an adult human.

77. Name the gland which is situated in abdominal cavity, below the diaphragm and has two lobes. ________________________________

78. The structural and functional unit of liver is Glisson's capsule, containing hepatic cells which are arranged in the form of cords.
(*Rewrite the statement by changing the underlined word.*)

____

79. Hepatic lobule is covered by a thin connective tissue sheath called the ________________________________

80. Complete the following flowchart.

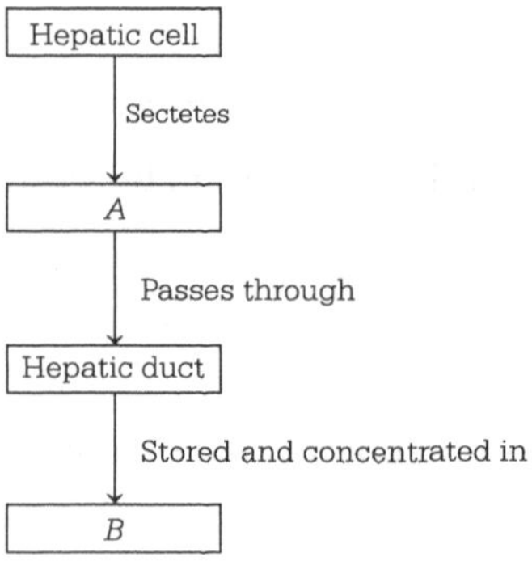

A. ________________________________

B. ________________________________

81. Refer to the diagram and identify the structure '*A*' in it.

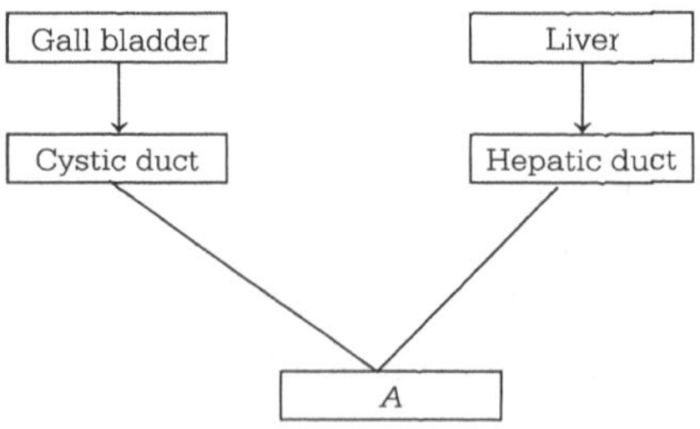

A. ________________________________

82. The ______________________ and the ______________________ open together into the duodenum as the common hepato-pancreatic duct.

83. Answer the following questions considering the given figure of gall bladder and pancreas.

(i) Identify the labelled parts as *A*, *B*, *C*, *D*, *E* and *F*.

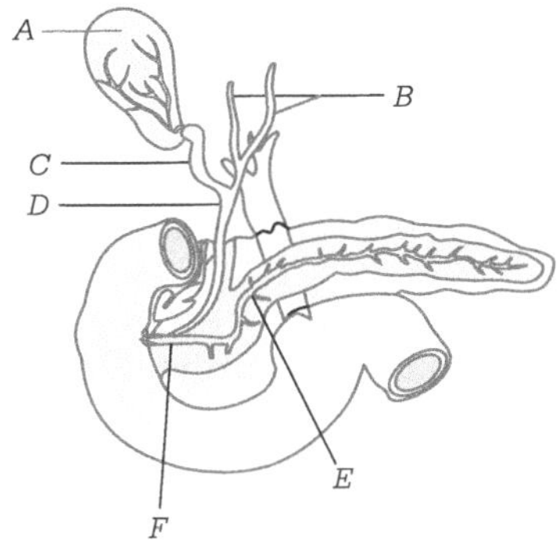

A. ____________________

B. ____________________

C. ____________________

D. ____________________

E. ____________________

F. ____________________

(ii) Where is sphincter of oddi is present in this figure?

____

84. Sphincter of Boyden surrounds the opening of bile duct before it is joined with ____________ (common hepatic duct/pancreatic duct).

85. Find the odd one out from following sequence related to gall bladder.

Birds, Human, Rat, Whale, Horse

____

86. The pancreas is an exocrine gland situated between the limbs of the C-shaped duodenum.

True ☐ False ☐

87. The ____________ portion of pancreas secretes an alkaline pancreatic juice containing enzymes and the ____________ portion secretes insulin and glucagon.

88. The ____________ (exocrine/endocrine) part of pancreas contains numerous acini or acinar cells.

Topic 2 : Digestion of Food

89. The ____________ performs two major functions, mastication of food and facilitation of swallowing.

90. The teeth and the tongue with the help of saliva masticate and mix up the food thoroughly.

True ☐ False ☐

91. Enzyme in saliva helps in lubricating and adhering the masticated food particles into a bolus.

(*Rewrite the statement by changing the underlined word.*)

____

92. The bolus is conveyed into the __________________ and then into the ______________ by swallowing or deglutition.

93. Name the muscular contraction of oesophagus which facilitates the bolus movement.

____

94. The buccal-oesophageal sphincter controls the passage of food into the stomach.

True ☐ False ☐

95. The saliva secreted into the oral cavity contains electrolytes (Na^+, K^+, Cl^-, HCO_3^-) and enzymes, gastric amylase and tributyrate.

(*Rewrite the statement by changing the underlined words.*)

____

96. Name the carbohydrate splitting enzyme present in saliva.

____

97. Food is digested in the oral cavity by the acylation action of the carbohydrate splitting enzyme.

(*Rewrite the statement by changing the underlined word.*)

____

98. Consider the following reaction and answer the following questions.

$$\text{Starch} \xrightarrow[\boxed{B}]{\boxed{A}} \text{Maltose}$$

(i) Identify enzyme, *A* and pH, *B*.

A. ____________________ B. ____________________

(ii) How much percentage of starch is hydrolysed by this reaction in buccal cavity? ________

(iii) What happed to the *A* enzyme if the pH of buccal cavity reduced to 2?

____

99. If you ingest some bacteria with food, which salivary enzyme acts on bacteria?

____

100. Complete the following flowchart.

Gastric Glands

Cardiac glands	*(A)*	**Pyloric glands**
• These are located in a narrow band around the *(B.)*	• These secrete HCl.	• These contain specialised cells called *(C,)* which secrete gastrin.

A. ________________ B. ________________ C. ________________

101. Match the following columns.

	Column I		Column II
A.	Mucus cells neck	1.	These secrete the proenzyme pepsinogen
B.	Parietal cells	2.	These cells secrete somatostatin, serotonin, histamine and gastric hormone
C.	Peptic cells	3.	These secrete alkaline mucus
D.	Argentaffin cell	4.	These secrete HCl and castle's intrinsic factor

A. ________________ B. ________________

C. ________________ D. ________________

102. Intrinsic factor is essential for absorption of vitamin ____________ (B_{12} /B_6).

103. The food mixes with the ____________ (acidic/basic) gastric juice of the stomach by the churning movements of its muscular wall and is called ____________ (bolus/chyme).

104. Write the products of the following reactions.

(i) Pepsinogen $\xrightarrow[\text{pH 1.8}]{\text{HCl}}$ ________________

(ii) Pepsinogen $\xrightarrow[\text{pH 7.8}]{\text{NaOH}}$ ________________

105. Pepsin is the ________________ enzyme of the stomach which converts proteins into ____________ and ____________.

106. The mucus and <u>carbonates</u> present in the gastric juice lubricate and protect the mucosal epithelium from excoriation by HCl.

(*Rewrite the statement by changing the underlined word.*)

____

107. Answer the following questions, considering the given equation.

$$\text{Casein} \xrightarrow[\text{Rennin}]{} \text{Paracasein} \xrightarrow{Ca^{+2}} A$$

(i) Rennin is a proteolytic enzyme.

True ☐ False ☐

(ii) Rennin is found in gastric juice of all mammals.

True ☐ False ☐

(iii) Identify compound *A*. ________________

(iv) Compound *A* helps in ____________ of milk.

108. The gastric juice contains small amount of ____________ and ____________.

109. How many of the following juices are secreted into the small intestine?

Bile juice, Gastric juice, Pancreatic juice, Intestinal juice

____

110. Pancreatic juice and bile are released into duodenum from _______________ (common hepatic duct/hepato-pancreatic duct).

111. How many of the following inactive enzymes are not found in pancreatic juice?

Trypsinogen, Chymotrypsinogen, Pepsinogen, Procarboxypeptidases, Amylases, Lipases, Nucleases, Prorennin

____

112. Identify the enzyme '*A*' in the following reaction.

$$\text{Trypsinogen} \xrightarrow{A} \text{Trypsin}$$

A. _______________

113. The bile released into the duodenum contains bile pigments, e.g. _______________ and _______________, bile salts, cholesterol and phospholipids.

114. Name the process in which fats break down into very small micelles by bile.

____

115. The pancreatic and intestinal juices in small intestine require _______________ (acidic/alkaline) medium to function property.

116. Complete the following analogy given below.

Gastric juice : Acidic : : _______________ : Alkaline

117. (i) Bile juice does not contain digestive enzyme.

True ☐ False ☐

(ii) Bile helps in fat emulsification and neutralising the acidity of gastric juice.

True ☐ False ☐

118. If gall bladder is removed, the bile flows into _______________

119. The secretion of the brush border cells of the mucosa and the secretion of Goblet cell constitutes _______________ (intestinal juice/pancreatic juice).

120. Name any four enzymes of succuss entericus.

____

121. The mucus and bicarbonates protect the intestinal wall from _______________ (base/acid) and provide _______________ (alkaline/acidic) medium for enzymatic activities.

122. Complete the following table.

	Nutrients	Enzymes for Digestion	Source of Enzyme
(a)	Carbohydrates	• Amylase	(A)
		• Disaccharidases	(B)
(b)	Proteins	• (C)	Chief cells
		• Trypsin, (D), carboxypeptidases	Pancreas
		• (E)	Small intestine
(c)	Fats	• Lipase	(F)
		• (G)	Liver

A. ________________ B. ________________ C. ________________

D. ________________ E. ________________ F. ________________

G. ________________

123. Answer the following questions considering the given reaction.

$$\left.\begin{array}{c}\text{Proteins}\\ \text{(A)}\\ \hline \text{Proteoses}\end{array}\right\} \xrightarrow[\text{(C)}]{\text{(B)/ Chymotrypsin}} \text{Dipeptides}$$

(i) Identify *A*, *B* and *C*.

A. ________________ B. ________________ C. ________________

(ii) This process takes place in ________________

124. Answer the following considering the given reaction.

$$\text{Polysaccharides (Starch)} \xrightarrow{A} \text{Disaccharides}$$

(i) Name the enzyme '*A*'.

A. ________________________________

(ii) What will happen to the above given reaction when medium is acidic?

____

125. During the conversion of polysaccharides to disaccharides ________________________ (hydrolysis/carboxylation) takes place.

126. Answer the following questions, considering the given reaction.

$$\text{Fats} \xrightarrow[A]{\text{Lipases}} \text{Diglycerides} \longrightarrow B$$

(i) Identify *A* and *B*.

A. ____________________ B. ____________________

(ii) What is the optimum pH for '*A*'? ________________________

127. Nucleases in the pancreatic juice act on ______________________ to form ________________ and ______________________.

128. Nucleases act upon nucleic acids in the______________ (acidic/alkaline) medium.

129. Identify the blanks *A*-*G* in the following reactions.

(i) Dipeptides $\xrightarrow{A}$ Amino acids

A. ______________________________

(ii) *B* $\xrightarrow{\text{Maltase}}$ Glucose + Glucose

B. ______________________________

(iii) Lactose $\xrightarrow{\text{Lactase}}$ Glucose + *C*

C. ______________________________

(iv) Sucrose $\xrightarrow{\text{Sucrase}}$ Glucose + *D*

D. ______________________________

(v) Nucleotides $\xrightarrow{\text{Nucleotidases}}$ Nucleosides $\xrightarrow{\text{Nucleosidases}}$ *E* + *F*

E. ______________________ F. ______________________

(vi) Di and monoglycerides $\xrightarrow{\text{Lipases}}$ Fatty acid + *G*

G. ______________________________

130. All of the given reactions in the above question occur in the ________________ region of the small intestine.

131. The products of duodenum region are absorbed in the large intestine.

True ☐ False ☐

132. The undigested and unabsorbed substances of food are passed on to the <u>small intestine</u>. (*Rewrite the statement by changing the underlined word.*)

____

133. Absorption of some water, ____________________ and ____________________ takes place in large intestine.

134. The large intestine secretes mucus which helps in adhering the undigested particles together and lubricate it for easy passage.

True ☐ False ☐

135. Faeces are the undigested, unabsorbed substances that enter into ____________________ (caecum/sigmoid colon) through ileo-caecal valve.

136. ______________________ valve prevents the back flow of faecal matter that is temporarily stored in the ______________ till defaecation.

137. Consider the following reactions and answer the questions that follows.

(i) Identify the pigment 'A' in the following reaction.

Bilirubin $\longrightarrow$ Urobilinogen $\longrightarrow A$

____

(ii) Where does pigment 'A' found? ____________________

138. The activities of the gastro-intestinal tract are under ____________________ and ____________________ control for proper coordination of different parts.

139. The sight, smell and the presence of food in the oral cavity do not stimulate the secretion of saliva.

True ☐ False ☐

140. What is effect of secretin on the following alimentary secretions?

(i) Bile secretion ____________________

(ii) Gastric secretion ____________________

Topic 3 : Absorption of Digested Products

141. ____________________ (Absorption/Egestion) is the process by which the end products of digestion pass through the intestinal mucosa into the blood or lymph.

142. Small amount of monosaccharides like ____________________, amino acids and some electrolyte like ____________________ are generally absorbed by simple diffusion.

143. Some fructose and amino acids are absorbed with the help of carrier proteins during osmosis.

True ☐ False ☐

144. ____________________ and ____________________ being insoluble, cannot be absorbed into the blood.

145. Consider the following flowchart of absorption of fat and answer the following questions.

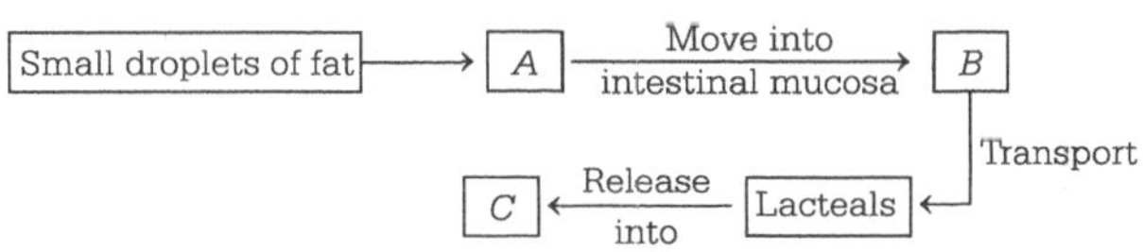

(i) Identify A, B and C.

A. ____________________ B. ____________________ C. ____________________

(ii) 'B' is small, protein coated ____________________

146. Complete the following analogy given below.

Glucose : Fructose : : ____________________ : Facilitated diffusion

147. Most of the amino acids are absorbed by ____________________, while some are absorbed by facilitated diffusion in ____________________ and jejunum.

148. By which process all of these components are absorbed into the intestinal cells?

Micelles, Fatty acids, Glycerides, Sterols, Fat soluble vitamins

____

149. Which of the following is/are absorbed in mouth?

Simple sugar, Alcohol, Drug

____

150. How many of the following are absorbed in stomach?

Water, Simple sugar, Fructose, Fatty Acids, Alcohol

____

151. The ______________ (small intestine/large intestine) is the principal organ for absorption of nutrients.

152. Box '*A*' and Box '*B*' have list of final products of digestion. Which box contains the digested food that is absorbed by small intestine?

Box '*A*'	Box '*B*'
Glucose, Fructose, Fatty acid, Glycerol, Amino acid, Nucleic acid, Vitamin-B_{12}	Water, Minerals, Drugs

____

153. Name the process by which absorbed substances reach the tissue so as to get utilised for various activities.

____

154. The energy requirements of animals and the energy content of food are expressed in the terms of ______________ .

155. Calorie and Joule are the unit of __________ (power/energy).

156. One kilo calorie is the amount of energy required to raise the temperature of _______ of water by ____________ °C.

157. The amount of heat liberated from complete hydrolysis of 1g food in a bomb calorimeter, in its gross calorific value.
(*Rewrite the statement by changing the underlined word.*)

____

158. Match the following columns.

	Column I (Macronutrients)		Column II (Gross calorific value (kcal/g)
A.	Carbohydrate	1.	9.45
B	Protein	2.	5.65
C.	Fats	3.	4.1

A. ____________ B. ____________ C. ____________

159. The digestive wastes that is solidified into coherent faeces in the rectum initiate a ______________ thus causing an urge for its removal.

160. The egestion of faeces to the outside through anal opening is a ______________ process and is carried out by a ______________ movement.

161. Complete the following analogy given below.

Starch : ____________________ : : Plants : Animals

Topic 4 : Disorders of Digestive System

162. Given below are the symptoms of vitamin deficiency diseases. Identify the vitamin.

Pellagra, Diarrhoea, Indigestion, Skin lesions, Vomiting

____

163. The inflammation of the intestinal tract is the most common ailment due to __________ or ______________ infections.

164. How many of the following parasites are responsible for the inflammation of the intestinal tract?

Tapeworm, Roundworm, Threadworm, Hookworm, Pinworm, Lice

____

165. Match the following columns.

	Column I (Disorders of digestive system)		**Column II** (Symptoms)
A.	Jaundice	1.	Food is not properly digested
B.	Vomiting	2.	Deposition of bile pigments
C.	Diarrhoea	3.	Bowel movements occur irregularly
D.	Constipation	4.	Ejection of stomach content through mouth
E.	Indigestion	5.	Increased liquidity of the faecal discharge

A.________________ B.________________ C.________________

D.________________ E.________________

166. The process of vomiting is controlled by the vomit centre in______________________ (cerebellum/medulla).

167. In ____________________, liver is affected and skin and eyes turn yellow.

168. Indigestion occurs due to inadequate enzyme secretion, anxiety, _________________, _______________________ and _______________________.

169. How many of the following diseases are related to Protein Energy Malnutrition (PEM)?

Marasmus, Indigestion, Constipation, Kwashiorkor

____

170. Marasmus is produced by a simultaneous deficiency of __________________________ (carbohydrates/proteins) and calories.

171. Due to protein deficiency, growth rate and body weight decline considerably, the growth and development of brain and mental faculties are impaired.

True ☐ False ☐

Answers

1. Carbohydrates, proteins, fats, vitamins and minerals.
2. False. Food provides energy and organic materials for growth and repair of tissues.
3. Metabolic
4. Biomacromolecules
5. Biomacromolecules
6. Digestion
7. Mechanical, biochemical
8. Alimentary canal and associated glands
9. *A*-Parotid gland, *B*-Oesophagus, *C*-Duodenum, *D*-Pancreas, *E*-Liver, *F*-Jejunum
10. Mouth, anus
11. Buccal cavity or oral cavity
12. Mouth → Buccal cavity → Oropharynx → Oesophagus → Stomach → Small intestine → Large intestine → Anus.
13. A-3, B-1, C-2
14. Teeth and muscular tongue
15. Velum palate, uvula
16. Thecodont
17. Deciduous
18. Diphyodont
19. (i) 32 (ii) 20
20. Incisors, Canines, Premolars, Molars
21. Heterodont
22. False. The arrangement of teeth in each half of the upper and lower jaw in the order I, C, Pm, M is represented by a dental formula.
23. $\frac{2102}{2102}$
24. (i) Molars are represented by 6, 7 and 8, while premolars are represented by 4 and 5.
    (ii) One (Incisors are 2 in number and canine is 1 in number in each half of each jaw).
    (iii) Upper molars
    (iv) *A*-Socket of jaw
25. Man is odd one out. All of them possess homodont dentition except man who possess heterodont dentition.
26. *A*-Thecodont, *B*-Amphibians, *C*-Pleurodont
27. (i) Enamel, mastication
    (ii) *A*-Enamel, *B*-Dentine, *C*-Pulp cavity, *D*-Root canal
28. Tongue
29. Frenulum
30. Papillae, taste buds
31. *A*-Sweet, *B*-Salty, *C*-Sour *D*-Bitter
32. Pharynx
33. 2(Oesophagus, trachea)
34. Epiglottis, glottis
35. Glottis
36. Oesophagus
37. 'J'-shaped
38. False. A muscular gastro-oesophageal sphincter regulates the opening of oesophagus into stomach.
39. The J-shaped structure (stomach) of alimentary canal is located in the upper left portion of the abdominal cavity.
40. (i) Cardiac, (ii) Fundic, (iii) Body, (main central region) (iv) Pyloric
41. (i) *A*-Fundus, *B*-Cardiac, *C*-Body *D*-Pyloric
    (ii) *B*-Cardiac
    (iii) *C*-Body
42. Pyloric, small
43. A-1, B-3, C-2
44. Duodenum → Jejunum → Ileum
45. Pyloric
46. Ileum
47. Large intestine
48. Caecum
49. False. A vestigial structure arises from caecum, which is a narrow finger-like tubular projection is known as vermiform appendix.
50. Colon
51. (i)-*A*, (ii)-*B*, (iii)-*C*, (iv)-*D*
52. Descending, anus
53. True
54. Serosa, muscularis, sub-mucosa and mucosa
55. Serosa, mesothelium
56. Muscularis
57. Oblique
58. Oesophagus is odd one out. All of them contain outer layer of serosa except oesophagus. It is not lined by serosa.
59. Longitudinal
60. Intestine
61. (i) *A*-Serosa, *B*-Muscularis, *C*-Sub-mucosa, *D*-Lumen
    (ii) *C*-Sub-mucosa
62. Brunner's glands
63. Mucosa, stomach, small intestine
64. Mucosa → Sub-mucosa → Circular muscularis → Longitudinal muscularis → Serosa
65. Microvilli
66. Lacteal
67. *A*-Villi, *B*-Lacteal, *C*-Crypts
68. Crypts of Lieberkuhn
69. (i) Mucosa, villi
    (ii) Goblet cell
70. Salivary glands, liver, pancreas
71. 3 (Parotid gland, sub-maxillary and sub-lingual gland)
72. Cheek, lower jaw, beneath tongue
73. Sub-lingual gland $<$ Sub mandibular gland $<$ Parotid gland
74. Parotid glands
75. False. Salivary glands situated just outside the buccal cavity secrete salivary juice into the buccal cavity.
76. Largest, 1.2 to 1.5
77. Liver
78. Hepatic lobule
79. Glisson's capsule
80. *A*-Bile juice, *B*-Gall bladder
81. *A*-Common bile duct
82. Common bile duct, pancreatic duct
83. (i) *A*-Gall bladder, *B*-Ducts from liver, *C*-Duct from gall bladder, *D*-Common bile duct, *E*-Pancreatic duct, *F*-Hepato-pancreatic duct

(ii) It is present at the junction of common bile duct (*D*), pancreatic duct (*E*) and guards the opening of common hepato-pancreatic duct (*F*).

84. Pancreatic duct

85. Human is odd one out. Gall bladder is absent in all of them except human.

86. False. The pancreas is a compound, i.e. exocrine and endocrine glands situated between the limbs of the C-shaped duodenum.

87. Exocrine, endocrine

88. Exocrine

89. Buccal cavity

90. True

91. Mucus

92. Pharynx, oesophagus

93. Peristalsis

94. False. The gastro-oesophageal sphincter controls the passage of food into the stomach.

95. Salivary amylase, lysozyme.

96. Salivary amylase

97. Hydrolytic action

98. (i) *A*-Salivary amylase, *B*-pH 6.8.
(ii) 30%
(iii) Enzyme *A* will not hydrolyse the starch because it acts on pH 6.8.

99. Lysozyme

100. *A*-Oxyntic glands, *B*-Cardiac orifice, *C*-G-cells

101. A-3, B-4, C-1, D-2

102. B_{12}

103. Acidic, chyme

104. (i) Pepsin
(ii) No product will be formed, because pepsinogen acts only on acidic medium.

105. Proteolytic, proteoses, peptones

106. Bicarbonates

107. (i) True
(ii) False. Rennin is found in gastric juice of the infants.
(iii) *A*–Calcium paracaseinate
(iv) Curdling

108. Lipases, amylase

109. 3 (Bile juice, pancreatic juice and intestinal juice).

110. Hepato-pancreatic duct

111. 2 (Prorennin, pepsinogen)

112. *A*–Enterokinase

113. Bilirubin, biliverdin

114. Emulsification of fats

115. Alkaline

116. Bile juice

117. (i) True
(ii) True

118. Intestine

119. Intestinal juice or Succus entericus

120. Disaccharides, dipeptidases, lipases, nucleosidases

121. Acid, alkaline

122. *A*-Salivary gland, *B*-Small intestine, *C*-Pepsin, *D*-Chymotrypsin, *E*-Aminopeptidases, *F*-Pancreas, *G*-Bile

123. (i) A-Peptones, B-Trypsin, C-Carboxypeptidase.
(ii) Small intestine

124. (i) *A*-Pancreatic amylase
(ii) The reaction will not takes place because pancreatic amylase act on alkaline medium.

125. Hydrolysis

126. (i) *A*-Bile, *B*-Monoglycerides,
(ii) pH (7.7-8.0)

127. Nucleic acid, nucleotides, nucleosides

128. Alkaline

129. (i) *A*-Dipeptidases, (ii) *B*-Maltose,
(iii) *C*-Galactose
(iv) *D*-Fructose, (v) *E*-Sugars *F*- N_2 Bases, (vi) *G*-Glycerol

130. Duodenum

131. False. The products of duodenum region are absorbed in the jejunum and ileum.

132. Large intestine

133. Minerals, drugs

134. True

135. Caecum

136. Ileo-caecal, rectum

137. (i) A-Stercobilin (ii) It is present in faeces, it gives colour to the faeces.

138. Neural, hormonal

139. False. The sight, smell and the presence of food in the oral cavity stimulate the secretion of saliva.

140. (i) It will increase in the presence of secretin.
(ii) It will decrease in the presence of secretin.

141. Absorption

142. Glucose, chloride ions

143. False. Some fructose and amino acids are absorbed with the help of carrier proteins during facilitated diffusion.

144. Fatty acids, glycerol

145. (i) *A*-Micelles, *B*-Chylomicrons, *C*-Blood stream
(ii) Fat globule

146. Active transport

147. Simple diffusion, ileum

148. Simple diffusion

149. Drug

150. 3(Water, simple sugar, alcohol)

151. Small intestine

152. Box '*A*'

153. Assimilation

154. Calories

155. Energy

156. 1 kg, 1°

157. Combustion

158. A-3, B-2, C-1

159. Neural reflex

160. Voluntary, peristaltic

161. Glycogen

162. Vitamin-B_3

163. Bacterial, viral

164. 5 (All of them except lice).

165. A-2, B-4, C-5, D-3, E-1

166. Medulla

167. Jaundice

168. Food poisoning, overeating, spicy food

169. 2 (Marasmus, kwashiorkor)

170. Proteins

171. True

CHAPTER SEVENTEEN

BREATHING AND EXCHANGE OF GASES

Topic 1 : Respiratory Organs

1. The process of exchange of O_2 from the atmosphere with CO_2 produced by the cells is called ____________________

2. Name animal group showing the following types of respiration.
   (i) Tracheal respiration (through trachea) ____________________
   (ii) Branchial respiration (through gills) ____________________
   (iii) Pulmonary respiration (through lungs) ____________________
   (iv) Cutaneous respiration (through moist skin) ____________________

3. Write the correct sequence of air passage in man through the structures given below.

Pharynx, Larynx, Trachea, Bronchi, Alveoli, Nasal passage, Internal nostril, External nostril, Bronchioles

   ____

4. Nasopharynx opens through the larynx region into the ____________________

5. Which portion of the human respiratory system is called sound box? ____________________

6. Larynx is present in between ____________________

7. Identify the structures *A*, *B*, *C* and *D* in the human respiratory system.

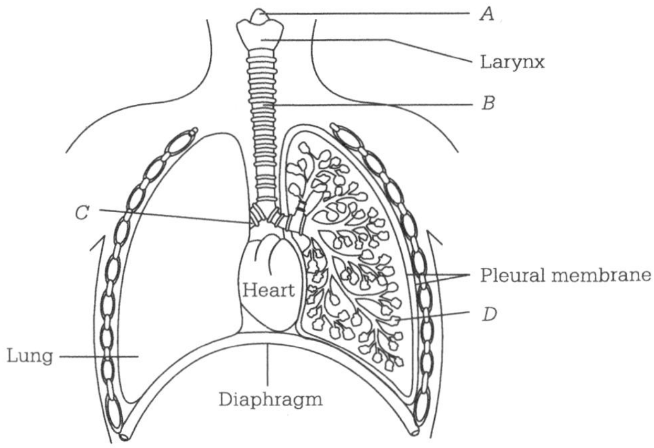

A. ______________________

B. ______________________

C. ______________________

D. ______________________

8. The process by which the vocal cords produce sounds is known as ____________

9. The cartilaginous flap which prevents the entry of food into the larynx during swallowing is ______________________

10. Trachea divides into right and left primary bronchi at ________ (5th/6th) thoracic vertebrae.

11. Incomplete cartilaginous rings support all of the following except

Trachea, Bronchi, Initial bronchioles, Alveoli

____

12. Lungs are comprised of network of bronchi, bronchioles and alveoli.

True ☐ False ☐

13. Each lung is enclosed by two membranes called the ______________

14. What is the function of pleural fluid? ______________________

15. Complete the following analogy.

The outer pleural membrane : Parietal pleura : : The inner pleural membrane :

____

16. The outer pleural membrane is in close contact with the ____________ lining whereas the inner pleural membrane is in contact with the ____________ surface.

17. Study the different parts given in the box and identify how many of them are related to conducting part (C) and respiratory/exchange part (R) of respiratory system?

Alveolar duct, Trachea, Bronchi, Larynx, Nasopharynx, External nostrils, Nasal chamber, Terminal bronchioles, Alveoli

____

18. Write the names of the structure mentioned as *A*, *B* and *C* in the diagram given below.

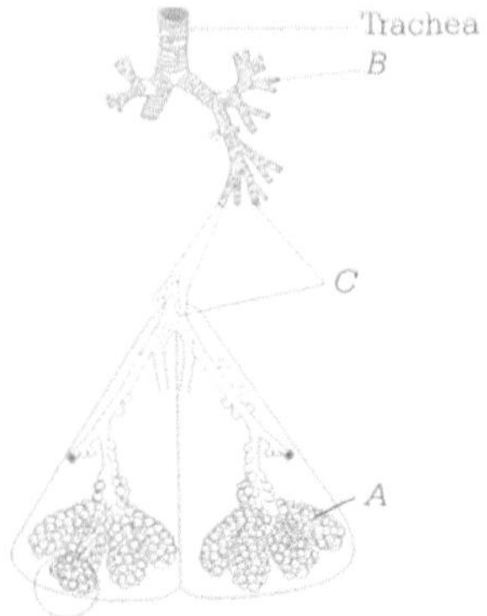

A. ______________________

B. ______________________

C. ______________________

19. Choose the words from the given box to fill in the blanks.

Diaphragm, Ribs, Vertebral column, Sternum

Thoracic chamber is formed dorsally by the ______________ ventrally by the ____________, laterally by the ______________ and on the lower side by the dome-shaped ____________.

20. Read the following functions and identify the part of respiratory system to which these functions belong.
   (i) Transport the atmospheric air to alveoli.
   (ii) Clear air from foreign particles.
   (iii) Humidifies and bring air to body temperature.

____

21. An instrument for measuring the air capacity of lungs is ______________ (Spirometer/Stethoscope).

22. The rate of breathing in a normal healthy human is <u>70-72 times/minutes.</u> (*Rewrite the statement by changing the underlined word.*)

____

23. Complete the missing words in the flowchart of process of respiration in humans.

Breathing or (A) ______________________ by which atmospheric air is drawn in and CO_2 rich alveolar air is released out.

↓

Diffusion of O_2 and CO_2 across (B) ______________

↓

Transport of gases by the (C) ____________

↓

(D) ____________ of O_2 and CO_2 between blood and tissues

↓

Utilisation of O_2 by the cells for (E) ______________ and resultant release of CO_2, i.e. (F) ______________

Topic 2 : Mechanism of Breathing

24. Breathing involves two phases, i.e. ______________________ during which atmospheric air drawn in and ______________________ by which the alveolar air is released out.

25. Inspiration and expiration are active processes.

True ☐ False ☐

26. The movement of air into and out of the lungs is carried by creating a <u>chemical reaction</u> between the lungs and the atmosphere.

(*Rewrite the statement by changing the underlined word.*)

____

27. Inspiration occurs when there is a __________________ (positive/negative) pressure in the lungs with respect to atmospheric pressure.

28. Negative pressure in lungs is achieved when intrapulmonary pressure is ____________ than the atmospheric pressure.

29. Expiration takes place when the ______________ pressure is greater than the atmospheric pressure.

30. Arrange the following steps of inspiration in correct order of their occurrence.

I. The contraction of external intercostal muscles raises the ribs and sternum.
II. Volume of thorax increases in the dorso-ventral axis.
III. Intrapulmonary pressure decreases.
IV. Diaphragm contraction.
V. Air rushes into the lungs.
VI. Volume of the thorax increases in the anterior-posterior axis.

____

31. Arrange the following steps of expiration in correct order of their occurrence.

I. Relaxation of external intercostal muscles and return of diaphragm and sternum to their normal position.
II. Air expelled from lungs.
III. Volume of thorax decreases.
IV. Intrapulmonary pressure increases.

____

32. During expiration, the diaphragm becomes ________ (concave/dome/oblique) shaped.

33. Additional muscles required for forceful breathing are

____

34. Write the names of the following muscles.

(i) Inspiratory muscles - contraction of these brings inspiration, e.g

____

(ii) Expiratory muscles - contraction of these bring expiration, e.g.

____

35. In the diagrams given below complete the labels for mechanism of breathing.

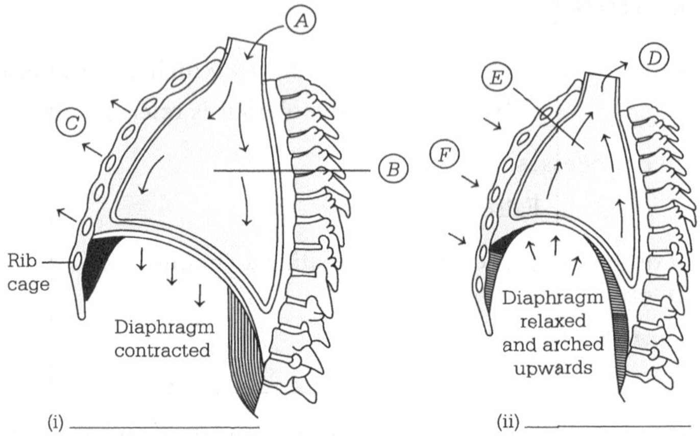

(i) ____________ (ii) ____________

A. ____________

B. ____________

C. ____________

D. ____________

E. ____________

F. ____________

Topic 3 : Respiratory Volumes and Capacities

36. Write the names of the respiratory volume against each statement.

(i) Additional volume of air, a person can inspire by a forcible inspiration.

____

(ii) Additional volume of air, a person can expire by a forcible expiration.

____

(iii) Volume of air remaining in the lungs even after a forcible expiration.

____

(iv) Volume of air inspired or expired during a normal respiration.

____

37. Write names of the respiratory capacities against each statement.

(i) Total volume of air accommodated in the lungs at the end of a forced inspiration.

____

(ii) Total volume of air a person can inspire after a normal expiration.

____

(iii) Total volume of air a person can expire after a normal inspiration.

____

(iv) Volume of air that will remain in the lungs after a normal expiration.

____

(v) The maximum volume of air a person can breathe in after a forced expiration.

____

38. Match the following columns.

	Column I		Column II
A.	Tidal volume	1.	2500-3000 mL of air
B.	Inspiratory reserve volume	2.	1000 mL of air
C.	Expiratory reserve volume	3.	500 mL of air
D.	Residual volume	4.	1200 mL of air

A .________________ B. ________________ C. ________________ D. ________________

39. Arrange the following in the order of increasing volume.

Tidal volume, Residual volume, ERV, Vital capacity

____

40. Different respiratory volumes are given below.

Tidal volume = 500 mL   Residual volume = 1000 mL
IRV = 2500 mL   ERV = 1000 mL

The Functional Residual Capacity (FRC) is ____________ mL

41. Match the following columns.

	Column I		Column II
A.	Total Lung Capacity (TLC)	1.	IRV + TV
B.	Expiratory Capacity (EC)	2.	FRC + TV + IRV
C.	Inspiratory Capacity (IC)	3.	ERV + TV
D.	Vital Capacity (VC)	4.	ERV + TV + IRV

A .________________ B. ________________ C. ________________ D. ________________

42. The volume of air from nostrils to terminal bronchioles that do not take part in exchange of gases due to incompetent tissue represents anatomical dead space.

True ☐   False ☐

43. In the given graph mark the following.

Inspiratory reserve volume, Vital capacity, Functional residual capacity, Residual volume, Total lung capacity

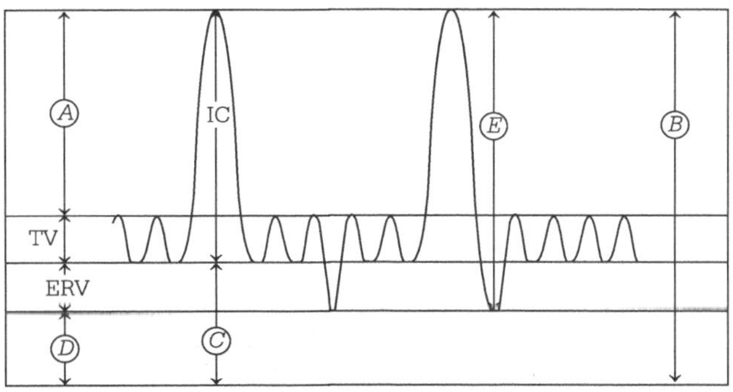

A .________________ B. ________________ C. ________________

D. ________________ E .________________

Topic 4 : Exchange of Gases

44. The exchange of gases in the alveoli of the lungs takes place by __________ (passive transport/active transport/simple diffusion).

45. What are the two important factors that can affect the rate of diffusion of gases in respiration?

____

46. Give the scientific term for the pressure exerted by an individual gas in a mixture of gases.

____

47. Identify *A*, *B*, *C* and *D* with respect to partial pressure of oxygen and CO_2 in different mediums.

Respiratory gas	Atmospheric air	Alveoli	Blood (Deoxygenated)	Blood (Oxygenated)	Tissues
O_2	159 mm Hg	(A)	40 mm Hg	(B)	40 mm Hg
CO_2	(C)	40 mm Hg	45 mm Hg	40 mm Hg	(D)

A. __________ B. __________ C. __________ D. __________

48. Complete the following sentences.

(i) The partial pressure of CO_2 is __________ (higher/lesser) in systemic veins than in systemic arteries.

(ii) The partial pressure of CO_2 is __________ (higher/lesser) in tissue spaces than in systemic arteries.

49. Identify *A* to *D* in the given diagram.

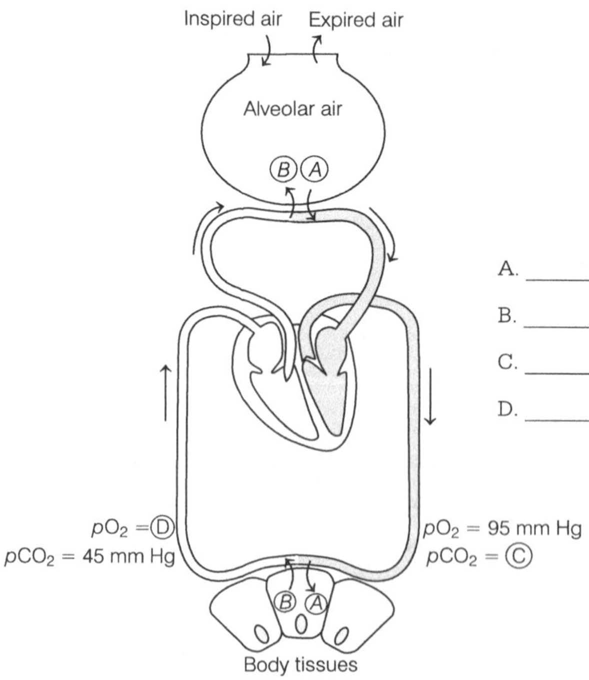

A. __________

B. __________

C. __________

D. __________

50. The solubility of CO_2 in blood is 20-25 times lesser than the solubility of O_2.

True ☐ False ☐

51. Given pie chart represents the approximate normal composition of alveolar air. What is *A*, *B* and *C*?

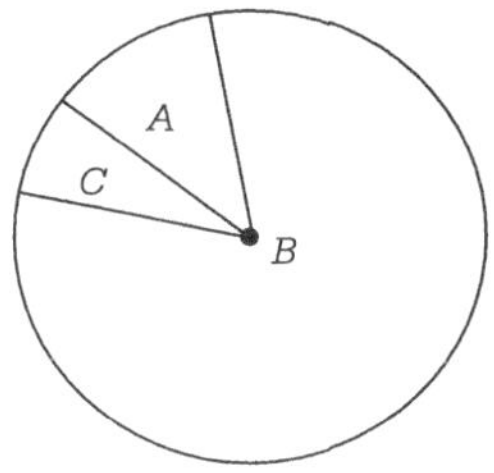

A. ____________________

B. ____________________

C. ____________________

52. In lungs, the air is separated from the venous blood through ____________________ and ____________________.

53. Given below the diagram of diffusion membrane made up of three major layers. Name them correctly.

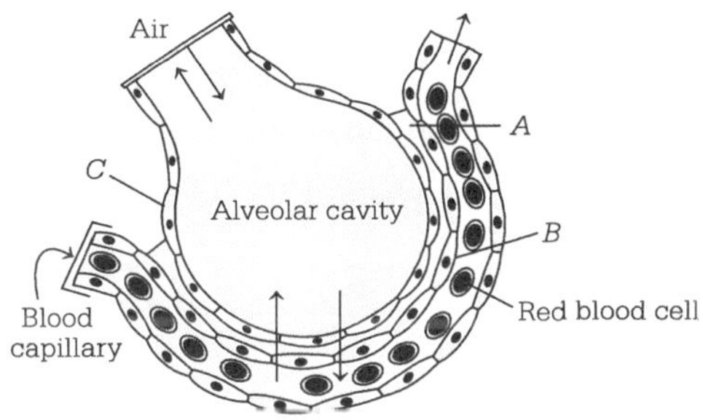

A. ____________________

B. ____________________

C. ____________________

54. The basement substance is bound by the basement membrane of ____________________ and ____________________.

Topic 5 : Transport of Gases

55. What percentage of O_2 is transported by RBCs and plasma, respectively in human body?

____

56. Blood carries the CO_2 in three form. The correct percentages of CO_2 in these forms are

(i) as carbaminohaemoglobin ____________________

(ii) as bicarbonates ____________________

(iii) dissolved form in plasma ____________________

57. Haemoglobin is a ______________ coloured ______________ containing pigment present in the ______________. Oxygen can bind with haemoglobin in a reversible manner to form ____________________. Each haemoglobin molecule can carry a maximum of ______________ molecules of O_2.

58. What are the four factors affecting the binding of haemoglobin with oxygen?

____

59. Complete the equation by stating the location for 'X' and 'Y'.

$$HbO_2 \underset{\text{Association (At 'Y')}}{\overset{\text{Dissociation (At 'X')}}{\rightleftharpoons}} Hb + O_2$$

X. ______________ Y. ______________

60. In a oxygen dissociation curve what is plotted on the

(i) x-axis ______________

(ii) y-axis ______________

61. The oxygen dissociation curve is ______________ in shape.

62. After taking a long deep breath we do not respire for few seconds due to more CO_2 in the blood.

True ☐ False ☐

63. Every 100 mL of oxygenated blood can deliver around 15 mL of oxygen to the tissues under normal physiological conditions.
(*Rewrite the statement by changing the underlined word.*)

____

64. The given graph shows an oxygen dissociation curve for haemoglobin.

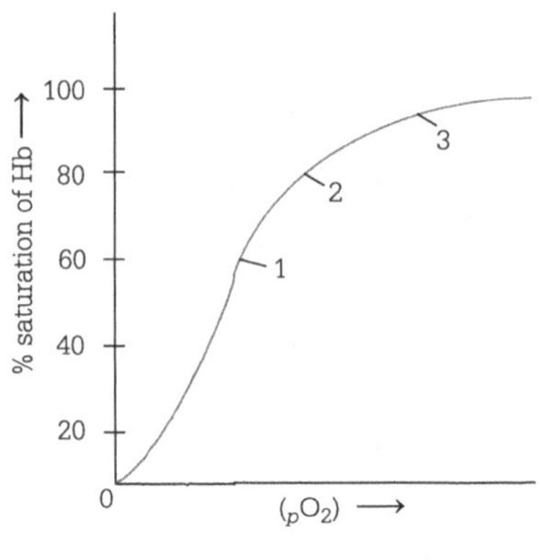

Locate left ventricle, pulmonary vein and vena cava in the given graph with respect to positions 1, 2 and 3.

____

65. Oxygen dissociation curve can be shifted from normal position depending on its various factor. Mention the conditions with respect to pO_2, pCO_2, H^+ ions pH and temperature when

(i) the curve is shifted to right ______________

(ii) the curve is shifted to left______________

66. Choose the odd one out for the factor favouring dissociation of oxyhaemoglobin to deliver O_2 to tissue.

$pO_2\uparrow$, $pCO_2\uparrow$, $H^+\uparrow$, Temperature $\uparrow$

____

67. Fetal haemoglobin has ________________ (same/higher/lower) affinity for oxygen than that of mother's haemoglobin during gestation.

68. The factor which favours the formation of HCO_3^- and H^+ in the blood is ____________ (high pCO_2/low pCO_2).

69. A high percentage of CO_2 is transported in dissolved state in comparison to O_2 because
(i) CO_2 has less solubility in plasma. (ii) CO_2 has high solubility in plasma.

____

70. CO_2 dissociates from carbaminohaemoglobin when

pO_2 : ______________ : : pCO_2 : ______________

71. In the given equation, what is X enzyme?

$$CO_2 + H_2O \overset{X}{\rightleftharpoons} H_2CO_3 \overset{X}{\rightleftharpoons} H^+ + HCO_3^-$$

X. ____

72. If 1g of Hb carries 1.34 mL of O_2 and 100 mL of blood has 15 g Hb then, how much O_2 will be carried approximately by 100 mL of arterial blood?

____

73. Identify the phenomena described in the statements.
(i) Exit of bicarbonate ions, considerably change ionic balance between the plasma and RBCs. To restore this ionic balance the Cl^- ions diffuse from the plasma into the RBCs.

____

(ii) Shifting of the oxygen haemoglobin dissociation curve to the right by increasing CO_2 partial pressure.

____

(iii) Binding of oxygen with haemoglobin tends to displace CO_2 from the blood.

____

74. Every 100 mL of deoxygenated blood delivers approximately ________ mL of CO_2 to the alveoli.

75. Write the correct sequence of CO_2 diffusion by choosing the terms from the box given below.

Capillary basement membrane, Alveolar epithelium, Alveolar epithelial basement membrane, Capillary endothelial membrane, Thin interstitial space

____

Topic 6 : Regulation of Respiration

76. Human beings have a significant ability to maintain and moderate the respiratory rhythm to suit the demands of the body tissues. This is achieved by ___________ system.

77. Respiratory process is regulated by certain specialised centres in the brain. Name them according to the description.
(i) Responsible for regulation of respiration or causes inspiration.

____

(ii) Reduce the inspiratory duration upon stimulation or limit inspiration.

____

78. Write the location of pneumotaxic centre and respiratory rhythm centre within human brain.

____

79. Fill in the blanks.

Respiratory Rhythm Centre

(A) respiratory group → It causes inspiration

(B) respiratory group → It causes either inspiration or expiration

A. ________________ B. ________________

80. Chemical regulation of respiration is done *via* ________________ area situated adjacent to the rhythm centre in the brain.

81. Receptors associated with aortic arch and carotid artery can recognise change in ________ and ________ concentration and send necessary signals to ________________ for remedial action.

82. Respiratory centre of brain is stimulated by __________ (O_2 / CO_2) content in ______________ (arterial/venous) blood.

83. The role of oxygen in the regulation of respiratory rhythm is ____________________

Topic 7 : Disorders of Respiratory System

84. Match the following columns.

	Column I		Column II
A.	Apnea	1.	Difficult breathing
B.	Dyspnea	2.	Absence of breathing
C.	Hypopnea	3.	Rapid breathing
D.	Hyperpnea	4.	Slow breathing

A.______________ B. ______________ C.______________ D. ______________

85. Choose the correct term from the box for the statements given below.

Hypoxia, Anoxia, Hiccups, Wheezing, Snoring, Yawning, Hypercapnia, Coughing

(i) Jerky incomplete inspiration ________________________

(ii) Breathing noisly ________________________

(iii) Condition of oxygen shortage in the tissue ________________________

(iv) A prolonged inspiration ________________________

(v) Rough rattling inspiratory noise during sleep ________________________

(vi) Excess carbon dioxide in the body fluids________________________

(vii) Absence of oxygen in inspired gases, arterial blood or tissue ________________________

(viii) Forceful sudden expiration ________________________

86. Given below are few respiratory disorders. How many of them are occupational respiratory disorders?

Coryza, SARS, Silicosis, Asbestosis, Emphysema

____

87. High percentage of CO_2 and very low percentage of O_2 may make a person unconscious due to _______________ (suffocation/asphyxia).

88. Name the pulmonary disease in which alveolar surface area involved in gas exchange is drastically reduced due to damage in the alveolar walls. _______________

89. _______________ is a chronic respiratory disorder caused mainly by cigarette smoking.

90. A respiratory disorder causing difficulty in breathing, wheezing due to inflammation of bronchi and bronchioles is_______________

91. Write the full form of SARS._______________

92. Name the causative agent for the following respiratory problems.

(i) Coryza (common cold)_______________

(ii) SARS _______________

(iii) Pneumonia_______________

(iv) Asthma _______________

Answers

1. Breathing
2. (i) Insects,
   (ii) Aquatic arthropods, molluscs and fishes
   (iii) Amphibians, reptiles, birds, mammals
   (iv) Earthworms, frogs
3. External nostril → Nasal passage → Internal nostril →Pharynx → Larynx → Trachea → Bronchi → Bronchioles → Alveoli
4. Trachea
5. Larynx
6. Epiglottis and trachea
7. *A* – Epiglottis, *B* – Trachea, *C* – Bronchus, *D* – Alveoli
8. Phonation
9. Epiglottis
10. 5th
11. Alveoli
12. True
13. Pleura
14. It reduces friction on the lung surface.
15. Visceral pleura
16. Thoracic, lung
17. C = 7; trachea, bronchi, larynx, nasopharynx, external nostrils, nasal chamber, terminal bronchiole.
    R = 2; alveoli and alveolar duct.
18. *A*–Alveolar sac, *B*–Secondary bronchus, *C*–Bronchioles
19. Vertebral column, sternum, ribs, diaphragm
20. Conducting part of respiratory system.
21. Spirometer
22. 12-16 times/minute.
23. *A*–Pulmonary ventilation, *B*–Alveolar membrane, *C*–Blood, *D*–Diffusion, *E*–Catabolic reaction, *F*–Cellular respiration
24. Inspiration, expiration
25. False. Inspiration is an active process whereas expiration is passive process.
26. Pressure gradient
27. Negative
28. Less
29. Intrapulmonary
30. IV → VI → I → II → III → V
31. I → III → IV → II
32. Dome
33. Internal intercostals and abdominal muscles
34. (i) Diaphragm muscle and external intercostal muscles.

(ii) Abdominal muscle and internal intercostal muscles.

35. (i) Inspiration ; *A*–Air entering the lungs, *B*–Volume of thorax increased, *C*–Ribs and sternum raised.
(ii) Expiration ; *D*–Air expelled from lungs, *E*–Volume of thorax decreased, *F*–Ribs and sternum returned to original position.

36. (i) Inspiratory Reserve Volume (IRV)
(ii) Expiratory Reserve Volume (ERV)
(iii) Residual Volume (RV)
(iv) Tidal Volume (TV)

37. (i) Total Lung Capacity (TLC)
(ii) Inspiratory Capacity (IC)
(iii) Expiratory Capacity (EC)
(iv) Functional Residual Capacity (FRC),
(v) Vital Capacity (VC).

38. A - 3, B - 1, C - 2, D - 4

39. $\underset{(500\,mL)}{\text{Tidal volume}} < \underset{(1000\,mL)}{\text{ERV}} < \underset{(1200\,mL)}{\text{RV}} < \underset{(4500\,mL)}{\text{Vital capacity}}$

40. $FRC = ERV + RV$
$= 1000 + 1000 = 2000$ mL

41. A - 2, B - 3, C - 1, D - 4

42. True

43. *A* – Inspiratory reserve volume,
B – Total lung capacity,
C – Functional residual capacity,
D – Residual volume,
E – Vital capacity

44. Simple diffusion

45. Solubility of the gases as well as the thickness of the membranes involved.

46. Partial pressure

47. *A* – 104mm Hg, *B* – 95mm Hg, *C* – 0.3mm Hg, *D* – 45mm Hg

48. (i) Higher
(ii) Higher

49. *A*–O_2, *B*–CO_2, *C*–40 mm Hg, D–40 mm Hg.

50. False. In comparison to solubility of O_2 in blood the solubility of CO_2 is 20-25 times higher.

51. *A*–14% oxygen, *B*–80% nitrogen, *C*–6% carbon dioxide

52. Squamous epithelium, endothelium of alveolar capillaries.

53. *A* – Basement substance,
B – Endothelium of blood capillary,
C – Squamous epithelium of alveolar wall (one celled thick).

54. Squamous epithelium, single layer endothelial cells of capillaries.

55. 97% and 3%

56. (i) 20-25%, (ii) 70%, (iii) 7%

57. Red, iron, RBCs, oxyhaemoglobin, four

58. Partial pressure of O_2(pO_2), partial pressure of CO_2(pCO_2), hydrogen ion concentration, temperature.

59. *X* is tissue, *Y* is alveoli in lungs.

60. (i) Partial pressure of oxygen (mm Hg)
(ii) Percentage saturation of haemoglobin with oxygen.

61. Sigmoid (or S-shaped).

62. False. After taking a long deep breath we do not respire for few seconds due to less CO_2 in the blood.

63. 5mL

64. 1. Vena cava (60%) because vena cava carry deoxygenated blood, but still some oxygen is there,
2. Left ventricle (80%),
3. Pulmonary vein (95-100%) Because pulmonary vein carry oxygenated blood.

65. (i) Shift to right ↓ in pO_2, ↑ in pCO_2, ↑ in H^+ inons, ↓ in pH, ↑ in body temperature.
(ii) Shift to left ↑ in pO_2, ↓ in pCO_2, ↓ in H^+ ions, ↑ in pH, ↓ in body temperature.

66. pO_2 ↑ odd one out. All other factors favour dissociation of oxyhaemoglobin.

67. Higher

68. High pCO_2

69. (ii) Because CO_2 has high solubility in plasma.

70. Increases, decreases

71. *X*-Carbonic anhydrase

72. 20 mL 1.34 mL of O_2 ×15 g of Hb = 20.1 mL or 20 mL approximately.

73. (i) Chloride shift or Hamburger's phenomenon
(ii) Bohr's effect,
(iii) Haldane effect

74. 4

75. Capillary endothelial membrane → Capillary basement membrane →Thin interstitial space → Alveolar epithelial basement membrane → Alveolar epithelium

76. Neural

77. (i) Respiratory rhythm centre
(ii) Pneumotaxic centre

78. Pons and medulla, respectively.

79. *A*–Dorsal, *B*–Ventral

80. Chemosensitive

81. CO_2, H^+, rhythm centre

82. CO_2, arterial

83. Insignificant

84. A - 2, B - 1, C - 4, D - 3

85. (i) Hiccups
(ii) Wheezing
(iii) Hypoxia
(iv) Yawning
(v) Snoring
(vi) Hypercapnia
(vii) Anoxia
(viii) Coughing

86. 2 (Silicosis and asbestosis)

87. Asphyxia

88. Emphysema

89. Emphysema

90. Asthma

91. Severe Acute Respiratory Syndrome

92. (i) Rhinovirus
(ii) Human corona virus
(iii) *Streptococcus pneumoniae*
(iv) Allergens

CHAPTER EIGHTEEN

BODY FLUIDS AND CIRCULATION

Topic 1 : Blood and Lymph

1. Fill in the blanks.
   (i) In simple organisms like sponges and coelenterates, ____________ circulates in body cavity from the surroundings to facilitate the exchange of O_2, CO_2, nutrients and waste products in cells.
   (ii) More complex organisms use special fluids within their bodies to transport such materials. The _________ is the most commonly found body fluid in most of the higher organisms including humans for this purpose.
2. What is the composition of blood?

   ____

3. Write the percentage of the following.
   (i) Plasma in blood ____
   (ii) Water in plasma ____
   (iii) Protein in plasma ____
4. Blood is slightly __________________ in nature with pH range ____________________ .
5. Plasma is a pink coloured, viscous fluid in blood.

   True ☐ False ☐

6. Choose the odd one out.

Albumin, Fibrinogen, Globulin, Cellulose

____

7. Match the following columns.

	Column I (Blood proteins)		Column II (Functions)
A.	Fibrinogen	1.	Defence mechanisms of the body
B.	Globulins	2.	Osmotic balance
C.	Albumins	3.	Clotting or coagulation of blood

A. _______ B. _______ C. _______

8. Name the minerals present in plasma. _______________

9. What is 'X' in the formula, Plasma – 'X' = Serum

X. _______________

10. Choose the odd one out.

Erythrocytes, Leucocytes, Thrombocytes, Plasma

____

11. The common names for erythrocytes, leucocytes and thrombocytes are _______, _______ and _______, respectively.

12. How many features are correct for erythrocytes?
(i) Least abundant of all the cells in blood.
(ii) Formed in the red bone marrow in adults.
(iii) Devoid of nucleus in most of the mammals.

____

13. What is the amount of haemoglobin present in 100 mL blood of human?

____

14. Mammalian RBCs are _______ (sickle/biconcave) in shape.

15. Choose the odd one out with respect to enzymes found in RBCs.

Glucose-6-phosphate dehydrogenase, Phosphofructokinase, Succinate dehydrogenase, Carbonic anhydrase

____

16. Oval and nucleated RBCs are present in which mammal? _______________

17. Give the scientific term for the following.
(i) Formation of erythrocytes. _______________
(ii) The process of formation of blood clot. _______________
(iii) The red coloured, iron containing complex protein in red blood cells. _______________

18. The graveyard of RBC's is kidney.
(*Rewrite the statement by changing the underlined word.*)

____

19. Give one significant role of RBCs in our body. ____________________

20. How many features are true for White Blood Cells (WBCs)?
(i) Lack haemoglobin
(ii) Lifespan ranges from a few hours to few days.
(iii) Enucleate cells.
(iv) Ranges from 6000 – 8000 mm^{-3} of blood in healthy human approximately.

____

21. Complete the flowchart depicting the types of leucocytes in human body.

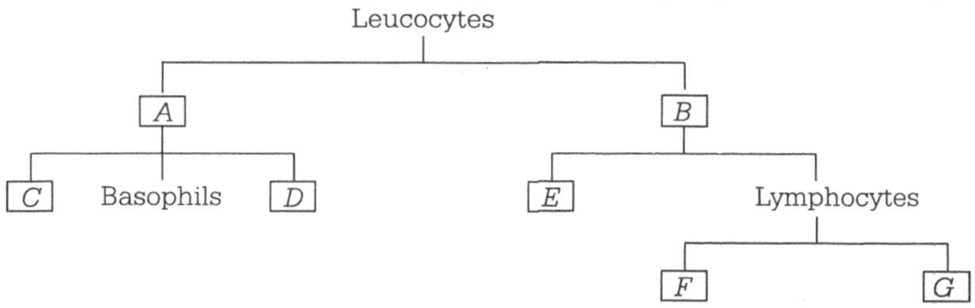

A. ____________ B. ____________ C. ____________ D.____________

E. ____________ F. ____________ G. ____________

22. Choose the odd one out.

Neutrophils, Basophils, Eosinophils, Monocytes

____

23. Complete the analogy with respect to WBCs.

Neutrophils : Most abundant in the body : : __________ : Least abundant in the body.

24. Give the scientific term for following sentence.
'Decrease in the number of leucocytes below 5000 mm^{-3} of blood'.

____

25. Match the following columns.

	Column I (Types of leucocytes/WBCs)		Column II (% of total WBCs)
A.	Neutrophils	1.	20-25%
B.	Basophils	2.	2-3%
C.	Monocytes	3	6-8%
D.	Eosinophils	4.	0.5 -1%
E.	Lymphocytes	5.	60-65%

A. ____________ B. ____________ C. ____________

D.____________ E. ____________

26. Match the following columns.

	Column I (WBCs)		Column II (Functions)
A.	Eosinophils	1.	Involved in inflammatory reactions
B.	Basophils	2.	Allergic reactions
C.	Neutrophils	3.	Responsible for immune response
D.	Lymphocytes	4.	Phagocytic cells
E.	Monocytes		

A. ______________ B. ______________ C. ______________

D.______________ E. ______________

27. How many of the given leucocytes can destroy foreign organisms entering the body?

Eosinophils, Basophils, Neutrophils, Monocytes, Lymphocytes

____

28. Identify the cells labelled as *A* to *E*.

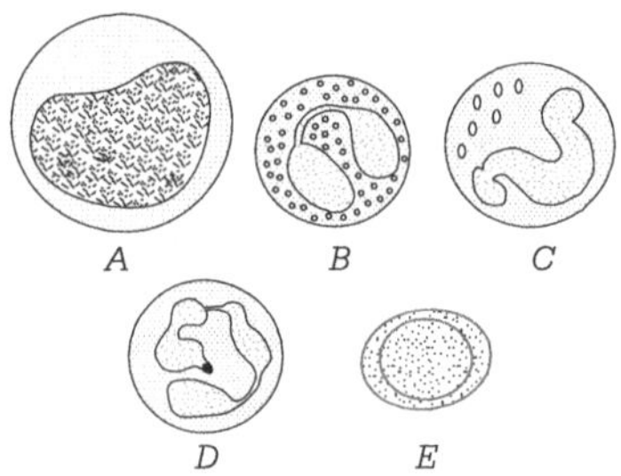

A. ______________

B.______________

C.______________

D.______________

E.______________

29. Match the following columns.

	Column I (Formed elements)		Column II (Number in/mm^3 of blood)
A.	Erythrocytes	1.	5–5·5 millions
B.	Leucocytes	2.	6000–8000
C.	Platelets	3.	1,50,000 – 3,50,000

A. ______________ B. ______________ C. ______________

30. Choose the odd one out with respect to secretions of basophils.

Histamine, Serotonin, Heparin, Antibodies

____

31. The largest leucocyte is neutrophil.
(*Rewrite the statement by changing the underlined word.*)

____

32. Name the two blood grouping systems used all over the world.

____

33. Write the scientific terms for the following.

(i) Chemicals that can induce immune response. ______________________

(ii) Proteins produced in response to antigens. ______________________

34. ABO blood grouping is based on the presence and absence of __________ (2/4) surface antigens.

35. Identify *W* to *Z* in the table given below.

Blood groups	Antigens on RBCs	Antibodies in plasma	Donor's group
A	(*W*)	anti-B	A, O
B	B	anti-A	(*Y*)
AB	A, B	(*X*)	AB, A, B, O
O	Nil	anti-A, B	(*Z*)

W. ______________ X. ______________ Y. ______________ Z. ______________

36. Name the blood group that is popularly known as

(i) universal donor ______________

(ii) universal recipient ______________

37. A sample of blood shows clumping with antiserum-A, but not with antiserum-B. The blood group would be ________ (A/AB/B).

38. Blood groups were discovered by Landsteiner.

True ☐ False ☐

39. Rh-factor is concerned with blood grouping. It derives its name from ______________ (monkey/chimpanzee).

40. It is estimated about ____________ (80%/20%) individuals in the world are Rh positive.

41. How many characteristics are true for erythroblastosis foetalis?

(i) Cause severe anaemia

(ii) Cause jaundice

(iii) Could be fatal to the foetus

____

42. In a Rh negative mother, erythroblastosis foetalis can be prevented by administering ________________ to the ________________ immediately after the delivery of the first Rh positive child.

43. What will happen if Rh(– ve) person donates blood to Rh(+ ve) person for the second time?

(i) Destruction of RBC in Rh(+ve) person.

(ii) Nothing will happen to Rh(+ve) person.

____

44. If a Rh^+ man marries a Rh^- woman then which of the following would happen?
(i) No problem with 1st pregnancy.
(ii) No problem during any pregnancy.
(iii) Problem would be expected in second pregnancy.

____

45. Blood exhibits ______________ mechanism in response to an injury or trauma. This is a mechanism to prevent excessive loss of blood from the body.

46. Name the dark reddish-brown scum formed at the site of a cut or an injury over a period of time.

____

47. $\underline{K^+}$ ions play a very important role in blood clotting.
(*Rewrite the statement by changing the underlined word.*)

____

48. Identify *A* and *B* for the given reactions with respect to blood clotting.

$$\text{Prothrombin} \xrightarrow{A} \text{Thrombin}$$

$$\text{Fibrinogen} \xrightarrow{B} \text{Fibrin}$$

A. ______________ B. ______________

49. Arrange the given steps of mechanism of blood coagulation in the correct sequence of occurrence.
(i) Conversion of prothrombin into thrombin
(ii) Coagulum formation
(iii) Thrombokinase formation
(iv) Site of injury release tissue thromboplastins
(v) Conversion of fibrinogen into fibrin

____

50. The proteolytic enzyme which causes lysis of fibrin during fibrinolysis is <u>thrombin</u>.
(*Rewrite the statement by changing the underlined word.*)

____

51. Match the following columns.

	Column I (Clotting factors)		Column II (Names)
A.	Factor I	1.	Ca^{2+}
B.	Factor IV	2.	Fibrinogen
C.	Factor IX	3.	Fibrin stabilising factor
D.	Factor XIII	4.	Christmas factor

A. ______________ B. ______________ C. ______________ D. ______________

52. Match the following columns.

	Column I (Clotting factors)		Column II (Names)
A.	Factor II	1.	Thromboplastin
B.	Factor III	2.	Prothrombin
C.	Factor VIII	3.	Hageman factor
D.	Factor XII	4.	Antihaemophilic globulins

A. ________ B. ________ C. ________ D. ________

53. Prothrombin, which helps in clotting of blood is released by the blood platelets and produced in liver.

True ☐ False ☐

54. Name the vitamin which takes part in blood clotting mechanism. ________

55. Complete the equation for lymph.

Lymph = Blood – (________ + ________)

56. Interstitial fluid or tissue fluid is collected and drained back to the major veins by an elaborate network of vessels called the ________ system. The fluid present in the system is called ________.

57. Choose the correct option for lymph.

(i) Colourless/Yellow-coloured fluid. ________

(ii) Lymphocytes present/Lymphocytes absent ________

(iii) Absorbs fats/Manufacture proteins________

58. Choose the odd one out with respect to lymph.

Hormones, Antibodies, Nutrients, RBCs

____

Topic 2 : Circulatory Pathways and Human Circulatory System

59. Name the animals group in which the following system is present.

(i) Open circulatory system ________

(ii) Closed circulatory system ________

60. In open circulatory system, blood pumped by the heart passes through large vessels into open spaces or body cavities called ________.

61. Advantage of closed circulatory system over open circulatory system includes which of the following?

(i) Closed system can direct blood to specific tissues.

(ii) Exchange of material occurs slowly.

(iii) Closed circulatory system can support higher levels of metabolic activity.

____

62. In closed circulatory system, all the components given in bracket are present except (heart, blood vessels, sinuses, blood).

____

63. Match the following columns.

Column I (Heart)	Column II (Animals)	Column III (Circulation types)

A.

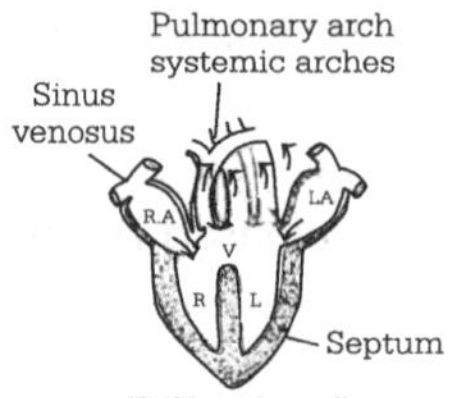

(3-Chambered)

1. Reptiles — K. Incomplete double circulation

B.

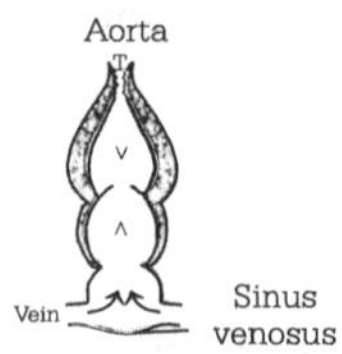

(2-Chambered)

2. Fishes — L. Single circulation

C.

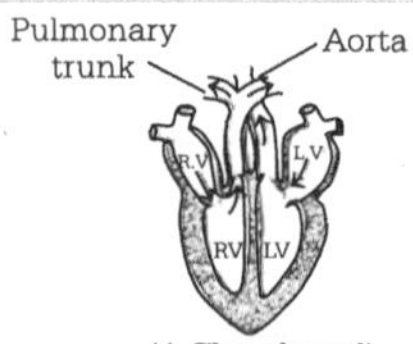

(4-Chambered)

3. Birds and mammals — M. Double circulation

D.

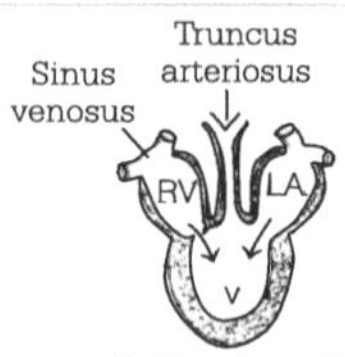

(3-Chambered)

4. Amphibians

A. ____________ B. ____________ C. ____________ D. ____________

64. In amphibians and reptiles, left atrium receives oxygenated blood from gills/lungs/skin.

True ☐ False ☐

65. In reptiles and amphibians, there is no clear cut separation of oxygenated and deoxygenated blood because they have ____________________.

66. Three major components of human circulatory system include

____

67. Choose the correct option for human heart.

(i) Origin : ectodermal/mesodermal ____________________

(ii) Situated in : thoracic/abdominal cavity ____________________

(iii) Position with respect to lungs : between/above________________

(iv) Slightly tilted : left/right ________________

68. The double-walled membranous bag that protects the heart and enclose pericardial fluid is ________________ .

69. Write the location of the following parts in human heart.

(i) Biscuspid valves ________________

(ii) Tricuspid valves ________________

(iii) Semilunar valves ________________

(iv) Chordae tendineae ________________

(v) Sino-Atrial (SA) node ________________

(vi) Atrio-Ventricular (AV) node ________________

70. Give two functions of heart valves.

(i) ________________

(ii) ________________

71. Ventricle is thick-walled as compared to atrium because

(i) it pumps blood to distant structures.

(ii) it does not contain SA and AV nodes.

____

72. The given figure represents a section of a human heart and write the labelled parts as *A* to *E*.

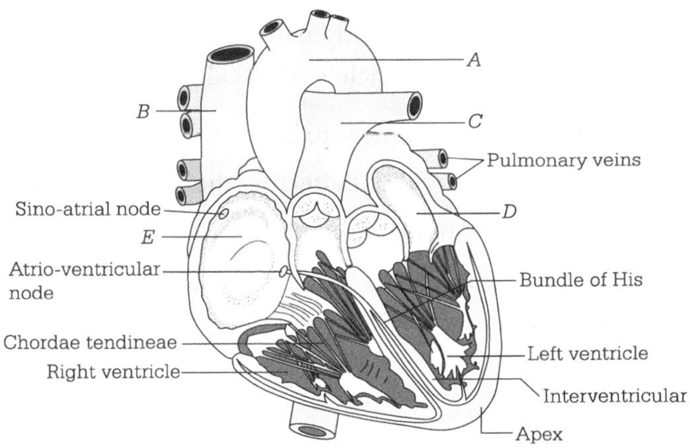

A. ________________

B. ________________

C. ________________

D. ________________

E. ________________

73. Arrange the following in correct conduction sequence in heart.

AV node, Bundle of His, Purkinje fibres, SA node, Heart muscles

____

74. Before the birth of a child, the major portion of blood from the right side by passes the pulmonary circulation through ____________ in between right and left auricles and ____________ in between pulmonary and systemic aorta.

75. What prevents bicuspid and tricuspid valves from collapsing back into atria during powerful ventricular contractions?

____

76. The papillary muscles are found in ____________.

77. In the human, name the septum which separates the following.

(i) Right and left atria ____________

(ii) Left and right ventricle ____________

(iii) Atrium and ventricle of same side ____________

78. Give scientific term for following.

Sino-atrial node and atrio-ventricular node. ____________

79. Complete the following analogy given below.
Pacemaker of heart : SA node : Pacesetter of heart : ____________

80. The SA node can generate the maximum number of action potential, i.e. ____________.

81. SA node is called the pacemaker of heart because

____

82. Artificial pacemaker is grafted at the place of ____________.

83. A bundle of muscle fibres called ____________ originates from AV node which consists of branches of fibres running along the ventricle's walls called ____________.

84. What will happen if the pacemaker becomes non-functional?

(i) Pacesetter would serve the similar functions without affecting the heart functioning.

(ii) Coordinated rhythmic movements would not occur in heart.

____

Topic 3 : Cardiac Cycle and ECG

85. Complete the flowchart of steps of cardiac cycle.

SA node generates action potential
↓
A
↓
B
↓
Atrial diastole
↓
C

A.____________

B.____________

C.____________

86. During systole of heart auricles and ventricles contract ________________ (simultaneously/separately).

87. Oxygenated blood is pumped into the aorta and deoxygenated blood is pumped into the pulmonary artery during ______________________ .

88. (i) Auricular diastole coincides

(ii) Tricuspid and bicuspid valves close

(iii) Semilunar valves open

When do the above events occur in a cardiac cycle?

____

89. Contraction of right ventricle pumps blood into _____________ artery.

90. During cardiac cycle, about ____________ (90%/50%) of ventricular filling occur prior to atrial contraction. ____________ (10%/50%) ventricular filling occur due to atrial contraction.

91. How many of the listed events occur during joint diastole?

(i) All 4 heart chambers are in relaxed state.

(ii) Tricuspid and bicuspid valves open.

(iii) Semilunar valves are closed.

(iv) Action potential is conducting.

(v) Blood flows into left and right ventricles from left and right atria.

____

92. The sequential event in the heart which is cyclically repeated is called ____________ .

93. Cardiac cycle consists of systole and diastole of both the atria and ventricles.

True ☐ False ☐

94. Write the numerical value of each of the following.

(i) Heart beats per minutes. ____________________________

(ii) Duration of the cardiac cycle. ____________________________

(iii) Number of cardiac cycles per minute. ____________________________

(iv) Stroke volume. ____________________________

(v) Cardiac output ____________________________

95. Give the scientific term for the following sentences.

(i) The volume of blood pumped out by each ventricle per minute. ________________

(ii) The volume of blood pumped by each ventricle during a cardiac cycle. ________________

96. Name the device that is

(i) used for hearing prominent heart sound. ______________________ .

(ii) used for measuring blood pressure. ______________________ .

97. Complete the following analogy with respect to a normal person.
The systolic pressure :: 120 mmHg : The diastolic pressure : ____________

98. Complete the equations.
(i) Systolic pressure − Diastolic pressure = ____________ pressure
(ii) Stroke volume (X) ____________ = Cardiac output

99. In a cardiac output of 5250 mL per minute, with 75 heartbeats per minute, the stroke volume is ____________.

100. The two prominent heart sounds are ________ and ____________.

101. Complete the following analogy given below.
____________ : First heart sound : : ____________ : Second heart sound

102. Name the heart sounds.
(i) Produced during closure of the semilunar valve. ____________
(ii) Produced during closure of tricuspid and bicuspid valves. ____________

103. Fill in the labels *A, B* and *C* in the given cardiac cycle.

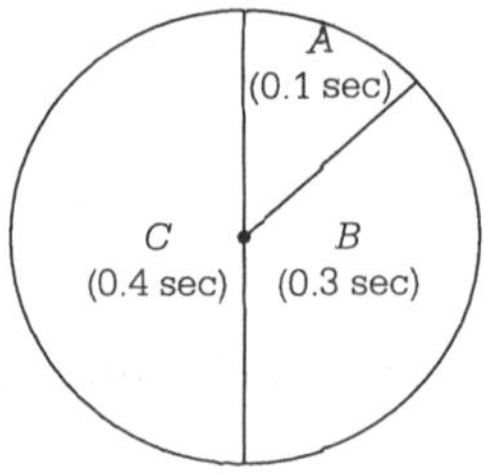

A. ____________
B. ____________
C. ____________

104. Give the full form of ECG. ____________

105. Name the machine used to obtain an ECG. ____________

106. What does ECG represent? (*Give answer in one line*)
____

107. Give the diagrammatic representation of a standard ECG.

____

108. Match the following columns.

	Column I		Column II
A.	P-wave	1.	Ventricular depolarisation followed by ventricular contraction
B.	QRS complex	2.	Atrial depolarisation followed by systole of both atria
C.	T-wave	3.	Ventricular repolarisation followed by ventricular relaxation

A. ____________ B. ____________ C. ____________

109. The electrocardiograph (monitoring machine) makes the sound _________pip_________pip________peeeeeeeeeeee as the patient goes into cardiac arrest.

True ☐ False ☐

110. By counting the number of <u>PQ</u> complexes that occur in a given time period, the heartbeat rate of a person can be determined.
(*Rewrite the statement by changing the underlined word.*)

____

111. To obtain a standard ECG, a patient is connected to the machine with three electrical leads at three locations, *viz.* ____

Topic 4 : Double Circulation and Regulation of Cardiac Activity

112. Give the scientific term for the following sentence.
'The tubes that conduct the blood in our body'.

____

113. Identify *A, B* and *C* in the given diagram.

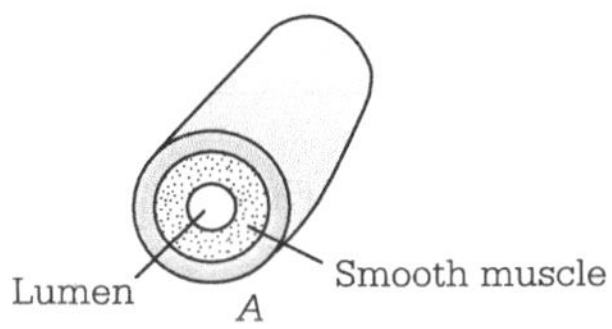

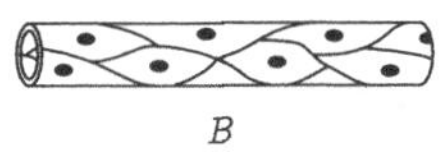

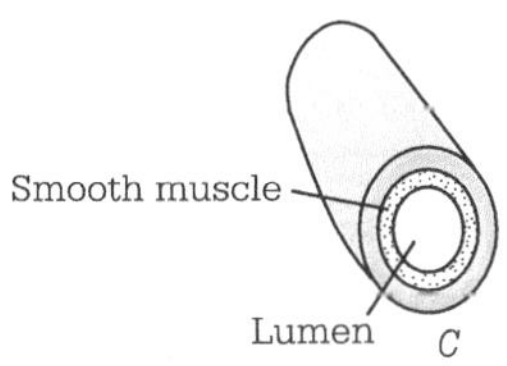

A. ____________________________________

B. ____________________________________

C. ____________________________________

114. Match the following columns.

	Column I (Layers of blood vessels)		Column I (Compositions)
A.	Tunica intima	1.	Fibrous connective tissue with collagen fibres
B.	Tunica media	2.	Smooth muscle fibres
C.	Tunica externa	3.	Squamous epithelium

A. ________________ B. ________________ C. ________________

115. Veins differ from artery in possessing ____.

116. Given below are differences between 'X' and 'Y'. Identify 'X' and 'Y'.

X	***Y***
Blood flow is fast.	Blood flow is slow.
Valves absent.	Valves present.
Walls are thick.	Walls are thin.

X. ________________ Y. ________________

117. What does double circulation mean?
(i) Blood passes through heart twice.
(ii) Heart pumps oxygenated blood, *via.* two major vessels.

____

118. Complete the equation.
Double circulation = ________________________ + ____________________________ .

119. Given below are two pathways of blood circulation. Which of the following represent pulmonary circulation and systemic circulation?

(i) Left ventricle $\xrightarrow[\text{blood}]{\text{Oxygenated}}$ Tissue $\xrightarrow[\text{blood}]{\text{Deoxygenated}}$ Right auricle

____

(ii) Right ventricle $\xrightarrow[\text{blood}]{\text{Deoxygenated}}$ Lungs $\xrightarrow[\text{blood}]{\text{Oxygenated}}$ Left auricle

____

120. Name two blood vessels possessing semilunar valves.

____

121. Name the type of circulation that exhibit the following features.
(i) Provides nutrients, oxygen and other essential substances to the tissues.
(ii) Take CO_2 and other harmful substances away for elimination.
(iii) Deoxygenated blood is collected by venules, veins and vena cava and emptied into right atrium.
(iv) Oxygenated blood entering aorta is carried by a network of arteries, arterioles and capillaries to the tissue.

____

122. The blood circulation which starts and ends into capillaries is renal circulation.
(*Rewrite the statement by changing the underlined word.*)

____

123. Vascular connection between the digestive tracts and liver is called

____.

124. The hepatic portal vein carries blood from ______________ to the ____________ before it is delivered to the systemic circulation.

125. Identify the parts labelled as *A, B, C, D* and *E* in the given schematic plan of blood circulation in human.

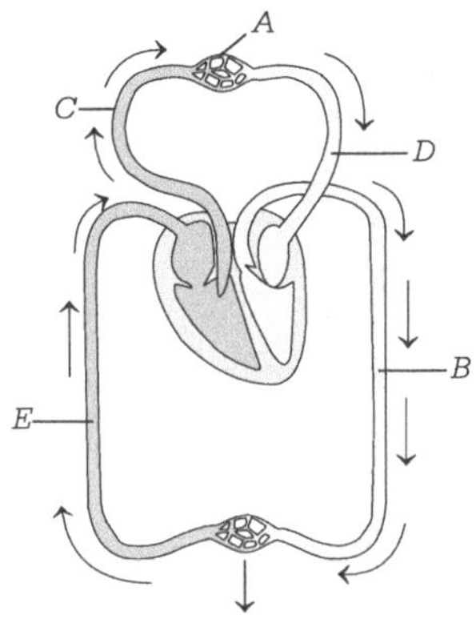

A. ____________________

B. ____________________

C. ____________________

D. ____________________

E. ____________________

126. Name the largest vein of the body. ____________________

127. The given features correspond to which blood vessel in the body?

(i) Narrowest of all blood vessels.

(ii) Permits rapid exchange of materials between blood and tissue.

(iii) Single layer of simple squamous epithelium.

____

128. What is the nature of blood passing through the blood vessels *A, B, C* and *D*?

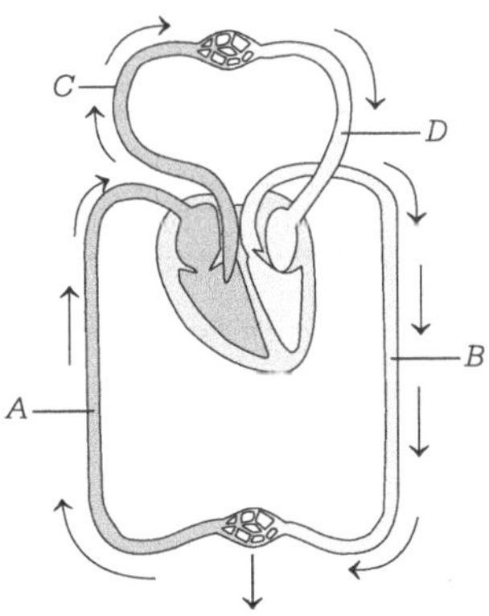

A. ____________________

B. ____________________

C. ____________________

D. ____________________

129. Match the following columns.

	Column I		Column II
A.	Superior vena cava	1.	Carries deoxygenated blood to lungs
B.	Inferior vena cava	2.	Carries oxygenated blood from lungs
C.	Pulmonary artery	3.	Brings deoxygenated blood from lower part of the body to right atrium
D.	Pulmonary vein	4.	Brings deoxygenated blood from upper part of the body to right atrium

A. __________ B. __________ C. __________ D. __________

130. Normal activities of the heart are regulated intrinsically by specialised muscles called __________, hence the heart is called __________.

131. Cardiac centre lies in which part of brain? ____________________

132. Cardiac centre can moderate the cardiac functions through somatic neural system. (*Rewrite the statement by changing the underlined word.*)

____

133. How can neural signal through the sympathetic nerves (part of ANS) increases cardiac output?

(i) By increasing ventricular systole and rate of heartbeat.

(ii) By decreasing ventricular systole and rate of heartbeat.

____

134. Parasympathetic neural signals decrease cardiac output by (increasing/decreasing) speed of conducting action potential.

135. Name the adrenal medullary hormones that can increase the cardiac output.

____

Topic 5 : Disorders of Circulatory System

136. Write the name of the scientific term given the following conditions.

(i) High blood pressure ____________________

(ii) Low blood pressure ____________________

(iii) Defect or disease of the heart from the birth ____________________

(iv) High rate of heartbeat ____________________

(v) Slow rate of heartbeat ____________________

(vi) Heart attack ____________________

(vii) Formation of clot in the coronary artery ____________________

(viii) Chest pain ____________________

(ix) Inadequate flow of blood to a part of the heart ____________________

137. Systolic or Pumping pressure in a normal human is ____________________

138. Which value indicates hypertension or high Blood pressure (BP) in humans?

120/ 80, 110/70, 130/80, 140/ 90

____

139. Given below are the conditions in a person which could lead to different circulatory system disorders. Write the correct disease/disorder name for each condition.

(i) Heart muscle is suddenly damaged by an inadequate blood supply.

____

(ii) Heart does not pump blood effectively due to lung congestion.

____

(iii) When enough oxygen is not reaching to the heart muscle. ____________________

(iv) When the heart stops beating. ____________________

(v) Deposition of calcium, fibrous tissue and lipids on the wall lining the lumen of large and medium-sized arteries.

____

140. Match the following columns.

	Column I (Abnormal ECG)		Column II (Heart problems)
A.	Flat T-wave	1.	Atherosclerotic heart disease
B.	Enlargement of P-wave	2.	Enlargement of atria
C.	S-T segment is elevated	3.	Myocardial infarction
D.	Enlarged Q and R waves		

A. ________ B. ________ C. ________ D. ________

Answers

1. (i) Water
(ii) Blood

2. Fluid matrix, plasma and formed elements.

3. (i) Plasma constitutes 55% of blood
(ii) 90-92% of water in plasma.
(iii) 6-8% of protein in plasma.

4. Alkaline, 7.3–7.5

5. False. Plasma is a straw coloured, viscous fluid in blood.

6. Cellulose is odd one out. Rest all are protein found in blood.

7. A-3, B-1, C-2

8. Na^+, Ca^{2+}, Mg^{2+}, HCO_3^-, Cl^-, etc.

9. '*X*'-Clotting factors

10. Plasma is odd one out. It is not a part of formed elements.

11. Red blood cells, white blood cells, platelets.

12. 2 (ii, iii). Erythrocytes are the most abundant of the cells in the blood.

13. 12-16 gms

14. Biconcave

15. Succinate dehydrogenase is not present in RBCs.

16. Camel and lama

17. (i) Erythropoiesis
(ii) Coagulation
(iii) Haemoglobin

18. Spleen

19. RBCs help in transport of respiratory gases (O_2 and CO_2).

20. 3(i, ii, iv). WBCs are nucleated

21. *A*-Granulocytes, *B*-Agranulocytes, *C*-Neutrophils, *D*-Eosinophils, *E*-Monocytes, *F*-T-lymphocytes, *G*-B-lymphocytes.

22. Monocytes is odd one out. Others are agranulocytes.

23. Basophils

24. Leucopenia

25. A-5, B-4, C-3, D-2, E-1.

26. A-2, B-1, C-4, D-3, E-4.

27. 3 (Neutrophils, lymphocytes and monocytes)

28. *A*-Monocyte, *B*-Eosinophil, *C*-Basophils, *D*-Neutrophil, *E*-Lymphocyte

29. A 1, B 2, C-3

30. Antibodies is odd one out. These are not released from basophils but are released from lymphocytes.

31. Monocyte

32. ABO blood grouping and Rh blood grouping

33. (i) Antigen
(ii) Antibodies

34. 2

35. *W*–A, *X*–Nil, *Y*–B,O, *Z*– O

36. (i) O blood group
(ii) AB blood group

37. A

38. True

39. Monkey

40. 80%

41. 3 (i, ii, iii), i.e. all of the given characteristics are true.

42. Anti Rh antibodies, mother

43. (ii) Nothing happens to Rh (+ve) person.

44. (i) and (iii), i.e. no problem in 1st pregnancy and problem would be expected in second pregnancy.

45. Coagulation or clotting

46. Clot or coagulum

47. Ca^{2+}

48. *A*-Thrombokinase, *B*-Thrombin

49. (iv) → (iii) → (i) → (v) → (ii)

50. Plasmin

51. A-2, B-1, C-4, D-3

52. A-2, B-1, C-4, D-3

53. True

54. Vitamin-K

55. RBCs + some proteins

56. Lymphatic, lymph

57. (i) Colourless
(ii) Lymphocytes present
(iii) Absorbs fats

58. RBCs is odd one out as they are not present in lymph.

59. (i) Arthropods and molluscs
(ii) Annelids and chordates

60. Sinuses

61. (i) and (iii). In closed circulatory system, exchange of material occurs slowly.

62. Sinuses are not present in closed circulatory system.

63. A-1– K, B-2–L, C-3–M, D-4–K

64. True

65. One ventricle

66. Heart, blood vessels, blood

67. (i) Mesodermal
(ii) Thoracic cavity
(iii) Between
(iv) Left

68. Pericardium

69. (i) Opening between left atrium and left ventricle.
(ii) Opening between right atrium and right ventricle.
(iii) Opening of the right and left ventricles into pulmonary artery and aorta, respectively.
(iv) Ventricles of heart.
(v) Right upper corner of the right atrium.
(vi) Lower left corner of the right atrium.

70. (i) Valves allow flow of blood in one direction, i.e. from atria to ventricle and from ventricles to pulmonary artery or aorta.
(ii) Valves prevent backflow of blood.

71. (i) Ventricles pump blood out of the heart to distant tissues therefore their walls are thicker than atria.

72. *A*-Aorta, *B*-Vena cava, *C*-Pulmonary artery, *D*-Left atrium, *E*-Right atrium

73. SA node → AV node → Bundle of His → Purkinje fibres → Heart muscles

74. Foramen ovale, ductus arteriosus

75. Chordae tendineae

76. Ventricles

77. (i) Interatrial septum
(ii) Interventricular septum
(iii) Atrio-ventricular septum

78. Nodal tissue

79. AV node

80. 70–75 min^{-1}

81. It initiates and maintains the rhythmic contractile activity of heart.

82. Sino-atrial (SA) node

83. Bundle of His, Purkinje fibres

84. (ii), i.e. Cardiac muscles will not undergo coordinated rhythmic movements.

85. *A*-Atrial systole. *B*-Ventricular systole, *C*-Joint diastole.

86. Separately

87. Ventricular systole

88. Ventricular systole occur

89. Pulmonary

90. 90%, 10%

91. 4, i.e. i, ii, iii, v (Action potential is not generated during joint diastole).

92. Cardiac cycle

93. True

94. (i) 72 times per minute
(ii) 0.8 S
(iii) 72 cycles per minute
(iv) 70 mL of blood
(v) 5000 mL or 5 L of blood (approximately)

95. (i) Cardiac output
(ii) Stroke volume

96. (i) Stethoscope
(ii) Sphygmomanometer

97. 80 mmHg

98. (i) Pulse
(ii) Heart rate

99. 70 mL
$5250 = \text{stroke volume} \times 75$
$\Rightarrow \text{Stroke volume} = \dfrac{5250}{75} = 70$

100. Lub, dub

101. Lub, dub

102. (i) Dub
(ii) Lub

103. *A*–Atrial systole, *B*–Ventricular systole, *C*–Ventricular diastole

104. Electrocardiogram

105. Electrocardiograph

106. ECG is graphical representation of the electrical activity of the heart during a cardiac cycle.

107.

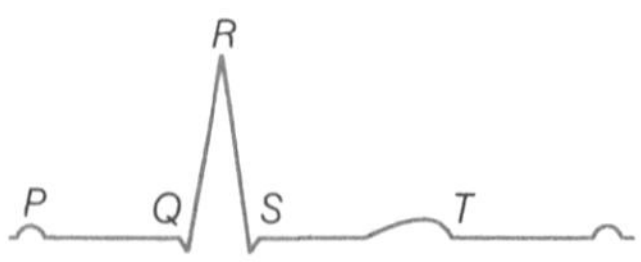

108. A-2, B-1, C-3

109. True

110. QRS

111. One to each wrist and to the left ankle

112. Blood vessels

113. *A*-Artery, *B*-Capillary, *C*-Vein

114. A-3, B-2, C-1

115. Thin tunica media

116. *X*-Arteries, *Y*-Veins

117. (i) It means that the blood passes through the heart twice for each circuit of the body.

118. Double circulation = Pulmonary circulation + Systemic circulation

119. (i) Systemic circulation
(ii) Pulmonary circulation

120. Aorta and pulmonary artery

121. Systemic circulation

122. Portal circulation

123. Hepatic portal system

124. Intestine, liver

125. *A*–Lungs, *B*–Dorsal aorta, *C*–Pulmonary artery, *D*–Pulmonary vein, *E*–Vena cava

126. Inferior vena cava

127. Capillaries

128. *A*–Deoxygenated, *B*–Oxygenated, *C*–Deoxygenated, *D*–Oxygenated

129. A-4, B-3, C-1, D-2

130. Nodal tissue, myogenic

131. Medulla oblongata

132. Autonomic nervous system

133. (i) By increasing the rate of heartbeat and strength of ventricular contraction.

134. Decreasing

135. Epinephrine

136. (i) Hypertension
(ii) Hypotension
(iii) Congenital heart disease
(iv) Tachycardia
(v) Bradycardia
(vi) Myocardial infarction
(vii) Coronary thrombosis
(viii) Angina pectoris
(ix) Myocardial ischemia

137. 120 mm Hg

138. 140/90

139. (i) Heart attack
(ii) Heart failure (Congestive heart failure)
(iii) Angina pectoris
(iv) Cardiac arrest
(v) Atherosclerosis (Coronary artery disease)

140. A-1, B-2, C-3, D-3

CHAPTER NINETEEN

EXCRETORY PRODUCTS AND THEIR ELIMINATION

Topic 1 : Human Excretory System

1. The process of excretion involves the removal of ______________________________ .

2. How many of the given molecules are excreted from the animal body?

Ammonia, Urea, Uric acid, Carbon dioxide, Oxygen, Water, Na^+, K^+, Cl^-, Phosphate, Sulphate

____

3. Ammonia and urea are waste products derived from the metabolic breakdown of ______________________ (carbohydrates/proteins).

4. Give common term for the following.

Ammonia, Urea, Uric acid

____

5. Complete the following analogy given below.

_________________ : Most toxic nitrogenous waste : : _________________ : Least toxic nitrogenous waste.

6. Identify *A, B* and *C* in the given table.

Type of animals	Excretory wastes	Representative animal groups
Ammonotelic	Ammonia	(*B*)
Ureotelic	Urea	(*C*)
Uricotelic	(*A*)	Reptiles and birds

A. ____________ B. ____________ C. ____________

7. Given below are animal species. Mention the type of excretory process shown by them.
   (i) Frog, Human, Alligators ____________
   (ii) Pigeon, Lizard, Cockroach ____________
   (iii) Rohu, *Clarias*, *Pila*, Prawn ____________

8. Ammonia is insoluble and is generally excreted by active transport through gill surfaces as ammonium ions.

   True ☐ False ☐

9. Kidneys do not play any significant role in the removal of ammonia.

   True ☐ False ☐

10. Ammonia is converted to urea in which part of the body? ____________

11. Terrestrial animals do not excrete ammonia to ____________ (conserve water/conserve energy)

12. Nitrogenous wastes which require large amount of water and least amount of water for their excretion are ________ and ________, respectively.

13. Excretion of nitrogenous waste in the form of pellet or paste is known as ammonotelism. (*Rewrite the statement by changing the underlined word.*)

    ____

14. Match the following columns.

	Column I (Excretory structures/organs)		Column II (Animals)
A.	Protonephridia	1.	Prawn
B.	Nephridia	2.	Cockroach
C.	Malpighian tubules	3.	Earthworm
D.	Green gland or Antennal gland	4.	Flatworms

A. ________ B. ________ C. ________ D. ________

15. Give the common excretory structure for the following.

    Platyhelminthes , Rotifers, Some annelids, Cephalochordates

    ____

16. How many of the listed parts of excretory system are found in humans?

    Kidneys, Ureters, Urinary bladder, Urethra.

    ____

17. Each kidney of an adult human measures
    (i) In length ________
    (ii) In width ________
    (iii) In thickness ________
    (iv) In weight ________

18. In humans, kidneys are reddish-brown and ________ shaped structure.

19. Kidneys are situated close to the dorsal inner wall of the abdominal cavity between the ________ and ________ vertebra.

20. Towards the centre of the inner concave surface of the kidney, a notch is present which is called ________ .

21. Name the described part in each statement.
    (i) Part of kidney which is the gateway for ureter, nerves and blood vessels. ________
    (ii) Inner to the hilum of kidney is a broad funnel-shaped space. ________
    (iii) Projections in renal pelvis. ________

22. Identify the following figure *X* and label its parts *A* to *E*.

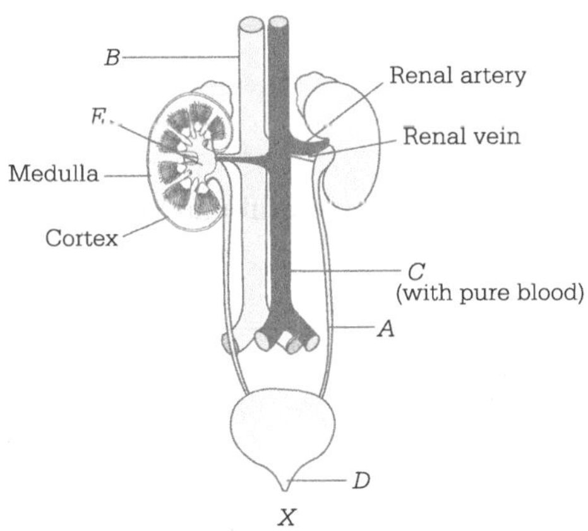

A. ________
B. ________
C. ________
D. ________
E. ________
X. ________

23. Recognise the figure *X* and label the parts *A* to *F*.

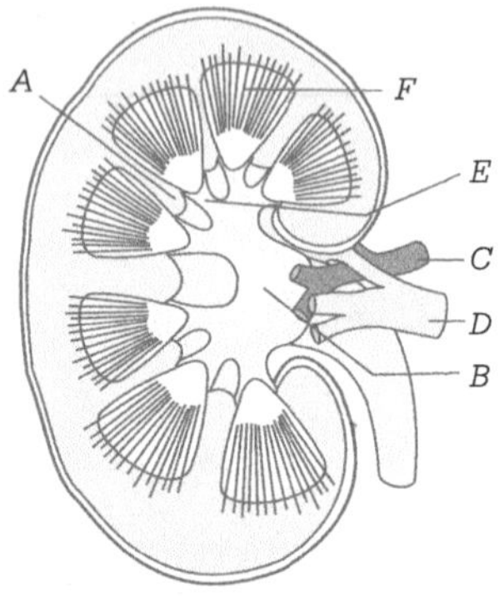

X

A. ____________________

B. ____________________

C. ____________________

D. ____________________

E. ____________________

F. ____________________

X. ____________________

24. The outer layer of human kidney is composed of a tough ____________ . Inside the kidney, the two zones, an outer ____________ and an inner ____________ is present.

25. Columns of Bertini in the kidneys of mammals are formed as extensions of ____________ (cortex in pelvis/ medulla in pelvis/cortex in medulla).

26. How many conical masses or medullary pyramids are present in kidney?

____

27. Arrange the given structures of mammalian kidney from outside capsule till the ureter.

Renal capsule, Renal pelvis, Ureter, Renal papilla, Minor calyx, Major calyx, Medullary pyramid, Cortex

____

28. The structural and functional unit of human kidney is ____________ .

29. Each kidney has nearly ____________ million complex tubular structures called nephrons.

30. Name the two major parts of a nephron. ____________________

31. Complete the flowchart depicting the components of nephron.

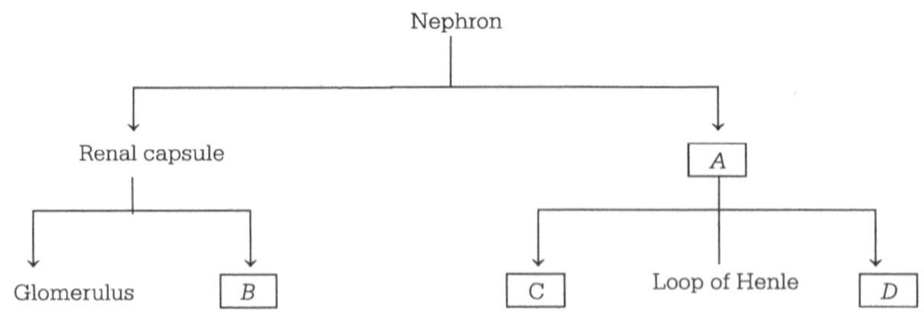

A. ____________ B. ____________ C. ____________ D. ____________

32. Complete the following analogy with respect to glomerulus.

Blood enters : : Afferent arteriole : : Blood leaves : : ____________________

33. Renal corpuscle is also known as ____________________.

34. A diagrammatic representation of a nephron showing blood vessels, ducts and tubule. How many of its parts are correctly labelled? Also write the correct labels for the incorrect ones.

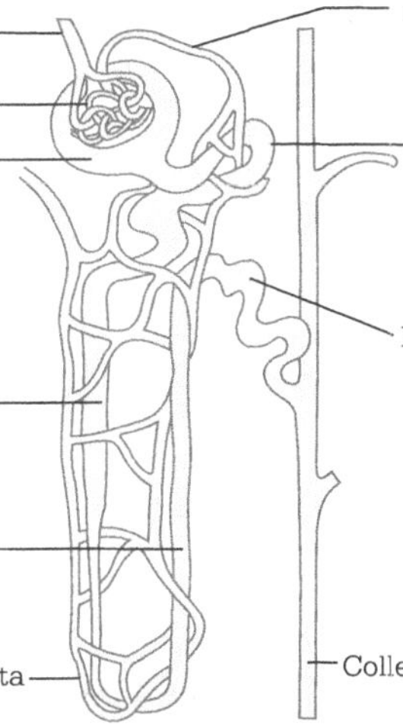

____

35. Give the full form of the following.

(i) PCT ____________________

(ii) DCT ____________________

36. Complete the flowchart depicting the flow of tubular fluid in the nephron.

Bowman's capsule
↓
A
↓
B
↓
C
↓
D
↓
Collecting duct

A. ____________________

B. ____________________

C. ____________________

D. ____________________

37. Complete the flowchart depicting the passage of urine in humans.

Collecting tubule
↓
A
↓
B
↓
C
↓
Outside the body

A. ____________________

B. ____________________

C. ____________________

38. Identify the structure based on the given description in a nephron.

(i) A tuft of capillaries formed by the afferent arteriole. ________________

(ii) A fine branch of renal artery. ________________

(iii) Double-walled cup-like structure enclosing glomerulus. ________________

(iv) A hairpin-shaped part of renal tubule having ascending and descending limb. ________

39. The following diagram represents the Malpighian body. Answer the questions based on it.

(i) Label parts *A* to *D*.

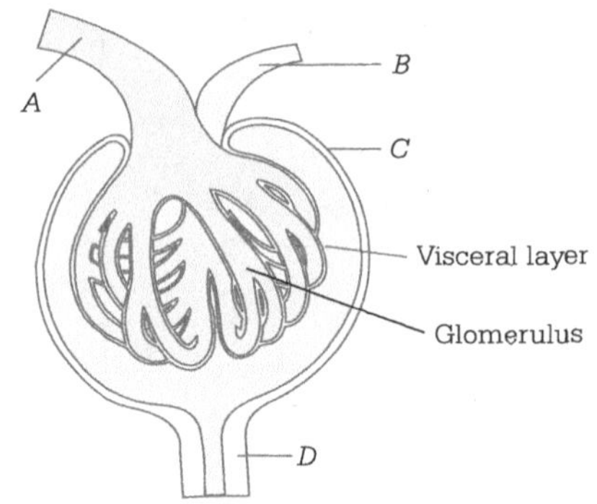

A. ________________

B. ________________

C. ________________

D. ________________

(ii) Name the special cells forming the visceral layer in the above diagram. ________________

(iii) What process takes place in the glomerulus?________________

40. Many collecting ducts converge and through medullary pyramids in the calyces open into the renal pelvis.

True ☐ False ☐

41. Name the two types of nephron present in human kidneys.

____

42. Choose the odd one out with respect to structure in cortical region of kidney.

PCT, DCT, Malpighian corpuscle, Loop of Henle

____

43. Name the type of nephron in which

(i) Loop of Henle is too short and extends very little into the medulla. ________________

(ii) Loop of Henle is very long and runs deep into the medulla. ________________

44. The efferent arteriole emerging from the glomerulus forms a fine capillary network around the renal tubule called ________________ capillaries.

45. Choose the odd one out with respect to structure found in cortical region.

Bowman's capsule, PCT, DCT, Vasa recta

____

46. Vasa recta run parallel to the ________________ .

47. Match the following columns.

Column I		Column II
A. Ureters	1.	Canal-like structure extending from the neck of the bladder to exterior
B. Urinary bladder	2.	Paired, narrow, tubular structures to carry urine from kidneys
C. Urethra	3.	Pear-shaped muscular sac-like structure in the pelvic cavity.

A. ________________ B. ________________ C. ________________

48. Sphincter muscles are present in which of the following structures?

Ureters, Urinary bladder, Urethra

____

Topic 2 : Urine Formation and Functions of Tubules

49. Urine formation involves three processes. Name them.

(i) ____

(ii) ____

(iii) ____

50. The part of nephron which performs the first step in urine formation is ______________.

51. On an average ________________ mL of blood is filtered by the kidneys per minute which constitute roughly ________________ (1/3rd 1/5th) of the blood pumped out the each ventricle of heart in a minute.

52. The glomerular capillary blood pressure causes filtration of blood through 3 layers, i.e.

(i) ____

(ii) ____

(iii) ____

53. $'X' = \dfrac{\text{Amount of filtrate formed by the kidney}}{\text{Minute}}$

What is 'X' in the above formula?

X. ____

54. Give the scientific term given to the epithelial cells of Bowman's capsule.

____

55. Filtration slits or slit pores present between podocytes allow filtration of blood, that filter almost all the constituents of the plasma except proteins.

True ☐ False ☐

56. In the nephron, the blood from the afferent arteriole filters into the lumen of the ________________________. This process is called as ultrafiltration.

57. Give the full forms of the following.

(i) JGA ____________________

(ii) GFR ____________________

(iii) ADH ____________________

(iv) ANF ____________________

58. GFR in a healthy individual is approximately

(i) __________ mL/minute

(ii) ____________ L /day

59. The kidney have built in mechanisms for the regulation of GFR. One such efficient mechanism is carried out by ____________________ (ANF/PNS/JGA).

60. JGA is a special sensitive region formed by cellular modification in the PCT and efferent arteriole at the location of their contact.

(*Rewrite the statement by changing the underlined words.*)

____

61. A fall in GFR can activate the ______________ to release ______________ hormone which can stimulate the glomerular blood flow and thereby the GFR back to normal.

62. Of the total filtrate, nearly how much (in per cent) of it is reabsorbed by the renal tubule?

____

63. Name the cells present in different segments of nephrons which perform reabsorption.

____

64. Choose the odd one out for the substance which are actively reabsorbed in the nephron.

Glucose, Amino acids, Na^+ ions, Nitrogenous wastes

____

65. During urine formation, the tubular cells secretes substances into the filtrate. How many of the given substances are secreted into the filtrate?

K^+, H^+, Ammonia, Water

____

66. Tubular secretion is an important step in urine formation because
(i) it maintains ionic and acid-base balance of body fluids.
(ii) it increases the amount of urine to be excreted out of the body.

____

67. Choose the odd one out with respect to tubules of the nephron.

PCT, Henle's loop, Collecting duct, Bowman's capsule

____

68. Match the following columns.

Column I	Column II
A. Proximal convoluted tubule	1. Formations of concentrated urine
B. Distal convoluted tubule	2. Reabsorption of large amount of water
C. Henle's loop	3. Reabsorption of 70-80% of electrolyte
D. Collecting duct	4. Ionic balance

A. __________ B. __________ C. __________ D. __________

69. Name the epithelial tissue lining PCT. __________

70. Given below are some substances in the PCT which are either reabsorbed or being filtered. Identify them.

HCO_3^-, NaCl, H_2O, Nutrients, K^+, H^+, NH_3^+

(i) Reabsorbed in the PCT. __________

(ii) Selective secretion into the filtrate. __________

71. If loop of Henle were absent from mammalian nephrons, the urine would be more __________ (dilute/concentrated).

72. Choose the correct option for the given statements about loop of Henle.

(i) Reabsorption is __________ (minimum/maximum) in this segment.

(ii) It plays __________ (significant/insignificant) role in maintaining high osmolarity of medullary interstitial fluid.

73. Write the correct name of limb of Henle's loop.

	Water	Electrolytes
Impermeable	*A*	*B*
Permeable	*C*	*D*

A. __________ B. __________ C. __________ D. __________

74. In the Henle's loop, as the concentrated filtrate pass upward, it gets diluted due to the passage of electrolytes to the __________.

75. Conditional reabsorption of Na^+ and water takes place in the __________ (DCT/PCT).

76. The long duct of collecting tubule extends from the cortex of the kidney to the inner parts of the medulla.

True ☐ False ☐

77. Identify the regions *A*, *B*, *C* and *D* in the given figure and choose the correct statement related to them.

(i) Large amount of water is reabsorbed from this region to produce concentrated urine.

(ii) It is capable of reabsorption of HCO_3^- and selective secretion of H^+ and K^+ ions to maintain pH and Na-K balance in the blood.

(iii) Vitamins, hormones, chloride ions, etc., are reabsorbed from the filtrate here.

(iv) In this tubule, the filtrate first becomes hypertonic to blood plasma and then becomes hypotonic to blood plasma.

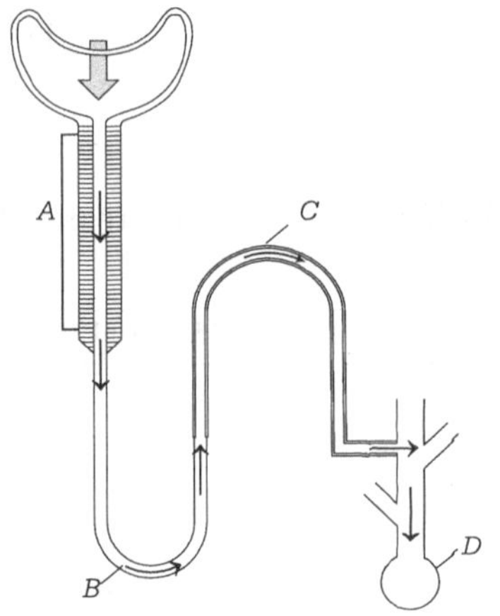

A. ____________

B. ____________

C. ____________

D. ____________

78. Give the scientific term for the process mentioned below as '*X*' and '*Y*'.

'*X*'	'*Y*'
It is absorption of selected material form the nephric filtrate into the blood of peritubular blood capillaries.	It is removal of selected material from the blood of the peritubular blood capillaries into the nephric filtrate.
It takes place by diffusion and active process.	It takes place by active transport.

X. ____________ Y. ____________

79. Collecting duct allows passage of small amounts of urea into the medullary interstitium to keep up the osmolarity.

True ☐ False ☐

80. Study the figure and write the correct names for *A* to *E* substances from the following.

H^+, H_2O, Urea, NaCl, HCO_3^-

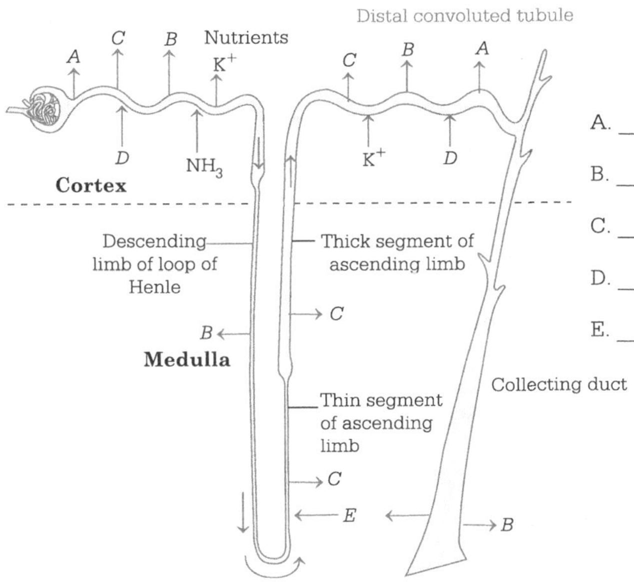

A. ____________

B. ____________

C. ____________

D. ____________

E. ____________

Topic 3 : Mechanism of Concentration of the Filtrate

81. Mammals have the ability to produce ____________ (hypotonic/hypertonic) urine.

82. Out of the following segments, which of the segments play an important role in counter current mechanism?

PCT, Vasa recta, DCT, Henle's loop, Collecting duct

____

83. The osmolarity (in mOsmol L^{-1}) in the kidney in

(i) outer cortex ____________

(ii) inner medulla ____________

84. How many factors help in maintaining an increased osmolarity towards the inner medullary interstitium?

(i) Counter-current pattern in vasa recta.

(ii) Counter-current pattern in Henle's loop.

(iii) Proximity between the Henle's loop and vasa recta.

____

85. Choose the odd one out with respect to the substances that help to maintain osmotic gradient between cortex and medulla regions.

Urea, Sodium Chloride, NH_3

____

86. Fill in the blanks.

(i) In counter-current mechanism, NaCl is transported by the ____________ limb of Henle's loop which is exchanged with the ____________ limb of vasa recta.

(ii) NaCl is returned to the interstitium by ascending portion of ____________ .

(iii) Small amount of urea enters the thin segment of ____________ limb of Henle's loop which is transported back to the interstitium by the ____________ .

87. The counter-current mechanism helps to

____

88. Human kidneys can produce urine nearly ____________ (twenty/five) times concentrated than the initial filtrate formed.

89. Under normal condition, which one is completely reabsorbed in the renal tubule?

Urea, Salts, Uric acid, Glucose, H_2O, Na^+ions, K^+ ions

____

90. In the given diagrammatic representation of counter-current mechanism, what is the osmolarity (mOsmol L^{-1}) at *P, Q, R* and *S* regions?

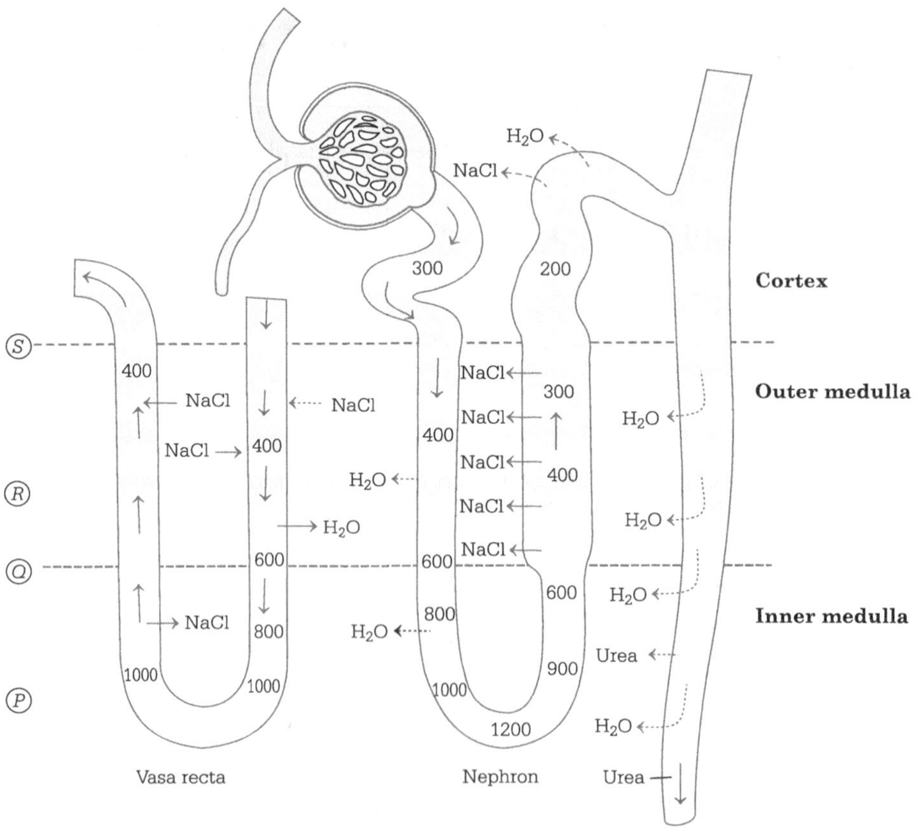

P. ________ Q. ________ R. ________ S. ________

91. Match the following columns.

	Column I		Column II
A.	Ultrafiltration	1.	Henle's Loop
B.	Concentration of urine	2.	Ureter
C.	Transport of urine	3.	Urinary bladder
D.	Storage of urine	4.	Malpighian corpuscle

A. ____________ B. ____________ C. ____________ D. ____________

Topic 4 : Regulation of Kidney Function

92. Choose the odd one out with respect to the parts monitoring the functioning of the kidneys.

JGA, Vasa recta, Hypothalamus, Heart

____

93. Osmoreceptors in the body are activated by changes in ____________ volume, ____________ volume and ____________ concentration.

94. Complete the sequence for the regulation of kidney function.

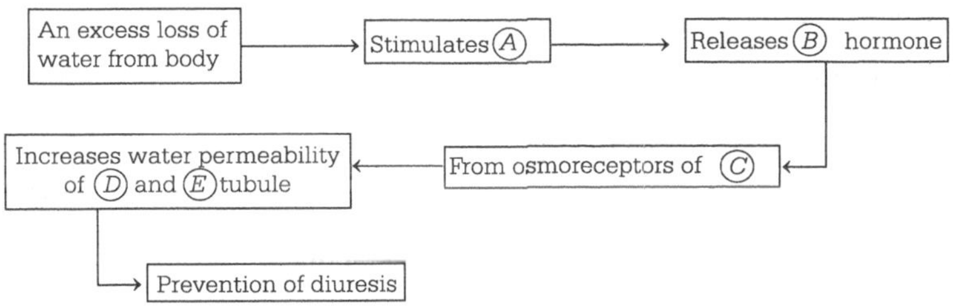

A. ____________ B. ____________ C. ____________

D. ____________ E. ____________

95. Osmoregulation is the function of ____________ (oxytocin/vasopressin).

96. An increase in body fluid volume can switch on the osmoreceptors and suppress ADH release.

True ☐ False ☐

97. Complete the flowchart.

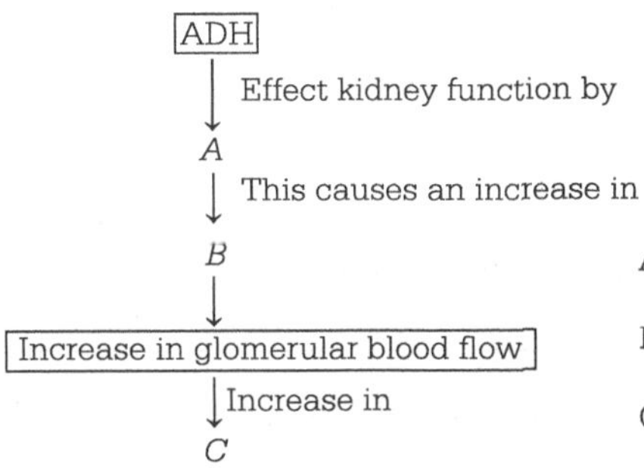

A. ____________

B. ____________

C. ____________

98. What cause the JG cells to release renin?

(i) Fall in GFR and blood pressure.

(ii) Increased GFR and blood pressure.

____

99. Complete the flowchart.

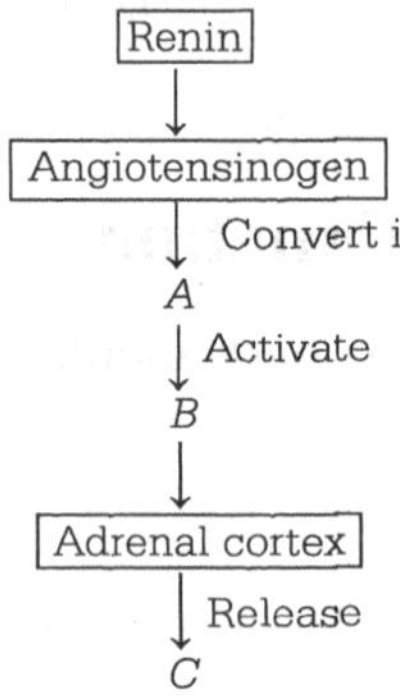

A. ____________________

B. ____________________

C. ____________________

100. Give the two functions of angiotensin-II.

____

101. Choose the correct option, if vasoconstrictions occur in the body.

(i) Glomerular blood pressure ____________________ (increase/decrease)

(ii) GFR ____________________ (increase/decrease)

(iii) Aldosterone secretion ____________________ (starts/stops)

(iv) ______________ (Reabsorption/Secretion) of Na^+ and water in the DCT occurs.

102. Give the full form of RAAS.

____

103. Renin-angiotensin mechanism controls ______________ (ultrafiltration/blood pressure).

104. Choose the odd one out in each of the following groups.

(i) Renal pelvis, Medullary pyramid, Renal cortex, Renal papilla. ______________

(ii) Afferent arteriole, Henle's loop, Vasa recta, Efferent arteriole. ______________

(iii) Glomerular filtration, Antidiuretic hormone, Hypertonic urine, Collecting duct. ________

105. <u>Renin</u> causes reabsorption of Na^+ and water from the distal parts of the tubule.
(*Rewrite the statement by changing the underlined word.*)

____

106. Angiotensinogen is a protein produced and secreted by <u>JG cells</u>.
(*Rewrite the statement by changing the underlined word.*)

____

107. Name the peptide hormone which opposses the regulation of RAAS.

____

108. Complete the flowchart.

Increased blood pressure
↓
ANF releases from the (A)
↓
Causes (B)
↓
Decreases blood pressure

A. ______________________

B. ______________________

109. Give a common term for following.

Angiotensinogen-II, ANF, ADH, Aldosterone

____

Topic 5 : Micturition

110. Give the scientific term used for the expulsion of urine from the urinary bladder.

____

111. Micturition occurs when ______________ (urethra relaxes/urethra contracts).

112. Given below is the outline of principal event of urination. Arrange them is ascending order.

(i) Stretch receptors on the wall of urinary bladder send signal to the CNS.

(ii) The bladder fills with urine and becomes distended.

(iii) Micturition.

(iv) CNS passes on motor messages to initiate the contraction of bladder.

____

113. During urination, contraction of smooth muscles of the bladder and relaxation of urethral sphincter causing the release of urine.

True ☐ False ☐

114. Name the specific smooth muscle of the bladder causing urination.

____

115. The neural mechanism that causes urination is called ______________ .

116. Urine is a light ____________ coloured watery fluid which is slightly ____________ in pH.

117. Answer the following questions in numbers only.

(i) The pH of a normal human urine. ______________

(ii) The amount of urine excreted by an adult per day in litres. ______________

(iii) The amount of urea excreted with urine per day (in gm) in an adult.______________

118. The colour of urine is caused by the pigment ______________ in it.

119. Urine contains 95% of ______________ and only 5% of ______________ in it.

120. There is no ______________ (vitamin/glucose) normally found in the urine.

Topic 6 : Role of Other Organs in Excretion

121. Give the common function of lungs, kidneys, liver and skin.

____

122. How much CO_2 is removed per minute by human lungs?

____

123. ______________ is the largest gland of the body, which functions as both secretory and excretory organs.

124. How many of the given substances are excreted from the liver.

Bile, Degraded steroid hormones, Vitamins, Drugs, Cholesterol, NaCl, Urea, Lactic acid

____

125. Bile comprises of two pigments, namely ______________ and ______________

126. Sweat glands and sebaceous glands are common in their nature as they both have. (*Give their common feature*) ______________

127. Complete the following analogy given below.

______________ : Secretes watery fluid with NaCl : : ______________ : Secretes sebum.

128. What is the composition of sweat?

____

129. Choose the odd one out with respect to composition of sebum.

Sterols, Lactic acid, Hydrocarbons, Waxes

____

130. Give the function of the following.

(i) Sweat ______________

(ii) Sebum ______________

Topic 7 : Disorders of the Excretory System

131. Given below are some abnormalities in the urine. Give the scientific term against each.

(i) Presence of excess urea in urine. ______________

(ii) Presence of abnormally high ketone bodies in urine. ______________

(iii) Presence of glucose in urine. ______________

(iv) Presence of albumin in urine. ______________

(v) Presence of blood cells in urine. ______________

(vi) Presence of haemoglobin in urine. ______________

(vii) Presence of WBCs or pus in the urine. ______________

132. Which symptom(s) is/are likely to be found in person suffering from diabetes mellitus?

Hematuria, Ketonuria, Glycosuria, Pyuria

____

133. Give common names for the kidney disorders mentioned below.

(i) Glomerulonephritis ______________________________

(ii) Renal calculi ______________________________

134. During diuresis in a person,

(i) volume of excreted urine increases.

(ii) volume of excreted urine decreases.

____

135. Give scientific term for following sentence.
An inflammation of the renal pelvis and medullary tissue of the kidney.

____

136. Choose the correct option for the following.

(i) During Bright's diseases, inflammation of the ____________ (glomerulus/PCT) occurs.

(ii) Chemical composition of renal calculi, besides uric acid is ______________________ (barium chloride/calcium oxalate).

(iii) Diabetes insipidus is characterised by ____________ (concentrated/dilute) urine.

(iv) Diabetes insipidus is caused due to the deficiency of (ADH/ANF). ______________

137. Uremia is highly harmful and may lead to ______________________________.

138. Name the process of removal of urea from the blood of a person with kidney failure.

____

139. What is the ultimate method in the correction of acute renal failure?

____

140. Dialysing unit (artificial kidney) contains a fluid which is almost same as plasma except that it has ______________________________

141. How many assumptions are correct with respect to kidney transplant?

(i) It is the ultimate method where drug or dialysis do not help.

(ii) Close relatives are often used as kidney donors to minimise risk of rejection.

(iii) Cyclosporin-A is used as immunosuppressive agent during kidney transplant.

____

142. Following are the steps of dialysis. Arrange them in proper order.

(i) Blood is passed into a vein.

(ii) Blood is mixed with heparin.

(iii) Blood is mixed with anti-heparin.

(iv) Blood is drained from a convenient artery.

(v) Blood is passed through coiled and porous cellophane tube bathing in dialysis fluid.
(vi) Removal of nitrogenous wastes from blood.

---

143. An X-ray of the lower abdomen shows a shadow in the region of the ureter suspected to be an ureteric calculus. A possible clinical symptom would be.

Anuria, Hematuria, Renal failure, Uremia

Answers

1. Metabolic wastes from the body.

2. 10 (All of them except oxygen). Oxygen is not excreted out.

3. Proteins

4. Nitrogenous wastes in animals.

5. Ammonia, uric acid

6. *A*-Uric acid, *B*-Aquatic amphibians, bony fishes and aquatic insects, *C*-Mammals, terrestrial amphibians and marine fishes.

7. (i) Ureotelism, (ii) Uricotelism (iii) Ammonotelism

8. False. Ammonia is readily soluble and is generally excreted by diffusion across body surfaces or through gill surfaces (in fish) as ammonium ions.

9. True

10. Liver

11. Conserve water

12. Ammonia, uric acid

13. Uricotelism

14. A-4, B-3, C-2, D-1

15. Protonephridia

16. 4, i.e. All of the given structures.

17. (i) 10-12 cm, (ii) 5-7 cm (iii) 2-3 cm, (iv) 120-170 g

18. Bean

19. T_{12} (last thoracic) L_3 (third lumbar)

20. Hilum

21. (i) Hilum
(ii) Renal pelvis
(iii) Calyces

22. The figure '*X*' is of human urinary system.
A-Ureter, *B*-Inferior vena cava, *C*-Dorsal aorta, *D*-Urethra, *E*-Pelvis

23. The figure '*X*' is the diagrammatic representation of LS of kidney in humans. *A*-Renal column, *B*-Renal pelvis, *C*-Renal artery, *D*-Renal vein, *E*-Calyx, *F*-Medullary pyramid

24. Renal capsule, cortex, medulla

25. Cortex in medulla

26. 15-16 approximately

27. Renal capsule → Cortex → Medullary pyramid → Renal papilla → Minor calyx → Major calyx → Renal pelvis → Ureter

28. Nephron

29. One

30. Glomerulus and renal tubule

31. *A*-Renal tubule, *B*-Bowman's capsule, *C*-Proximal Convoluted Tubule (PCT), *D*-Distal Convoluted Tubule (DCT).

32. Efferent arteriole

33. Malpighian body

34. 6 parts are correctly labelled, *viz. C, D, E, F, I, J* (*A*-Efferent arteriole, *B*-Afferent arteriole, *G*-PCT, *H*-DCT)

35. (i) PCT - Proximal Convoluted Tubule
(ii) DCT - Distal Convoluted Tubule

36. *A*-Proximal convoluted tubule, *B*-Descending limb of loop of Henle, *C*-Ascending limb of loop of Henle, *D*-Distal convoluted tubule.

37. *A*-Ureter, *B*-Urinary bladder, *C*-Urethra

38. (i) Glomerulus
(ii) Afferent arteriole
(iii) Bowman's capsule
(iv) Henle's loop

39. (i) *A*-Afferent arteriole
B-Efferent arteriole
C-Bowman's capsule
D-Proximal convoluted tubule
(ii) Podocytes
(iii) Ultrafiltration

40. True

41. Cortical nephron and juxta medullary nephron.

42. Loop of Henle is odd one out as it dips into the medulla region of kidney.

43. (i) Cortical nephron
(ii) Juxtamedullary nephron

44. Peritubular

45. Vasa recta is odd one out as it is absent in cortical region.

46. Henle's loop.

47. A-2, B-3, C-1

48. Urinary bladder

49. (i) Ultrafiltration
(ii) Reabsorption
(iii) Secretion

50. Glomerulus

51. 1100-1200, 1/5th

52. (i) The endothelium of glomerular blood vessels
(ii) A basement membrane
(iii) The epithelium of Bowman's capsule.

53. Glomerular Filtration Rate (GFR)

54. Podocytes

55. True

56. Bowman's capsule

57. (i) Juxta Glomerular Apparatus
(ii) Glomerular Filtration Rate

(iii) Antidiuretic hormone
(iv) Atrial Natriuretic Factor

58. (i) 125 (ii) 180

59. JGA

60. Distal Convoluted Tubule (DCT) and afferent arteriole.

61. JG cells, renin

62. About 99% of the filtrate has to be reabsorbed by the renal tubule.

63. Tubular epithelial cells

64. Nitrogenous wastes is odd one out as these are reabsorbed passively in the nephron.

65. 3 (K^+, H^+, ammonia)

66. (i) It helps in the maintenance of ionic and acid-base balance of the body fluids.

67. Bowman's capsule is odd one out as it is not a tubule.

68. A-3, B-4, C-1, D-2

69. Simple cuboidal brush-bordered epithelium.

70. (i) HCO_3^-, NaCl, H_2O, nutrients
(ii) H^+, NH_3^+, K^+

71. Dilute

72. (i) Minimum
(ii) Significant

73. *A*-Ascending limb
B-Descending limb
C- Descending limb
D-Ascending limb

74. Medullary fluid

75. DCT

76. True

77. *A*-PCT-(iii), *B*-Henle's loop-(iv), *C*-DCT-(ii), *D*-Collecting duct-(i)

78. '*X*' is tubular reabsorption, '*Y*' is tubular secretion.

79. True

80. *A*–HCO_3^-, *B*–H_2O, *C*–NaCl, *D*–H^+, *E*–Urea

81. Hypertonic

82. Vasa recta, Henle's loop

83. (i) 300
(ii) 1200

84. 3, i.e. All of them.

85. NH_3 is the odd one out.

86. (i) Ascending, descending,
(ii) Vasa recta
(iii) Ascending, collecting duct

87. Counter-current mechanism helps to maintain a concentration gradient in the medullary interstitium.

88. Five

89. Glucose

90. *P*-1200 mOsmol L^{-1}, *Q*-900 mOsmol L^{-1},
R-600 mOsmol L^{-1},
S-300 mOsmol L^{-1}

91. A-4, B-1, C-2, D-3

92. Vasa recta is odd one out.

93. Blood, body fluid, ionic

94. *A*-Hypothalamus, *B*-ADH, *C*-Neurohypophysis, *D*-DCT, *E*-Collecting duct

95. Vasopressin

96. False. An increase in body fluid volume can switch off the osmoreceptors and suppress ADH release.

97. *A*-Constriction of blood vessels, *B*-Blood pressure, *C*-GFR

98. (i) A fall in Glomerular Filtration Rate (GFR) or glomerular blood flow and blood pressure.

99. *A*-Angiotensin-I, *B*-Angiotensin-II, *C*-Aldosterone

100. Vasoconstriction and release of aldosterone.

101. (i) Increases, (ii) Increases
(iii) Starts,, (iv) Reabsorption

102. Renin-Angiotensin-Aldosterone System.

103. Blood pressure

104. (i) Renal cortex is odd one out as other structures are found in medulla region.
(ii) Henle's loop is odd one out as it is filtration duct, while other three are blood vessels.
(iii) Glomerular filtration is odd one out as rest options are concerned with kidney function regulation.

105. Aldosterone

106. Liver cells

107. Atrial-Natriuretic Factor (ANF)

108. *A*- Heart (atria), *B*-Vasodilation

109. Hormones

110. Micturition

111. Urethra relaxes

112. (ii) → (i) → (iv) → (iii)

113. True

114. Detrusor muscle

115. Micturition reflex

116. Yellow, acidic

117. (i) 6.0, (ii) 1-1.5 L/day
(iii) 25-30 g/day

118. Urochrome

119. Water, other substances

120. Glucose

121. Help in the elimination of excretory wastes.

122. 200 mL/minute

123. Liver

124. 7 (All of them except NaCl). NaCl is not secreted or excreted from liver.

125. Bilirubin, biliverdin

126. Secretory glands on skin

127. Sweat gland, sebaceous gland

128. Water, NaCl, small amount of urea, lactic acid, etc.

129. Lactic acid is odd one out as it is not a part of sebum.

130. (i) It provides cooling effect on the body surface and also helps in removal of wastes.
(ii) It provides protective oily covering for the skin.

131. (i) Uremia
(ii) Ketonuria
(iii) Glycosuria
(iv) Albuminuria
(v) Hematuria
(vi) Hemoglobinuria
(vii) Pyuria

132. Ketonuria, glycosuria

133. (i) Bright's disease
(ii) Kidney stones

134. (i) Volume of excreted urine increases.

135. Pyelonephritis

136. (i) Glomerulus
(ii) Calcium oxalate
(iii) Dilute
(iv) ADH

137. Kidney failure

138. Hemodialysis

139. Kidney transplantation

140. No urea

141. 3, i.e. All of the given assumptions are correct.

142. (iv) → (ii) → (v) → (vi) → (iii) → (i)

143. Anuria and hematuria

CHAPTER TWENTY

LOCOMOTION AND MOVEMENT

Topic 1 : Types of Movement, Muscle and Mechanism of Muscle Contraction

1. Movement is one of the significant features of both plants and animals.

   True ☐ False ☐

2. Name a unicellular organism which shows movement by streaming of protoplasm.

   ____

3. How many of the listed structures in human body show movement?

Tongue, Ear pinna, Eyelids, Limbs, Jaws

   ____

4. Some movements result in change of place or location. Such voluntary movements are called ______________ .

5. All movements are locomotion, but all locomotions are not movements.

   True ☐ False ☐

6. Complete the following analogy with respect to locomotory structures in different animals.

   Cilia : Flagella : Tentacles : : ______________ : *Euglena* : ______________ .

7. Locomotion performed by animals vary with their ______________ .

8. The purpose of locomotion in living beings is to
   (i) search of food, shelter, mate and climatic conditions.
   (ii) derive metabolic energy for survival.

   ____

9. Name the three types of movements exhibited by cells in human body.

   ____

10. Which movement is responsible for swimming of spermatozoa and maintenance of water current in the canal system of sponges?____________________

11. Choose the odd one out with respect to the type of movement.

(i) Macrophages, Leucocytes, Microfilaments, Jaws ____________________

(ii) Tongue, Hands, Feet, Basophils ____________________

12. Ciliary movement in humans occurs in ________________ organs which are lined by ________________ epithelium.

13. Given below are locations where ciliary movement occurs in human body. Write the functions of this movement.

(i) Trachea ____________________

(ii) Fallopian tube ____________________

14. Name the property of muscle effectively used for locomotion.

____

15. Locomotion requires a perfect coordinated activity of ____________, ____________ and ____________ systems.

16. Origin of muscle is ________________ (ectodermal/endodermal/mesodermal).

17. How much of the body weight is contributed by muscles in humans?

____

18. Muscles exhibit four special properties. Name them.

____

19. The criteria used for classification of muscles is

(i) location, appearance, nature of activity regulation.

(ii) number of myofibrils and their appearance.

____

20. Based on which criteria does muscles are classified as skeletal, visceral and cardiac?

____

21. Smooth muscles is a name given to ____________ muscle.

22. Lactic acid can accumulate in ____________ muscles upon excess use.

23. Complete the following table by filling blanks *A-H*.

Skeletal muscles	Visceral muscles	Cardiac muscles
Associated with skeletal components of body	(A)	(B)
(C)	(D)	Striated
Voluntary	(E)	(F)
(G)	Performs various functions like transport of food through digestive tract	(H)

A. ________ B. ________ C. ________ D. ________

E. ________ F. ________ G. ________ H. ________

24. Name the muscle present in the following structures.

Alimentary canal, Stomach, Reproductive tract

____

25. Write two functions performed by visceral muscles.

(i) ______________________________

(ii) ______________________________

26. Skeletal muscle is made up of a number of muscle bundles known as ____________.

27. A layer surrounding each muscle bundle is called ____________, which is made up of ____________ tissue.

28. Muscle fibres are basically muscle cells present inside each muscle bundle.

True ☐ False ☐

29. Give the scientific terms for the following components of muscle fibre.

(i) Plasma membrane ____________

(ii) Cytoplasm ____________

(iii) Endoplasmic reticulum ____________

30. Anatomical unit of the skeletal muscle is muscle bundle.
(*Rewrite the statement by changing the underlined word.*)

____

31. Muscle fibre is a syncitium due to the presence of ____________.

32. The storehouse of Ca^{2+} ions in muscle fibre is ____________ (sarcolemma/sarcoplasmic reticulum).

33. The characteristic feature of muscle fibre is the presence of large number of myofibril in the sarcoplasm.

True ☐ False ☐

34. In the diagrammatic cross-sectional view of a muscle, label the parts *A* to *D*.

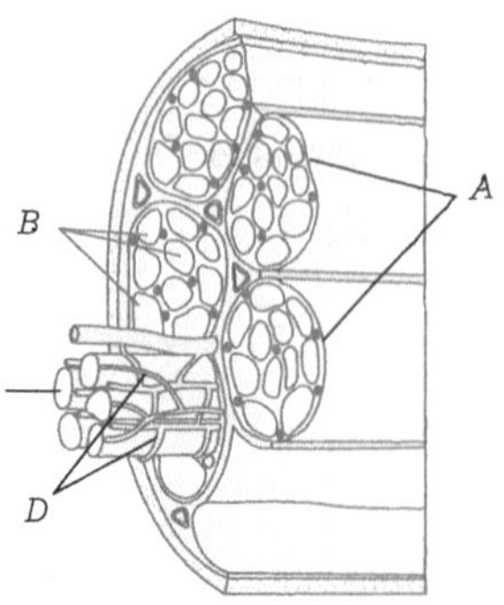

A. ____________

B. ____________

C. ____________

D. ____________

35. Each myofibril has alternate dark and light bands. The light bands contain ____________ protein whereas dark bands contain ____________ protein.

36. The functional unit of contraction of muscle fibre is ________________ .

37. How many features are true for a sarcomere?

(i) These are the sections of myofibrils joined end to end all along the muscle fibre.
(ii) It is the area between two M-lines.
(iii) It comprises of one full A-band and half I-band.

____

38. Match the following columns.

Column I	**Column II**
A. I-band	1. Light band of actin filament
B. A-band	2. Dark band of myosin filament
C. M-line	3. Thin fibrous membrane present in the middle of A-band
D. Z-line	4. It is central part of thick filament not overlapped by thin filament
E. H-zone	5. Elastic fibre bisecting I-band

A. ____________ B. ____________ C. ____________
D. ____________ E. ____________

39. Actin filaments are called ________ filaments whereas myosin filaments are called ________ filaments.

40. The portion of the myofibril between two successive Z-lines is considered as the functional unit of contraction and is called ____________ (sarcosome/sarcomere).

41. Give the full form of the following.

(i) I-band ____________________
(ii) A-band ____________________
(iii) H-zone ____________________

42. In a resting stage, the edges of thin filaments on either side of the thick filaments partially overlap the free ends of the thick filaments thus, leaving the central part of the thick filament.

True ☐ False ☐

43. Label the parts *A* to *G* in the diagrammatic representation of anatomy of a muscle fibre showing a sarcomere.

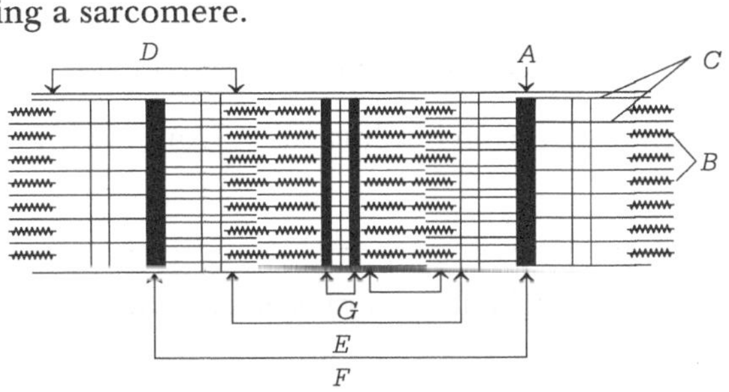

A. ____________
B. ____________
C. ____________
D. ____________
E. ____________
F. ____________
G. ____________

44. Myofilaments consist of mainly four types of proteins, i.e. ____________, ____________, ____________ and ____________.

45. Name the following based on the description given below.

(i) Two contractile proteins which help to shorten muscle fibre.

____

(ii) Two regulatory proteins which determine when the muscle should contract.

____

46. Each actin (thin) filament is made up of two ____________ actins that are helically wound to each other.

47. ____________ actin is a polymer of monomeric ____________ actin.

48. Tropomyosin masks the binding sites for myosin on actin filaments during resting state.

True ☐ False ☐

49. Identify *A, B* and *C* in the given diagram of actin (thin) filament.

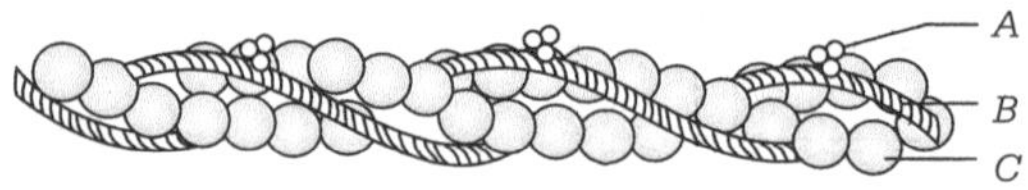

A. ____________

B. ____________

C. ____________

50. Myosin filament is a polymer of monomeric protein called ____________.

51. Identify *A, B, C* and *D* in the given diagram of myosin monomer.

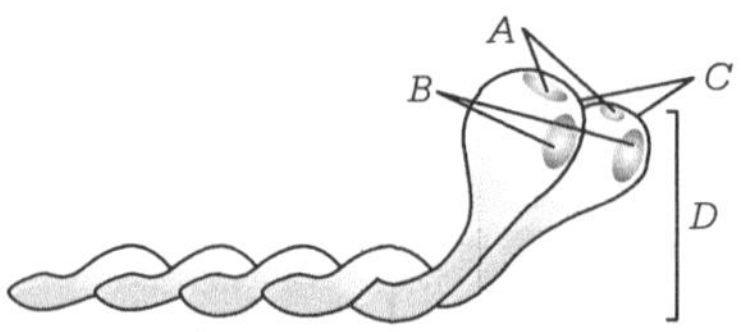

A. ____________

B. ____________

C. ____________

D. ____________

52. Give the full form of the following.

(i) HMM ____________

(ii) LMM ____________

53. The HMM component project outward at regular distance and angle from the surface of polymerised myosin filament and form a ____________.

54. The globular head of HMM is an active ATPase enzyme and it contains.

ATP binding sites, Actin binding sites, Myosin binding sites, Troponin binding sites, Calcium binding sites

____

55. Name the theory that best explain the mechanism of muscle contraction.

____

56. Contraction of muscle fibres takes place by the sliding of the thin filaments over the thick filaments.

True ☐ False ☐

57. Muscle contraction is initiated by a signal sent by CNS, *via.* ________________ neurons.

58. Complete the following equation.
Motor unit = Motor neuron + ____________________

59. Give the scientific term for the following sentence.

The junction between a motor neurons and the sarcolemma of the muscle fibre.

____.

60. When a neural signal reaches neuromuscular junction and releases neurotransmitters ___________________________ is generated.

61. Given below is the sequence of events which causes muscle contraction. Arrange them correctly in the order of their occurrence.

I. Myosin binds with active sites on actin (Cross-bridge formation).

II. Nerve impulse.

III. Release of Ca^{2+} from sarcoplasmic reticulum.

IV. The acetylcholine is released.

V. Sarcomere shorten.

VI. Action potential generated in sarcolemma.

VII. Spread of impulses over sarcolemma on T-tubules.

____

62. During contraction, ___________ bands get shorten whereas ____________ bands retain the length.

63. How many of listed events help in relaxation of muscle fibre?

Hydrolysis of ATP, Pumping of Ca^{2+} ions into sarcoplasmic cisternae, Masking of actin filaments, Z-lines at original position

____

64. Identify *A*, *B*, *C* and *D* in the given diagram of muscle contraction,

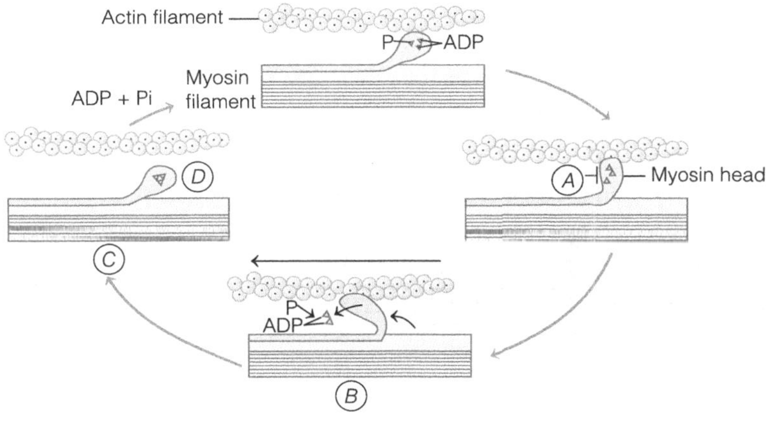

A. ________________

B. ________________

C. ________________

D. ________________

65. Fill in the blanks by choosing the correct pair from the box.

Sarcolemma and sarcoplasm, Myosin head and actin filament Calcium ions and troponin.

(i) Muscle contraction is initiated by the binding of ________________ to ________________.

(ii) Cross-bridge is formed between ________________ and ________________ during muscle contraction.

(iii) Action potential causes calcium ions to diffuse from the ________________ into ________________.

66. Give the scientific term used for the reduction in the force of contraction of muscle after prolonged stimulation. ________________

67. Repeated activation of the muscles lead to lactic acid accumulation thus, causing fatigue due to aerobic breakdown of glucose.
(*Rewrite the statement by changing the underlined word.*)

____

68. A red coloured oxygen storing pigment in muscle is ________________ (myoglobin/ haemoglobin).

69. Complete the following analogy given below.
Aerobic muscle : Anerobic muscle : : Red muscle fibres : ________________.

70. Complete the table by writing more or less in the blanks.

	Characteristics	Red muscle fibres	White muscle fibres
A.	Myoglobin content	________	________
B.	Number of mitochondria	________	________
C.	Number of sarcoplasmic reticulum	________	________
D.	Amount of oxygen stored	________	________

71. Identify the state of the sarcomere in the diagram *A*, *B* and *C* given below.

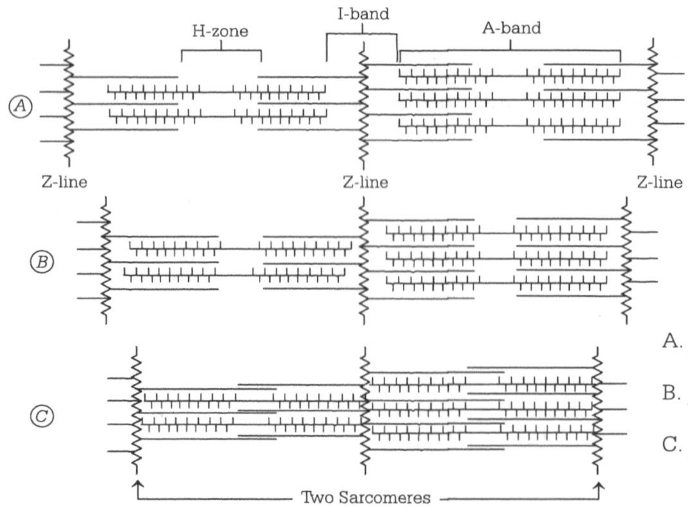

A. ________________

B. ________________

C. ________________

Topic 2 : Skeletal System

72. Which type of skeleton includes hard parts inside the body?

____

73. Choose the correct example for exoskeleton and endoskeleton.

Scales, Feathers, Hairs, Nails, Spicules, Cartilages, Bones

(i) Endoskeleton _______________

(ii) Exoskeleton _______________

74. Complete the following analogy given below.

Bones : Cartilage : : Calcium salts : _______________.

75. _______________ has slightly pliable matrix whereas _______________ has a very hard matrix.

76. Write the number of bones present in axial and appendicular skeleton.

Human Skeletal System

Axial system/skeleton — A. _______________

Appendicular system/skeleton — B. _______________

77. Write the common term for the following.

(i) Skull, Vertebral column, Sternum, Ribs. _______________

(ii) Cranial bones, Facial bones, Hyoid, Ear ossicles. _______________

78. (i) In the given diagram of human skull, identify the bones labelled as *A*-*H*.

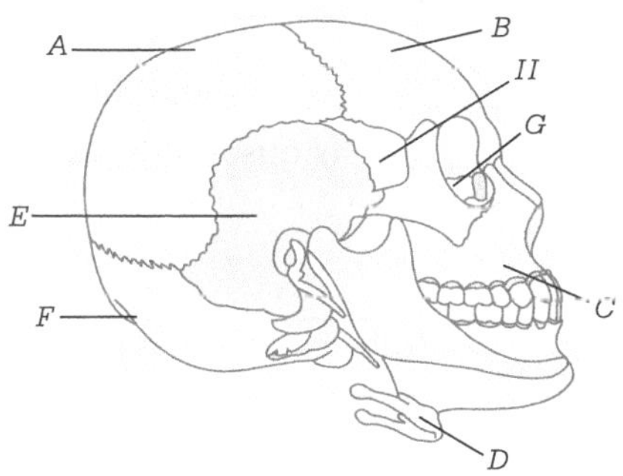

A. _______________

B. _______________

C. _______________

D. _______________

E. _______________

F. _______________

G. _______________

H. _______________

(ii) How many of the bones shown in diagram form cranial bones?

____

79. Choose the odd one out from the following.

(i) Frontal, Temporal, Maxillae, Sphenoid _______________

(ii) Vomer, Nasals, Mandible, Ethmoid _______________

(iii) Cervical, Thoracic, Zygomatic, Coccyx _______________

80. Write the number of bones in the following.

(i) Cranial _______________ (ii) Facial _______________

(iii) Hyoid _______________ (iv) Middle ear _______________

81. How many of the listed bones are paired in humans and how many are unpaired?

Frontal, Parietal, Temporal, Occipital, Sphenoid, Ethmoid

Paired ____________________

Unpaired ____________________

82. Name the hard protective covering for the brain.____________________

83. Write the location of hyoid bone.____________________

84. Ear ossicles are present in the ____________________ part of ear.

85. Choose the odd one out with respect to ear ossicle.

Malleus, Hyoid, Incus, Stapes

____

86. Write the specific shape of each bone.

	Bones		Shapes
(i)	Malleus	–	________
(ii)	Incus	–	________
(iii)	Stapes	–	________
(iv)	Hyoid	–	________

87. The skull in human articulates with the superior region of the vertebral column with the help of two occipital condyles, thus it is known as ____________________ skull.

88. The skull bones fit together by ____________________ and the air spaces within these bones are known as ____________________ (sinuses/sutures).

89. Name all the facial bones in human.

____

90. Name two facial bones that found singly. ____________________

91. Except for ____________________, all skull bones are joined together by sutures.

92. The main framework of the trunk is constituted by ____________________ . It provides places for the attachment of the ribs and bones of pelvis.

93. Our vertebral column is formed by ____________ serially arranged units called vertebrae and is dorsally placed.

94. Human vertebral formula is $C_7 T_5 L_{12} S_4 C_4$.

True ☐ False ☐

95. Each vertebrae has a central hollow portion through which spinal cord passes. This hollow portion of vertebrae is called ____________________ .

96. Examine the figure of vertebral column (right lateral view) and identify *A, B, C, D, E* and *F*.

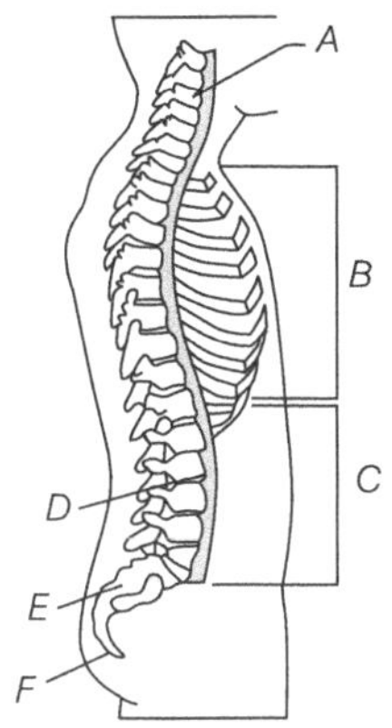

A. ____________________

B. ____________________

C. ____________________

D. ____________________

E. ____________________

F. ____________________

97. __________ and __________ are fused bones of vertebral column.

98. Complete the following analogy given below.

First vertebrae : Second vertebrae. : : Atlas : ____________________

99. What are the functions of vertebral column in humans?

(i) ____________________

(ii) ____________________

(iii) ____________________

100. Identify the bone based on the description.

(i) Flat in shape (ii) Present in the ventral midline of thorax

(iii) True ribs are attached to it

____

101. Give the scientific name for the following.

(i) Breast bone ____________________

(ii) False ribs ____________________

102. Each typical rib in human is a thin flat bone connected ______________ to the vertebral column and ______________ to the sternum.

103. Typical ribs are ______________ (monocephalic/bicephalic/tricephalic).

104. Match the following columns.

	Column I (Ribs)		Column II (Numbers)		Column III (Locations)
A.	True ribs	1.	First 7 pairs	P.	Attached to vertebral column dorsally and to seventh rib ventrally
B.	False ribs	2.	8th, 9th and 10th pairs	Q.	Do not connect ventrally
C.	Floating ribs	3.	11th and 12th pairs	R.	Attached to vertebral column dorsally and to sternum ventrally

A. ______________ B. ______________ C. ______________

105. Name the cartilage which helps to join sternum with ribs. ____________________

106. Complete the equation.

____________________ + Ribs + ____________________ = Ribcage

107. Identify *A*, *B* and *C* in the given figure of ribcage.

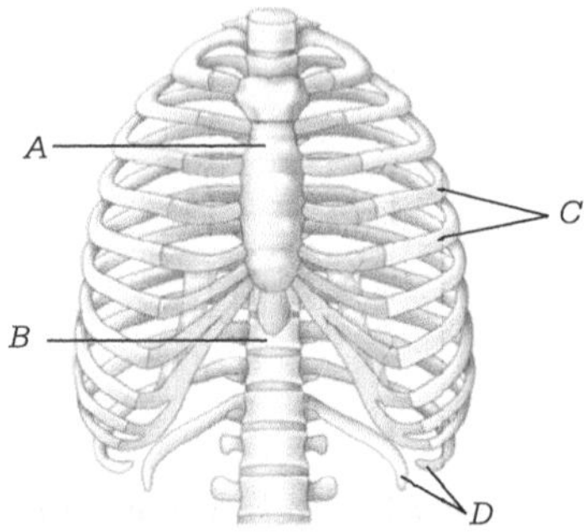

A. ____________________

B. ____________________

C. ____________________

D. ____________________

108. Write the number of bones in the given flowchart.

Appendicular skeleton (126 bones)
- Limbs (A)
- Girdle (B)
  - Pectoral girdle (C)
  - Pelvic girdle (D)

A. ______________ B. ______________ C. ______________ D. ______________

109. Each limb (upper to lower) consists of how many bones? ____________________

110. Give common names for the following.

(i) Femur, Tibia, Fibula, Tarsals. ____________________

(ii) Humerus, Ulna, Radius, Carpals. ____________________

111. Complete the following based on the number of given bones in human body.

(i) Phalanges = ________ × 2 = ________

(ii) Carpals = ________ × 2 = ________

(iii) Metacarpals = ________ × 2 = ________

(iv) Tarsals = ________ × 2 = ________

(v) Metatarsals = ________ × 2 = ________

112. Give the scientific names of the following bones.

(i) Wrist bones ____________________

(ii) Palm bones ____________________

(iii) Digits ____________________

(iv) Thigh bone ____________________

(v) Ankle bone ____________________

113. Write the phalangeal/digital formula for human hand/foot. ____________

114. Complete the following analogy given below.

Longest bone : Smallest bone : : ____________ : ____________

115. ____________ is a cup-shaped bone that covers the knee ventrally.

116. Give one common function of pectoral girdle and pelvic girdle.

____

117. Consider the following flowchart and identify *A*-*E*.

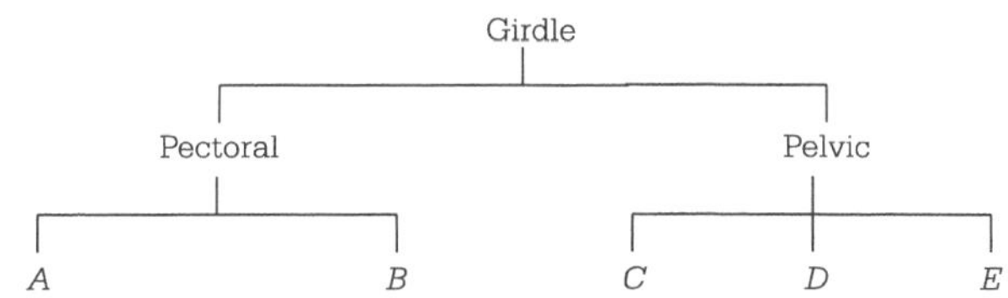

A. ____________ B. ____________ C. ____________ D. ____________

E. ____________

118. In the given figure of right pectoral girdle and upper arm, how many labels are correctly named? Also rewrite the incorrect ones correctly.

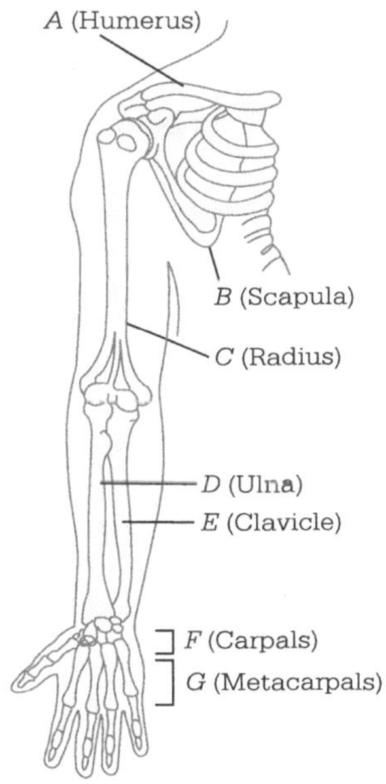

____

119. ____________ (Clavicle/Scapula) bone is situated in the dorsal part of the thorax between the second and seventh ribs.

120. Give the scientific names given to each of the following.

(i) Collar bone ____________

(ii) Shoulder blade ____________

121. Scapula has a slightly elevated ridge called ______________ which projects as a flat expanded process called the ______________________. The latter articulates with the clavicle.

122. For articulation of head of humerus, a depression is found in scapula called as <u>occipital cavity</u>.
(*Rewrite the statement by changing the underlined word.*)

____

123. How many curvatures are present in clavicle? ______________________

124. Pelvic girdle consists of ____________ (three/two) coxal bones. Each coxal bone is formed by the fusion of __________ (three/two) bones.

125. Each coxal bone is formed by the fusion of three bones namely,

____

126. Complete the following analogy given below.

Glenoid cavity : ________________ : : Pectoral girdle : Pelvic girdle

127. Two halves of pelvic girdle articulate ventrally at a fibrocartilaginous joint called ______________________ (Sutures/Pubic symphysis).

128. Name the cavity through which the femur bone articulates. ____________________

129. Identify *A, B, C* and *D* in the figure given below.

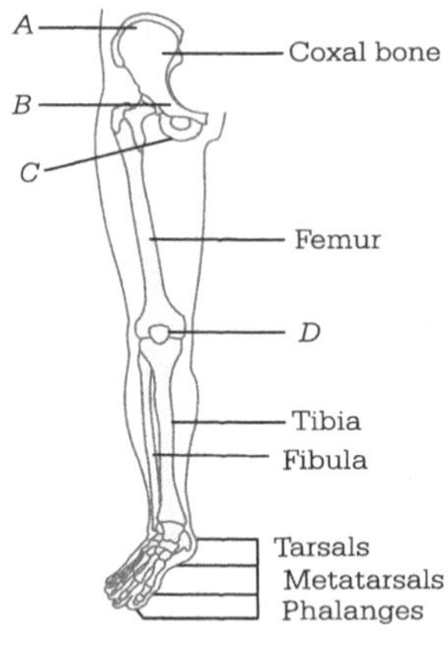

A. ______________________

B. ______________________

C. ______________________

D. ______________________

130. Match the following columns.

	Column I		Column II
A.	Scapula	1.	Thin flat bones
B.	Sternum	2.	Long slender bone
C.	Clavicle	3.	Dagger-shaped bone
D.	Ribs	4.	Large triangular flat bone

A. ____________ B. ______________ C. ______________ D. ______________

131. Bones are classified according to their shape and size as long-bones, short bones, flat bones, irregular bones and sesamoid bones. Write the common bone name given to each category.

(i) Skull bones, Sternum, Ribs ____________________

(ii) Carpals, Tarsals ____________________

(iii) Humerus, Radius, Ulna, Femur, Tibia ____________________

(iv) Ear Ossicles, Vertebrae ____________________

(v) Patella, Fabella, Pisiform ____________________

Topic 3 : Joints

132. How many statements are correct for joints?

(i) These are essential for all types of movements involving bony parts.

(ii) They connect bones and cartilages only.

(iii) These are structural arrangement of tissues.

____

133. Force generated by the muscle is used to carryout movement through joints. In this case, joint acts as point of ____________________ (effort/fulcrum/resistance).

134. Name different types of joints found in human body.

____

135. Match the following columns.

	Column I		Column II
A.	Cartilaginous joints	1.	Immovable joints
B.	Fibrous joints	2.	Imperfectly movable or slightly movable joints
C.	Synovial joints	3.	Perfectly movable or freely movable joints

A. __________ B. __________ C. __________

136. Sutures are fixed, immovable or fibrous joints in which articulating bones are held together by white fibrous tissue.

True ☐ False ☐

137. Complete the flowchart.

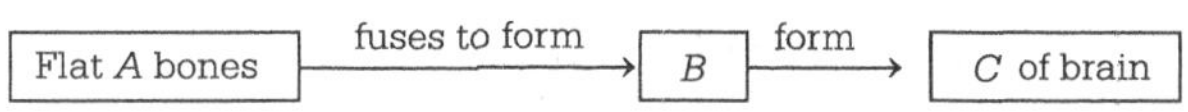

A. __________ B. __________ C. __________

138. Give an example of cartilaginous joint. ____________________

139. Synovial joints in humans are characterised by the presence of fluid-filled ____________________ cavity between the articulating surfaces of the two bones.

140. Articular cartilage of synovial joint is ____________________ (fibro/elastic/hyaline) cartilage.

141. Cartilaginous joints help in locomotory movements.

True ☐ False ☐

142. Name the fluid rich in albumin and hyaluronic acid found in synovial joints.

____

143. How many types of synovial joints are present in human body? ______________

144. Name the common term given to these joints.

Knee joint, Elbow joint, Interphalangeal joint

____

145. Give one example of each type of joint.

(i) Ball and socket joint ______________

(ii) Pivot joint ______________

(iii) Gliding joint ______________

146. Match the following columns.

	Column I		Column II
A.	Hip joint	1.	Ball and socket joint
B.	Between radius and ulna	2.	Pivot joint
C.	Between metacarpals and phalanges	3.	Ellipsoid joint
D.	Between carpals and metacarpals of thumb	4.	Saddle joint
E.	Between tarsals	5.	Gliding joint

A. ______________ B. ______________ C. ______________ D. ______________

E. ______________

Topic 4 : Disorders of Muscular and Sketetal System

147. Choose the odd one out with respect to the symptoms of Myasthenia gravis.

Fatigue, Tremors, Paralysis of skeletal muscle, Weakening of muscles

____

148. Lack of which protein in a person causes muscular dystrophy? ______________

149. Arthritis is a disease caused by degradation of bone.
(*Rewrite the statement by changing the underlined word.*)

____

150. Choose the odd one out with respect to arthritis.

Rheumatoid arthritis, Gout, Osteoarthritis, Osteoporosis

____

151. Match the following columns.

	Column I (Disorders)		Column II (Features)
A.	Myasthenia gravis	1.	Degeneration of skeletal muscles
B.	Muscular dystrophy	2.	Neuromuscular junction affected
C.	Tetany	3.	Low Ca^{2+} in body fluid
D.	Arthritis	4.	Inflammation of joints

A. ______________ B. ______________ C. ______________ D. ______________

152. A person suffering from arthritis may experience how many of the listed symptoms?

Swelling in joint, Pain in joint, Sweating, Softness in bones, Fusion of joint, Tissue degeneration

____

153. An abnormal tissue growth that occurs on the surface of the articular cartilage in rheumatoid arthritis is called as __________ (rheumatoid factor/pannus).

154. Name the type of arthritis based on the description given below.

(i) Inflammation of the synovial membrane in synovial joint. __________

(ii) Sodium urate or uric acid crystal accumulate in joints. __________

(iii) Articular cartilage at the synovial joint slowly degenerates and bone ends become exposed. __________

155. Match the following columns.

	Column I		Column II
A.	Gouty arthritis (Gout)	1.	Autoimmune disease
B.	Rheumatoid arthritis	2.	Progressive degenerative disorder
C.	Osteoarthritis	3.	Hereditary disease

A. __________ B. __________ C. __________

156. Rheumatoid arthritis is diagnosed by the presence of ______ (IgM/IgA/IgE) in blood.

157. Name an age related disorder characterised by decreased bone mass and increased chances of fracture.__________

158. Choose the correct option for osteoporosis in each statement.

(i) Osteoporosis occurs more in __________ (men/women).

(ii) Decreased level of __________ (adrenaline/oestrogen) in common cause.

(iii) It occurs due to __________ (high level/deficiency) of calcium in the body.

159. Osteomalacia is a deficiency disease, sometimes called adult rickets.

True ☐ False ☐

160. Deficiency of vitamin __________ causes osteomalacia.

161. __________ is a symptom characterised by rapid spasms in muscles due to hypocalcemia.

Answers

1. True
2. *Amoeba*
3. 4(All of them except ear pinna) Ear pinna does not show movement.
4. Locomotion
5. False. All locomotions are movements, but all movements are not locomotion.
6. *Paramoecium, Hydra*
7. Habitats
8. (i) Locomotion is performed for search of food, shelter, mate, suitable breeding grounds, favourable climatic conditions or to escape from enemies and predators.
9. Amoeboid, ciliary and muscular
10. Flagellar movement
11. (i) Jaws is odd one out. Rest show amoeboid movement.
    (ii) Basophils is odd one out. Rest show muscular movement.
12. Internal tubular, ciliated.
13. (i) Helps in removing dust particles and same of the foreign substance inhaled along with atmospheric air
    (ii) Passage of ova and fertilised zygote.
14. Contractile property
15. Muscular, skeletal and neural
16. Mesodermal
17. 40-50%
18. Excitability, contractility, extensibility and elasticity.
19. (i) Location, appearance and nature of regulation of their activities.
20. Based on their location
21. Visceral
22. Skeletal
23. *A*-Located in the inner walls of hollow visceral organs, *B*-Muscles of heart, C-Striated, *D*-Non-striated, *E*-Involuntary, *F*-Involuntary, *G*-Primarily involved in locomotory actions, *H*-Pumping of heart
24. Visceral or smooth muscle
25. (i) Transportation of food through the digestive tract
    (ii) Transport of gametes through the genital tract.
26. Fascicle
27. Fascia, collagenous connective
28. True
29. (i) Sarcolemma
    (ii) Sarcoplasm
    (iii) Sarcoplasmic reticulum
30. Muscle fibre
31. Many nuclei
32. Sarcoplasmic reticulum
33. True
34. *A* - Fascicles (muscle bundles)
    B - Muscle fibres (muscle cells)
    C - Sarcolemma
    D - Blood capillaries
35. Actin, myosin
36. Sarcomere
37. 2 statements, i.e. (i) and (iii) are true. Statement (ii) is false because sarcomere is the area between two Z-lines
38. A-1, B-2, C-3, D-5, E-4
39. Thin, thick
40. Sarcomere
41. (i) Isotropic band
    (ii) Anisotropic band
    (iii) Hensen zone
42. True
43. *A* - Z-line, *B* - Myosin filament, *C* - Actin filaments, *D* - I-band (light band),
    E - A-band, *F* - Sarcomere,
    G - H-zone
44. Myosin, actin, troponin, tropomyosin
45. (i) Actin and myosin
    (ii) Troponin and tropomyosin
46. 'F' (filamentous)
47. Filamentous, globular
48. False. Troponin masks the active binding sites for myosin on actin filaments during resting state.
49. *A*–Troponin complex,
    B–Tropomyosin,
    C–F-actin (HMM)
50. Meromyosin
51. *A*–Actin binding site, *B*–ATP binding site, *C*–Myosin head, *D*–Cross arm (HMM)
52. (i) Heavy Meromyosin
    (ii) Light Meromyosin
53. Cross arm
54. ATP binding sites, actin binding sites
55. Sliding filament theory
56. True
57. Motor
58. Muscle fibre
59. Neuromuscular junction or motor end plate.
60. An action potential.
61. II $\rightarrow$ IV $\rightarrow$ VI $\rightarrow$ VII $\rightarrow$ III $\rightarrow$ I $\rightarrow$ V
62. I, A
63. 3 (Pumping of Ca^{2+} ions into sarcoplasmic cisternae, masking of actin filaments, Z-line returns to original position).
64. *A*-Cross-bridge, *B*-Sliding/ rotation, *C*-Breaking of cross bridge, *D*-ATP
65. (i) Calcium ions and troponin
    (ii) Myosin head and actin filament.
    (iii) Sarcolemma and sarcoplasm
66. Muscle fatigue
67. Anaerobic breakdown of a glycogen
68. Myoglobin
69. White muscle fibres
70. A-Red - more, White - less,
    B-Red - more, White - less,
    C-Red - less, White - more,
    D- Red - more, White - less
71. *A* - Relaxed state of sarcomere, *B* - Contracting state of sarcomere, *C* - Maximally contracted state of sarcomere.
72. Endoskeleton
73. (i) Endoskeleton – Spicules, cartilages and bones.
    (ii) Exoskeleton – Scales, feathers, hair and nails.
74. Chondroitin salts
75. Cartilage, bone
76. A-80, B-126
77. (i) Axial skeleton
    (ii) Skull bones
78. (i) *A*-Parietal, *B*-Frontal, *C*-Maxilla, *D*-Hyoid, *E*-Temporal, *F*-Occipital, *G*-Ethmoid, *H*-Sphenoid
    (ii) Only 6 labels represent cranial bones

A-Parietal, *B*-Frontal, *E*-Temporal, *F*-Occipital, *G*-Ethmoid, *H*-Sphenoid

79. (i) Maxillae is odd one out as it is facial bone (Rest are cranial bones).
(ii) Ethmoid is odd one out as it is cranial bone (Rest are facial bones).
(iii) Zygomatic is odd one out as it is facial bone (Rest are bones of vertebral column).

80. (i) 8 bones
(ii) 14 bones
(iii) 1 bone
(iv) 3 bones × 2 pairs each = 6 bones

81. Paired – 2 (Parietal and temporal)
Unpaired-4 (Frontal, occipital, sphenoid, ethmoid)

82. Cranium

83. At the base of the buccal cavity

84. Middle

85. Hyoid is odd one out. It is not an ear ossicle.

86. (i) Hammer-shaped
(ii) Anvil-shaped
(iii) Stirrup- shaped
(iv) U-shaped

87. Dicondylic skull

88. Sutures, sinuses

89. Nasals, maxillae, zygomatic, mandible, lacrimal, palatines, inferior nasals and vomer

90. Mandible and vomer

91. Mandible

92. Vertebral column

93. 26

94. False. Human vertebral formula is $C_7T_{12}L_5S_{1(5)}Co_{1(4)}$

95. Neural canal

96. *A*-Cervical vertebra, *B*-Thoracic vertebra, *C*-Lumbar vertebra, *D*-Intervertebral disc, *E*-Sacrum, *F*-Coccyx

97. Sacrum, coccyx

98. Axis

99. (i) Protects the spinal cord
(ii) Supports the head
(iii) Provides surface as an attachment for ribs and musculature of back

100. Sternum

101. (i) Sternum
(ii) Vertebrochondral ribs

102. Dorsally, ventrally

103. Bicephalic

104. A-1-R, B-2-P, C-3-Q

105. Hyaline cartilage

106. Thoracic vertebrae, sternum

107. *A*-Sternum, *B*-Vertebral column , *C*-Ribs, D-Floating ribs

108. A-120, B-6, C-4, D-2

109. 30 bones

110. (i) Hindlimb bones
(ii) Forelimb bones

111. (i) Phalanges = 14 × 2 = 28
(ii) Carpals = 8 × 2 = 16
(iii) Metacarpals = 5 × 2 = 10
(iv) Tarsals = 7 × 2 = 14
(v) Metatarsals = 5 × 2 = 10

112. (i) Carpals
(ii) Metacarpal
(iii) Phalanges
(iv) Femur
(v) Tarsals

113. 2, 3, 3, 3, 3

114. Femur, stapes

115. Patella

116. Help in the articulation of limbs.

117. *A*-Clavicle, *B*-Scapula, *C*-Ilium, *D*-Ischium, *E*-Pubis

118. 3 are correctly labelled, i.e. *B, F, G*. Rest are *A*-Clavicle, *C*- Humerus, *D*-Radius, *E*-Ulna

119. Scapula

120. (i) Clavicle
(ii) Scapula

121. Spine, acromion process

122. Glenoid cavity

123. Two

124. Two, three

125. Ilium, ischium and pubis

126. Acetabulum

127. Pubic symphysis

128. Acetabulum

129. *A*-Ilium, *B*-Pubis, *C*-Ischium, *D*-Patella

130. A-4, B-3, C-2, D-1

131. (i) Flat bones
(ii) Short bones
(iii) Long bones
(iv) Irregular bones
(v) Sesamoid bones

132. 2 statements (i) and (iii) are correct.
Statement (ii) is incorrect because joints are contact between bones or between bones and cartilages.

133. Fulcrum

134. Joints are of three types, i.e. fibrous, cartilaginous and synovial.

135. A-2, B-1, C-3

136. True

137. *A*-skull, *B*-Sutures, *C*-Cranium

138. Joints between the adjacent vertebrae in the vertebral column.

139. Synovial

140. Hyaline

141. False. Synovial joints help in locomotory movements.

142. Synovial fluid

143. 6 (Ball and socket, hinge, saddle, pivot, gliding, ellipsoid)

144. Hinge joint

145. (i) Between humerus and pectoral girdle
(ii) Between atlas and axis
(iii) Between carpals

146. A-1, B-2, C-3, D-4, E-5

147. Tremors is odd one out as they are not a symptom of Myasthenia gravis.

148. Dystrophin

149. Inflammation of joints

150. Osteoporosis is odd one out as it is not a type of arthritis.

151. A-2, B-1, C-3, D-4

152. 4 (Swelling in joint, pain in joint, fusion of joint, tissue degeneration). Softness in bones and sweating are not the symptoms of arthritis.

153. Pannus

154. (i) Rheumatoid arthritis
(ii) Gout
(iii) Osteoarthritis

155. A-3, B-1, C-2

156. IgM

157. Osteoporosis

158. (i) Women , (ii) Oestrogen
(iii) Deficiency

159. True

160. Vitamin-D

161. Tetany

CHAPTER TWENTY ONE

NEURAL CONTROL AND COORDINATION

Topic 1 : Human Neural System

1. The function of organ and organ system in our body must be coordinated to maintain ______________________.

2. Name the process of interaction and complementation of functions of two or more organs.

   ____

3. The increased supply of oxygen necessitates an increase in the rate of ______ increased blood flow and increased heartbeat.

4. ______ system and ______ system coordinate and integrate all the activities of the organs.

5. The endocrine system provides chemical integration through enzymes.

   (Rewrite the statement by changing the underlined word.)

   ____

6. The endocrine system provides an organised network of point-to-point connections for a quick coordination.

   True ☐ False ☐

7. Complete the following analogy given below.

   Endocrine system : Nervous system : : Hormone : ______________

8. Name the specialised cells which detect, receive and transmit different kinds of stimuli and which makes neural system. ______________________

9. Presence of brain with ganglia and better organised neural system is found in ______ (*Hydra* /insects)

   ____

10. Write the two parts of the human neural system.

    ____

11. The CNS includes the PNS and ANS and is the site of information processing and control. (*Rewrite the statement by changing the underlined word.)*

____

12. The PNS comprises of all the nerves of the body associated with brain and __________.

13. Write the two nerve fibres of PNS. ______________________________

14. Afferent nerve fibres transmit impulses from ______________ to ______________.

15. Efferent fibres transmit regulatory signals from the CNS to the brain. (*Rewrite the statement by changing the underlined word.*)

____

16. The PNS includes somatic neural system and ______________________.

17. The system that transmits impulse from the CNS to skeletal muscles is______________.

18. Somatic neural system relays impulses from CNS to voluntary organs.

True ☐ False ☐

19. The system that transmits impulse from the CNS to the involuntary organs and smooth muscles of the body ______________________________.

20. Write the two parts of autonomic neural system.

____

21. The flowchart given below is the functional organisation of the human nervous system. Identify *A, B, C, D* and *E* in the following flowchart.

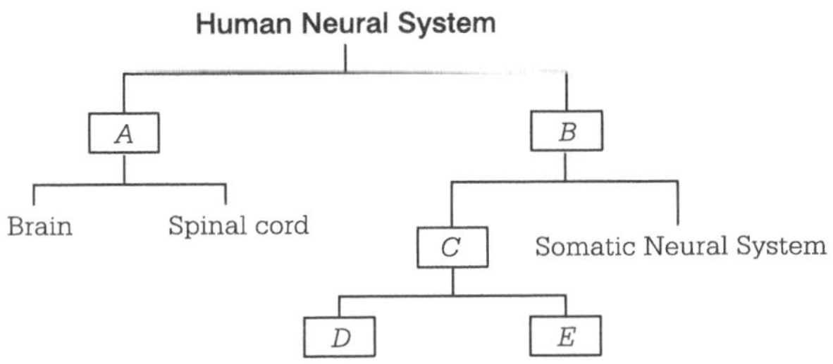

A. ______________________

B. ______________________

C. ______________________

D. ______________________

E. ______________________

22. Visceral nervous system is a part of ____________________ and comprises of____________________ and plexuses. It constitutes the connection between ______________ and the viscera.

Topic 2 : Neuron as Structural and Functional Unit of Neural System

23. Write the three major parts of a neuron. ______________________________

24. ____________________ are the short fibres which branch and project out of the cell body.

25. The cluster of rough endoplasmic reticulum in cell body and dendrites are called

______________________________________.

26. Dendrites and ____________________ of neuron transmit impulses towards and away from the cell body.

27. Name the bulb-like structure present in the distal end of axon branches.

____

28. Write a technical name for the following.
(i) Neuron with one axon and two or more dendrites. _______________
(ii) Neuron with one axon and a cell body. _______________

29. Identify the correct pair.
(i) Bipolar neuron–Cornea (ii) Multipolar neuron–Cerebral cortex
(iii) Unipolar neuron–Embryonic stage

____

30. Name the neuron with one axon and one dendrite.

____

31. Myelinated nerve fibres are enveloped by the _______________ cells.

32. Small sacs called synaptic cleft contain neurotransmitters.
(*Rewrite the statement by changing the underlined word.*)

____

33. Name the cell wraps around the axons of motor and sensory neurons and are involved in neurogeneration.

____

34. Periodic gap in insulating sheath on axon that facilitate rapid nerve conduction are called _______________ .

35. Match the following columns.

Column I	Column II
A. Spinal nerves	1. Myelinated nerve fibre
B. ANS	2. Non-myelinated nerve fibre
C. Cranial nerves	
D. Somatic neural system	

A. _______________ B. _______________ C. _______________ D. _______________

36. Consider the following figure and identify *A-G*.

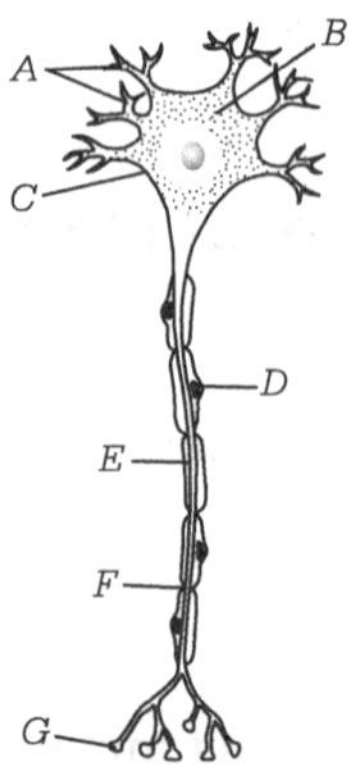

A. _______________
B. _______________
C. _______________
D. _______________
E. _______________
F. _______________
G. _______________

37. The ion channels present on the neuronal membrane are ______________________ (permeable/semipermeable/selectively permeable).

38. In the resting phase, axonal membrane is comparatively more permeable to _________ and impermeable to ______________.

39. During resting phase, the membrane is impermeable to negatively charged proteins. Consequently inside the axoplasm, there is high concentration of ____________________ and low concentration of _______________ .

40. Maintenance of the ionic gradients across the resting membrane is done by

____ .

41. For the maintenance of ionic gradients across the resting membrane, the sodium-potassium pump transports _________________ outwards for ___________________ into the cell.

42. The electrical potential difference across the polarised plasma membrane when concentration of K^+ is high inside the axon and low outside, it is known as action potential. (*Rewrite the statement by changing the underlined word.*)

____

43. When a stimulus is applied at a site on the polarised membrane, the membrane at that site becomes freely permeable to ____________________________ .

44. During the propagation of a nerve impulse, the action potential results from the movement of K^+ ions from intracellular fluid to extracellular fluid.

True ☐ False ☐

45. During the transmission of nerve impulse through a nerve fibre, the potential on the inner side of the plasma membrane has which type of electric charge?

____

46. A wave of action potential is termed as ______________________________________ .

47. A nerve impulse is transmitted from one neuron to another through the junctions called

____________________________ .

48. Name the junction between pre-synaptic and post-synaptic neurons.

____

49. Write the technical term for the gap between a pre and post-synaptic neurons.

____

50. The occurrence of synaptic cleft is usually found for electrical synapses.

True ☐ False ☐

51. Electrical synapse transmission is always faster than chemical synapse.

True ☐ False ☐

52. Write a general term for the chemicals that are involved in the transmission of impulses at the synapses. ______

53. The axon-terminals contain ______ vesicles filled with the neurotransmitters.

54. The neurotransmitters bind to their specific ______ present on the post-synaptic membrane.

55. Complete the following sequence.

Synaptic vesicles → Pre-synaptic membrane → ______ → Synaptic cleft → ______ → Post-synaptic neuron.

56. Complete the following flowchart of transmission of impulses by identifying *A*, *B*, and *C*.

(*A*) of pre-synaptic membrane → Rupturing of synaptic vesicle → Propagation of post-synaptic (*B*) → (*C*) of transmitter.

A. ______ B. ______ C. ______

57. Consider the following figure. Identify *A, B, C, D, E* and *F*.

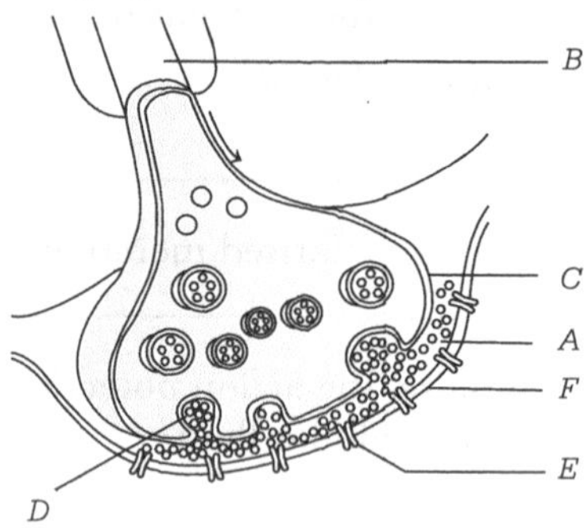

A. ______

B. ______

C. ______

D. ______

E. ______

F. ______

58. Jumping of action potential at nodes of Ranvier is known as ______.

59. Write any two examples of neurotransmitter. ______

Topic 3 : Central Nervous System and Reflex Arc

60. Spinal cord acts as 'command and control system' of the body.

True ☐ False ☐

61. The brain is covered by a layers of protective coverings called ______.

62. Find the odd one out from the following .

Duramater, Arachoid, Piamater, Sphenoid

______

63. Duramater maintains contact with the brain tissue.

True ☐ False ☐

64. Piamater is the toughest among all the meninges.
(*Rewrite the statement by changing the underlined word.*)

______

65. Find the odd one out from the following parts of brain.

 Cerebrum, Cerebellum, Thalamus, Hypothalamus

 ____

66. Name the nerve fibre that connects cerebral hemispheres.

 ____

67. The layer of cells which cover cerebral hemisphere is called <u>corpus cortex</u>.
 (Rewrite the statement by changing the underlined word.)

 ____

68. Cerebral cortex is referred to as ______________ which consists of motor, sensory and associated areas.

69. Association areas are responsible for intersensory associations, memmory and communications.

 True ☐ False ☐

70. The ______________ which acts as an insulation gives white appearance.

71. Myelin sheath in the CNS is synthesised by ______________.

72. Write the part of the forebrain that is the coordinating centre for sensory and motor signaling. ______________

73. The inner parts of cerebral hemispheres and a group of associated deep structures form a complex structure called the ______________ system.

74. Find the odd one out from the following words. .

 Pons, Amygdalus, Hippocampus

 ____

75. Alzheimer's disease in human is associated in the deficiency of ______________.

76. Name the brain disease caused due to accumulation of amyloid β-peptide.

 ____

77. Midbrain is located between thalamus/hypothalamus of the forebrain and ______________ of the hindbrain.

78. Name the canal that passes through the midbrain. ______________

79. Write the technical name for the lobes on the dorsal portion of midbrain.

 ____

80. The ______________ and ______________ form the brain stem.

81. Find the odd one out from the following words.

 Pons, Thalamus, Medulla, Cerebellum

 ____

82. Match the following columns.

	Column I (Parts of brain)		Column II (Features)
A.	Hypothalamus	1.	Neurosecretory cells
B.	Midbrain	2.	Centres to control cardio-vascular reflexes
C.	Pons	3.	Corpora- quadrigemina
D.	Medulla	4.	Fibres interconnecting different region of brain

A. ________ B. ________ C. ________ D.________

83. Complete the following analogy given below.

Wernicke's area : Broca's area : ________ : Production of speech

84. Name the part of the hindbrain that is connected to the spinal cord.

____

85. ________ has very convoluted surface to provide additional space for the neurons.

86. Consider the following figure, identify *A*, *B*, *C*, *D* and *E*.

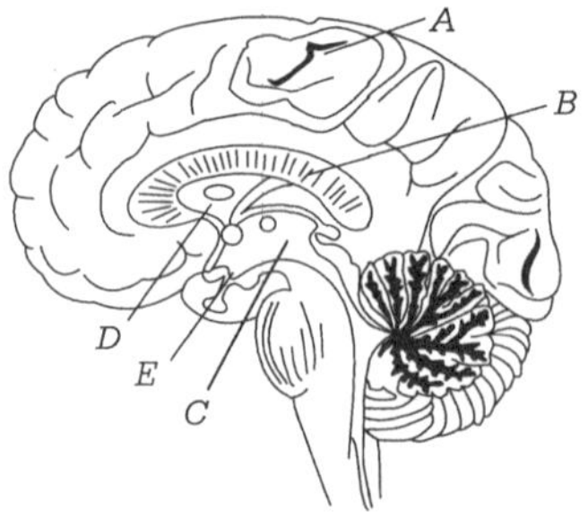

A. ________

B. ________

C. ________

D. ________

E. ________

87. Consider the following figure and identify the incorrect labellings.

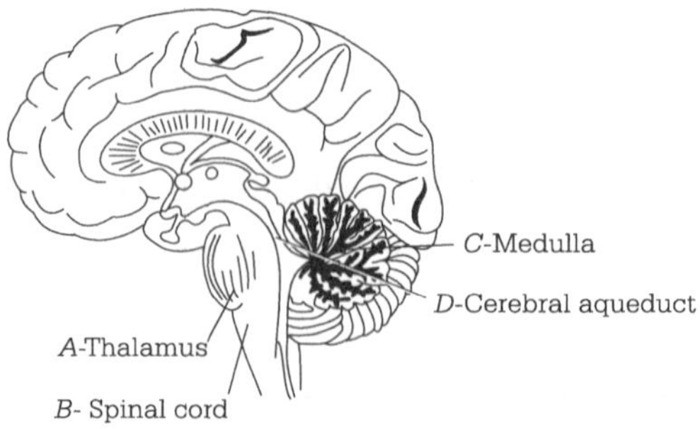

____

88. Name the part of the hindbrain that controls respiration, gastric secretions.________

89. Brain stem includes cerebral aqueduct.

True ☐ False ☐

90. Match the following columns.

	Column I (Parts of brain)		Column II (Functions)
A.	Cerebrum	1.	Controls the pituitary
B.	Cerebellum	2.	Controls vision and hearing
C.	Hypothalamus	3.	Maintains body posture
D.	Midbrain	4.	Site of intelligence

A.________ B. ________ C. ________ D. ________

91. Write the scientific term for the action of sudden withdrawal of a body part when comes in contact with extremely hot object. ________

92. The reflex pathway comprises atleast one ________ neuron (receptor) and one neuron (exciter).

93. The ________ neuron receives signal from a sensory organ and transmits the impulse into the CNS.

94. Complete the following sentence using the words given below in the box.

Afferent, PNS, CNS, Efferent, Reflex action, Reflex arc

The ________ neuron carries signals from ________ to the effector.

The stimulus and response form a ________ .

95. Arrange the following in the sequence of reflex arc.

Grey matter, Effectors, Motor fibres, Sensory fibres, Receptors

____

96. Identify the parts labelled as *A* to *E* from the representation of reflex action showing knee-jerk reflex.

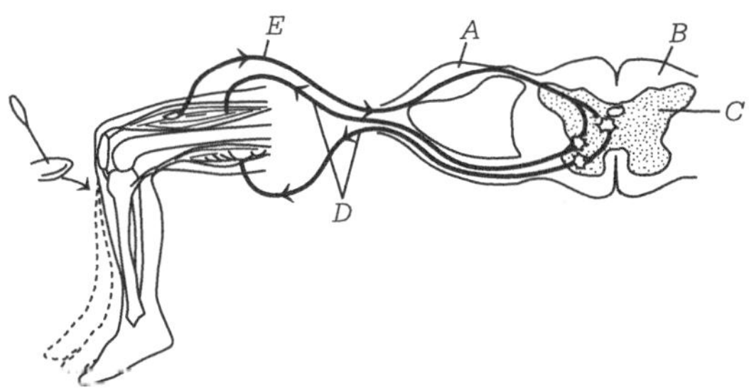

A. ________

B. ________

C. ________

D. ________

E. ________

97. The reflex arc is shown in the diagram below. Write the correct label for the pathway of this reflex arc.

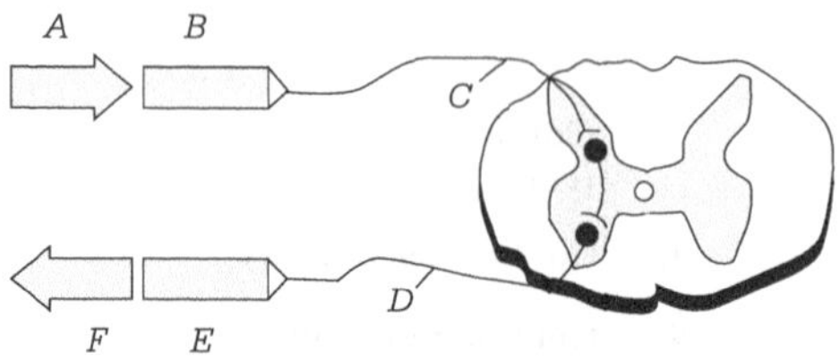

A. ______
B. ______
C. ______
D. ______
E. ______
F. ______

98. How many of the following is not a reflex action?

Salivation, Sweating, Withdrawal of hand when pinched by needle

______

99. Given below are the different components of reflex arc.

I. Effector neuron — II. Dorsal nerve root
III. CNS — IV. Afferent neuron
V. Sensory receptor — VI. Effector organ

Arrange these in correct order of action potential that follows a sensory receptor stimulation.

______

Topic 4 : Sense Organ : Eye

100. Name the sockets of the skull where eyes are located. ______

101. Find the odd one out.

Sclera, Choroid, Orbit, Labyrinth

______

102. Name the three layers of eyeball. ______

103. The front transparent part of the sclera is called ______.

104. The external layer of the eyeball composed of a dense connective is called ______.

105. Cornea is the posterior position of sclera layer.

True ☐ False ☐

106. Absence of blood circulation is a characteristic of ______ of an eye.

107. The ______ layer becomes thick in the anterior part to form ______ body.

108. Write the pigmented opaque structure formed by ciliary body.

______

109. The eyeball contains a transparent crystalline iris, held in place by ligaments.
(*Rewrite the statement by changing the underlined word.*)

______

110. The apperture surrounding by the iris is called ______.

111. The diameter of the pupil is regulated by the muscle fibre of iris.

True ☐ False ☐

112. Arrange these layers from outside to inside.

Bipolar cells, Photoreceptor cells, Ganglion cells

____

113. Name the two types of photoreceptor cells.____________________

114. Consider the following figure of an eye. Identify *A* to *H*.

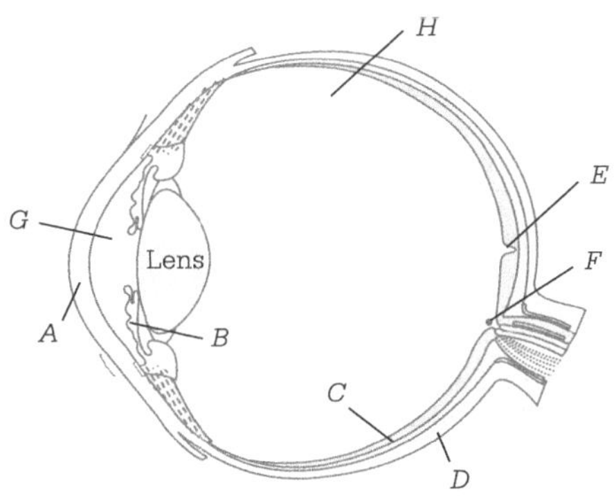

A. ____________________

B. ____________________

C. ____________________

D. ____________________

E. ____________________

F. ____________________

G. ____________________

H. ____________________

115. Name the light sensitive proteins in the photoreceptor cells. ____________________

116. Complete the following analogy given below.

Cones : Phototopic : : ______________ : Scotopic.

117. Name the purplish-red protein present in the rods. ____________________

118. Visual purple contains a derivative of vitamin-C.

True ☐ False ☐

119. The three types of cones respond to red, green and black lights.

True ☐ False ☐

120. Match the following columns.

	Column I (Parts of an eye)		Column II (Features)
A.	Sclera	1.	Visible coloured portion of eye
B.	Choroid	2.	Compound of dermal connective tissue
C.	Pupil	3.	Central opening of the iris
D.	Iris	4.	Position of retina where only the cones are densely packed
E.	Fovea	5.	Contains many blood vessels and looks bluish in colour

A. __________ B. __________ C. __________ D. __________

E. __________

121. Yellow spot is the region where the photoreceptor cells are absent.

True ☐ False ☐

122. Name the point medial to and slightly above the posterior pole of the eyeball where the retinal blood vessel enter. ______________

123. The black pigment in the eye which reduces the internal reflection is ______________.

124. The yellowish pigmented spot lateral to blind spot is called ______________.

125. Name the central pit in the yellowish pigmented spot. ______________

126. ______________ is the point where the visual acuity is the greatest.

127. The space between the cornea and the lens is called the ______________ chamber.

128. The space between the lens and the retina is called ______________ chamber.

129. Name the transparent gel in the vitreous chamber. ______________

130. Match the following columns.

	Column I (Features)		Column II (Parts of an eye)
A.	External layer of eyeball	1.	Choroid
B.	Inner layer of eyeball	2.	Sclera
C.	Middle layer of eyeball	3.	Retina

A. ______________ B. ______________ C. ______________

131. Name the protein of which photosensitive compounds are composed of

______________.

132. Photopigments are composed of ______________, is an aldehyde of vitamin-A.

133. Complete the following sentence using the words given below in the box.

Association, Permeability, Semiportability, Dissociation, Retinal, Opsin

Light induces ______________ of the retinal from opsin resulting in changes in the structure of ______________. This causes membrane ______________ changes.

134. Complete the following sentence using he words given below in the box.

Chemoreceptor, Ganglion, Unipolar, Bipolar, Photoreceptor, Muscle

Potential difference generated in the ______________ cell produce a signal that generates action potential in the ______________ cells through ______________ cells.

135. Action potentials are transmitted by the optic nerves to the ______________ area of the brain, where neural impulses are analysed.

136. The visual image is formed on the retina, where it is recognised based on earlier memory and experience.

True ☐ False ☐

Topic 5 : Sense Organ : Ear

137. Write the two sensory functions of ear.

____

138. Find the odd one out from the following words.

Pinna, Auditory meatus, Labyrinth, Tympanic membrane

____

139. Outer ear consists of _______________ and _______________.

140. The _______________ collects the vibrations in the air which produce sound.

141. Write the common name for tympanic membrane. _______________

142. Wax secreting sebaceous glands are present in the middle ear.

True ☐ False ☐

143. The tympanic membrane is composed of _______________ (endothelial/connective/muscular) tissue.

144. Fenestra ovalis is the opening of tympanic cavity.

True ☐ False ☐

145. Name the three ossicles of middle ear. _______________

146. The ear ossicles increase the efficiency of transmission of sound waves to the inner ear.

True ☐ False ☐

147. Complete the following analogy given below.

Oval window of cochlea : Stapes : : _______________ : Malleus.

148. Name the central part of the tympanic membrane.

____

149. Name the structure that connects the middle ear cavity with the pharynx.

____

150. Cochlea helps in equalising the pressures on either sides of the eardrum.
(*Rewrite the statement by changing the underlined word.*)

____

151. The fluid-filled inner ear is called _______________.

152. Write the two parts of the inner ear. _______________.

153. Membranous labyrinth lies inside the channels of bony labyrinths.

True ☐ False ☐

154. The membranous labyrinth is surrounded by a fluid _______________ and is filled by fluid _______________.

155. The coiled portion of the labyrinth is called _______________.

156. Bony labyrinth is divided into upper ______________ and lower ______________.

157. Scala media is the space within cochlea.

True ☐ False ☐

158. Match the following columns.

	Column I (Membranes of cochlea)		Column II (Parts of bony labyrinth)		Column III (Termination points)
A.	Reissner's	1.	Scala tympani	(i)	Oval window
B.	Basilar	2.	Scala vestibuli	(ii)	Round window

A. ______________ B. ______________

159. Name the structure located on the basilar membrane which contains hair cells, acting as auditory receptors. ______________

160. The hair cells are present on the external side of organ of Corti.

True ☐ False ☐

161. Match the following columns.

	Column I (Parts of an ear)		Column II (Features)
A.	Vestibule	1.	Evaluation of rotational motion
B.	Cochlea	2.	Responds to gravity and head movement
C.	Endolymph	3.	Spiral organ of Corti
D.	Semicircular canal	4.	Fluid within organ of Corti

A. ______________ B. ______________ C. ______________ D. ______________

162. The Reissner's membrane is found in the middle ear.

True ☐ False ☐

163. Diagrammatic view of ear is given below. Identify *A* to *H*.

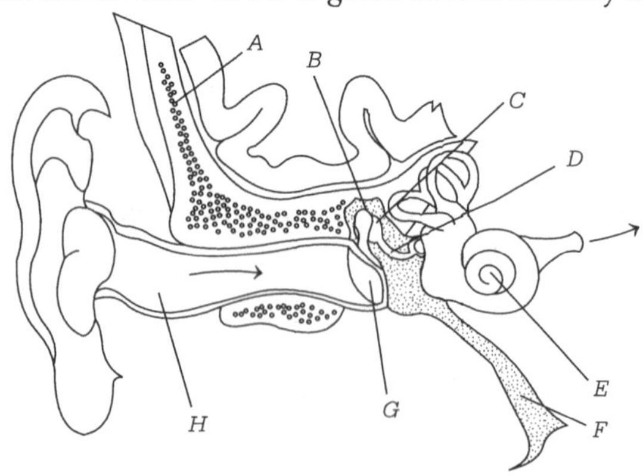

A. ______________
B. ______________
C. ______________
D. ______________
E. ______________
F. ______________
G. ______________
H. ______________

164. Complete the sentence using the words given below in the box.

Dorsal, Apical, Efferent, Stereo cilia, Afferent, Stereo fimbriate

The basal end of the hair cells is in close contact with ______________ nerve fibres. A large number of processes called ______________ are projected from the ______________ part of each hair cell.

165. Consider the following figure. Identify *A*, *B*, *C*, *D*, *E* and *F*.

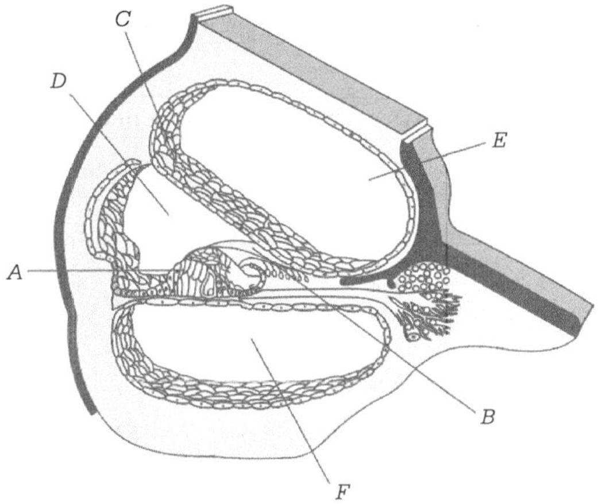

A. ______________________

B. ______________________

C. ______________________

D. ______________________

E. ______________________

F. ______________________

166. The thin elastic membrane above the rows of the hair cells is called ______________ membrane.

167. Name the system located above the cochlea of the inner ear.

____

168. The vestibular apparatus is composed of three ______________ canals and the ______________ organ.

169. The otolith organ consists of utricle and saccule.

True ☐ False ☐

170. The swollen base of semicircular canals in the vestibular apparatus is called ________.

171. The projecting ridge with hair cells in the ampulla is called ______________.

172. Name the projecting ridge in saccule and utricle.______________

173. Write the two structures of vestibular apparatus that act as specific receptors for maintenance of balance of body and posture.______________

174. The movements of the basilar membrane bend the hair cells, processing them against ______________ membrane.

175. The impulses are transmitted by the afferent fibres to the ______________ cortex of the brain.

176. Name the brain lobe which gets affected due to hearing impairment.

____

177. Consider the following flowchart of travelling of vibration into the ear. Identify the missing parts.

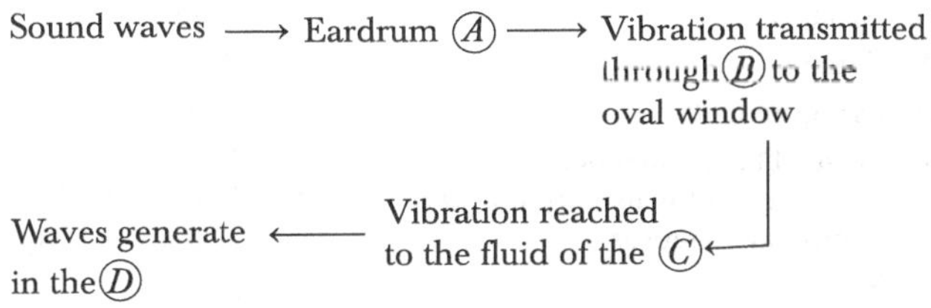

A. ______________________

B. ______________________

C. ______________________

D. ______________________

178. Complete the following sequence of mechanism of hearing.

External ear → *A* → Ear ossicles → Oval window → *B* → Lymph → Tectorial membrane → Afferent neurons → Auditory nerves → *C*

A. ________________ B. ________________ C. ________________

179. The macula of utricle responds to vertical movements of the head.

True ☐ False ☐

Answers

1. Homeostasis

2. Coordination

3. Respiration

4. Neural, endocrine

5. Hormones

6. False. The neural system provides an organised network of point-to-point connection for a quick coordination.

7. Neurotransmitter

8. Neurons

9. Insects

10. Central Neural System (CNS) and Peripheral Neural System (PNS).

11. Brain and spinal cord

12. Spinal cord

13. Afferent and efferent fibres

14. Tissues/Organs, CNS

15. Peripheral tissues/organs

16. Autonomic neural system

17. Somatic neural system

18. True

19. Autonomic neural system

20. Sympathetic neural system and parasympathetic neural system.

21. *A*–Central neural system, *B*–Peripheral neural system, *C*–Autonomic neural system, *D*–Sympathetic nervous system, *E*–Parasympathetic nervous system

22. Peripheral nervous system, nerves, ganglia, CNS

23. Cell body, dendrites and axon

24. Dendrites

25. Nissl's granules

26. Axon

27. Synaptic knob

28. (i) Multipolar neuron
(ii) Unipolar neuron

29. (ii) and (iii) are correct pair.
(i) is incorrect and can be corrected as
Bipolar neuron-Retina of an eye.

30. Bipolar

31. Schwann

32. Synaptic vesicles

33. Schwann cells

34. Nodes of Ranvier

35. A-1, B-2, C-1, D-2

36. *A*-Dendrites, *B*-Nissl's granules, *C*-Cell body, *D*-Schwann cells, *E*-Axon, *F*-Node of Ranvier, *G*-Synaptic knob

37. Selectively permeable

38. Potassium ion (K^+), Sodium ion (Na^+).

39. Potassium, sodium

40. Active transport of ions

41. $3Na^+$, $2K^+$

42. Resting potential

43. Na^+

44. False. During the propagation of a nerve impulse, the action potential results form the movement of Na^+ ions from extracellular fluid to intracellular fluid.

45. First negative then positive and again back to negative.

46. Nerve impulse

47. Synapse

48. Synapse

49. Synaptic cleft

50. False. The occurrence of synaptic cleft is usually found for chemical synapses.

51. True

52. Neurotransmitters

53. Synaptic

54. Receptors

55. Neurotransmitters, Receptors on post-synaptic membrane.

56. *A*– Depolarisation, *B*–Action potential, *C*–Hydrolysis

57. *A*-Synaptic cleft, *B*-Axon, *C*-Pre-synaptic membrane, *D*-Neurotransmitter, *E*-Receptor, *F*-Post-synaptic membrane

58. Saltatory conduction

59. Acetylcholine and epinephrine

60. False. Brain acts as 'command and control system' of the body.

61. Meninges

62. Sphenoid is odd one out. Rest others are the layer of cranial meninges.

63. False. Piamater maintains contact with the brain tissue.

64. Duramater

65. Cerebellum is odd one out. Rest of the options are the parts of forebrain.

66. Corpus callosum

67. Cerebral cortex

68. Grey matter

69. True

70. Myelin sheath

71. Oligodendrocytes

72. Thalamus

73. Limbic

74. Pons is odd one out. Rest of the options are parts of the limbic system.

75. Acetylcholine

76. Alzheimer's disease

77. Pons

78. Cerebral aqueduct
79. Corpora quadrigemina
80. Midbrain, hindbrain
81. Thalamus is odd one out. Rest of the options are the parts of hindbrain.
82. A-1, B-3, C-4, D-2
83. Language development
84. Medulla
85. Cerebellum
86. *A*-Cerebral hemisphere, *B*-Corpus callosum, *C*-Midbrain, *D* -Thalamus, *E*-Hypothalamus
87. *A, B* and *C* are labelled incorrectly and can be corrected as
A-Pons, *B*-Medulla,
C-Cerebellum
88. Medulla
89. False. Brain stem includes midbrain, pons and medulla oblongata.
90. A-4, B-3, C-1, D-2
91. Reflex action
92. Afferent, efferent
93. Afferent
94. Efferent, CNS, reflex arc
95. Receptors → Sensory fibres → Grey matter → Motor fibres → Effectors
96. *A*–Dorsal root ganglion, *B*–White matter, *C*–Grey matter, *D*–Efferent pathway, *E*–Afferent pathway
97. *A*-Stimulus, *B*-Receptor, *C*-Sensorys nerve, *D*-Motor nerve, *E*-Effector, *F*-Response
98. 1 (Sweating)
99. V → IV → II → III → I → VI
100. Orbits
101. Labyrinth is odd one out. Rest of the options are the parts of an eye.
102. Sclera, choroid and retina
103. Cornea
104. Sclera
105. False. Cornea is the anterior position of sclera layer.
106. Cornea
107. Choroid, ciliary
108. Iris
109. Lens
110. Pupil
111. True
112. Photoreceptor cells → Bipolar cells → Ganglion cells
113. Rods and cones
114. *A*-Cornea, *B*-Iris, *C*-Retina, *D*-Sclera, *E*-Fovea, *F*-Blind spot, *G*-Aqueous chamber, *H*-Vitreous chamber
115. Photopigments
116. Rods
117. Rhodopsin
118. False. Visual purple contains a derivative of vitamin-A.
119. False. The three types of cones respond to red, green and blue lights.
120. A-2, B-5, C-3, D-1, E-4
121. False. Blind spot is the region where the photoreceptor cells are absent.
122. Blind spot
123. Retina
124. Macula lutea
125. Fovea
126. Fovea
127. Aqueous
128. Vitreous
129. Vitreous humor
130. A-2, B-3, C-1,
131. Opsin
132. Retinal
133. Dissociation, opsin, permeability
134. Photoreceptor, ganglion, bipolar
135. Visual cortex
136. True
137. Hearing and maintenance of body balance.
138. Labyrinth is odd one out. Rest of the options are the part of outer ear, while labyrinth is a part of inner ear.
139. Pinna, external auditory meatus (canal)
140. Pinna
141. Eardrum
142. False. Wax secreting sebaceous glands are present in the pinna and the meatus.
143. Connective
144. False. Fenestra ovalis is an intersection between middle ear with the inner ear.
145. Malleus, incus and stapes
146. True
147. Tympanic membrane
148. Umbo
149. Eustachian tube
150. Eustachian tube
151. Labyrinth
152. Bony labyrinth and membranous labyrinth
153. True
154. Perilymph, endolymph
155. Cochlea
156. Scala vestibuli, scala tympani
157. True
158. A-2-(i), B-1-(ii)
159. Organ of Corti
160. False. The hair cells are present on the internal side of organ of Corti.
161. A-2, B-3, C-4, D-1
162. False. The Reissner's membrane is found in the inner ear.
163. *A*–Temporal bone, *B*–Malleus, *C*–Incus, *D*–Stapes, *E*–Cochlea, *F*–Eustachian tube, *G*–Tympanic membrane, *H*–External auditory canal
164. Afferent, stereo cilia, apical
165. *A*-Organ of Corti, *B*-Tectorial membrane, *C*-Reissner's membrane, *D*-Scala media, *E*-Scala vestibuli, F-Scala tympani.
166. Tectorial
167. Vestibular apparatus
168. Semicircular, otolith
169. True
170. Ampulla
171. Crista ampullaris
172. Macula
173. Crista and macula
174. Tectorial
175. Auditory
176. Temporal lobe
177. *A*–Vibrates, *B*–Ear ossicles, *C*–Cochlea, *D*–Lymph
178. *A*-Eardrum, *B*-Fluid of cochlea/Cochlea, *C*-Auditory cortex
179. True

CHAPTER TWENTY TWO

CHEMICAL COORDINATION AND INTEGRATION

Topic 1 : Human Endocrine System

1. Name the substance, that integrates and coordinates the all metabolic activities within the organism. ______________________

2. The ______________ and the ______________ system jointly coordinate and regulate the physiological functions in the body.

3. Endocrine glands are also known as __________ glands.

4. Enzyme is a chemical produced by endocrine glands and released into the blood. (*Rewrite the statement by changing the underlined word.*)

   ____

5. Name the chemical which is non-nutrient, acts as intercellular messengers and is produced in trace amount.

   ____

6. The ______________ and ____________ producing diffused tissues/cells located in different parts of our body constitute the endocrine system.

7. (i) Consider the following sequence of endocrine gland and rearrange the correct sequence from top to bottom.

Thymus → Pituitary → Adrenal → Ovary → Pancreas

   ____

(ii) Consider the given diagram and answer the following questions.

(a) Identify glands *A* to *F*.

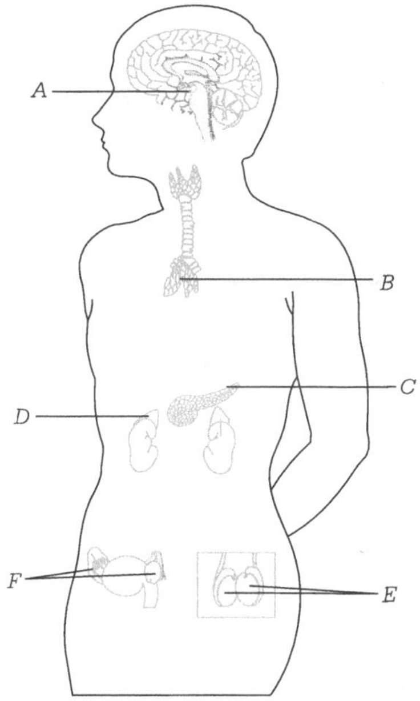

A. ______________

B. ______________

C. ______________

D. ______________

E. ______________

F. ______________

(b) Which of the following gland function as both exocrine and endocrine?

____

(c) Which of the following gland controls female secondary sexual character?

____

8. Name any four organs other than endocrine system that produce hormones.

____

Topic 2 : Hypothalamus

9. Hypothalamus contains several groups of neurosecretory cells called ______________ (chroma/nuclei) which produce hormones.

10. The hormones released by hypothalamus regulate the synthesis and secretion of pineal gland.

True ☐ False ☐

11. (i) Name the two types of hormones released by hypothalamus.

____

(ii) ______________ hormones are hypothalamic hormones that stimulate the secretion of pituitary hormone.

(iii) ______________ are hypothalamic hormones that inhibit the secretion of pituitary hormone.

(iv) The hypothalamic releasing and inhibitory hormones reach the pituitary gland through portal circulation and regulate the functions of ______________ pituitary.

12. Complete the following analogy.

Releasing hormone : Inhibiting hormone : : Gonadotropin : ______________

13. Which hormone from the hypothalamus inhibits the release of growth hormone from the pituitary.

____

14. Name the part of pituitary gland that is under the direct neural regulation of the hypothalamus.

____

Topic 3 : The Pituitary Gland

15. Pituitary gland is lodged in the cavity of the sphenoid bone, called neural lobe. (*Rewrite the statement by changing the underlined words.*)

____

16. Pituitary gland is anatomically divided into _______________ and _______________ .

17. Write the names of the two portions of adenohypophysis.

____

18. Identify the odd one out from the following hormones.

Prolactin, Thyroid stimulating hormone, Melanocyte stimulating hormone, Luteinizing hormone

____

19. How many of the given hormones are released from the portion *A* in the figure of pituitary gland?

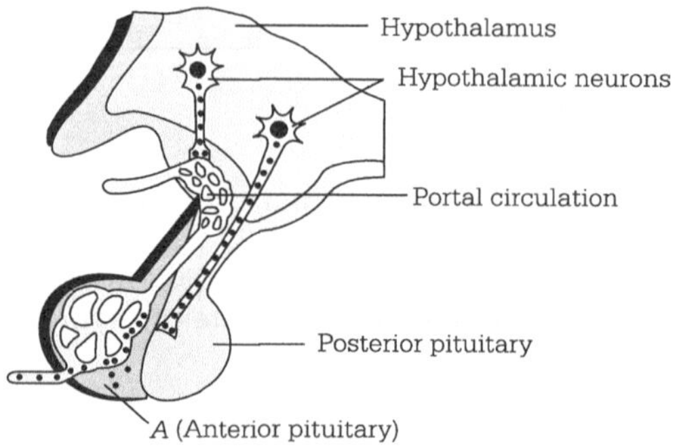

(i) Growth hormone
(ii) Prolactin hormone
(iii) Melanocyte stimulating hormone
(iv) Oxytocin

____

20. Give full form of the following abbreviation of hormones.

(i) GnRH ____________________

(ii) GH ____________________

(iii) PRL ____________________

(iv) TSH ____________________

(v) ACTH ____________________

(vi) LH ____________________

(vii) FSH ____________________

(viii) MSH ____________________

21. Which portion of pituitary gland releases oxytocin and vasopressin hormones.

____

22. Neurohypophysis is also known as ____________________.

23. The synthesis site of vasopressin and oxytocin hormones is ____________________ (neurohypophysis/hypothalamus).

24. Complete the following flowchart.

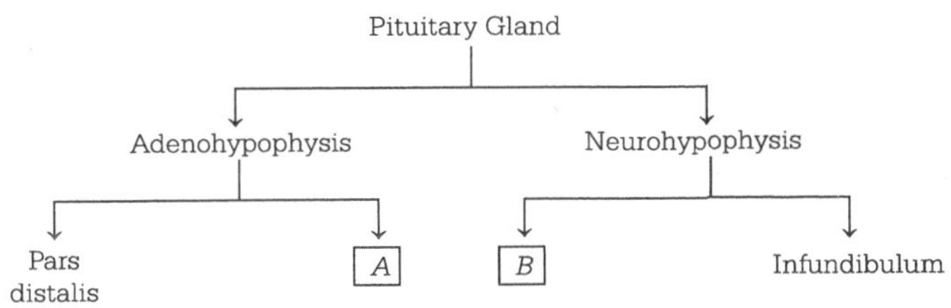

A. ____________________

B. ____________________

25. (i) The WWE fighter 'The Great Khali' suffered from gigantism, it is also known as

____________________.

(ii) Deficiency of adenohypophysis secretion leads to a condition called

____________________.

(iii) ____________________ (Hyperpituitarism/Hypopituitarism) causes disorders in the acidophil cells developing gigantism and acromegaly.

(iv) Complete the following analogy.

Hypopituitarism : Hyperpituitarism : : Simmond's disease : ____________________

26. Match the following columns.

	Column I (Hormones)		Column II (Chemical nature)		Column III (Functions)
(i)	Prolactin (PRL)	(a)	Polypeptide	1.	Secretion of glucocorticoids
(ii)	Thyroid (TSH)	(b)	Protein	2.	Regulates growth of mammary gland
(iii)	ACTH	(c)	Glycoprotein	3.	Stimulate gonadal activity
(iv)	FSH			4.	Secretion of thyroid hormone

(i) ____________ (ii) ____________ (iii) ____________ (iv) ____________

27. Name the hormones named as gonadotropins.

____

28. Luteinizing hormone stimulates the synthesis and secretion of ____________ (androgens/glucocorticoids) from testis.

29. In males spermatogenesis is regulated by ____________ and ____________.

30. Complete the following actions of gonadotropins in females.
   (i) LH induces ovulation of ____________ follicles.
   (ii) FSH development of the ____________ follicles.

31. Melanin containing cells are known as ____________.

32. The darkening of the skin in humans is due to ____________ (TSH/MSH).

33. Oxytocin acts on the smooth muscles of our body and stimulates their contraction.
   True ☐ False ☐

34. Which hormone is given to induce labor pain during pregnancy?____________

35. <u>Vasopressin</u> stimulates milk ejection from the mammary gland.
   (*Rewrite the statement by changing the underlined word.*)

____

36. Vasopressin stimulates resorption of water and electrolytes by the ____________ of kidney and reduces loss of water through ____________.

37. Which hormone is also known as anti-diuretic hormone?

____

38. In which of the following condition blood becomes diluted?
   (i) ADH is released.
   (ii) ADH is not released.

____

39. Give one word for the following conditions.
   (i) A condition in which ADH is deficient. ____________
   (ii) A condition in which ADH is hyperactive. ____________

Topic 4 : Pineal and Thyroid Glands

40. Name the gland which is present on the dorsal side of forebrain.

____

41. Melatonin regulates ____________ rhythm of body, maintain ____________ cycle, regulates body ____________ and influence menstrual cycle.

42. Name the hormone released by pineal which help in jet lag condition.

____

43. The thyroid gland is composed of two lobes which are located on either side of the <u>larynx</u>. (*Rewrite the statement by changing the underlined word.*)

____

44. The thyroid gland is composed of __________ and ______________ .

45. The thyroid gland secretes thyroxine or ________________, triiodothyronine and ______________________.

46. Write the full form of the following abbreviation.

(i) T_3 ____________________

(ii) T_4 ____________________

(iii) ADH ____________________

(iv) TCT ____________________

47. Identify the labelled parts *A*, *B* and *C* in the diagram given below.

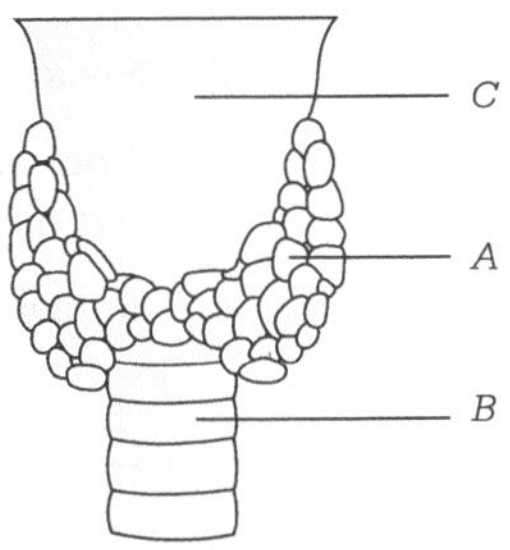

A. ______________________

B. ______________________

C. ______________________

48. Dietary iodine is converted into iodide and absorbed in the blood from where it is actively transported into the thyroid gland.

True ☐ False ☐

49. Consider the following reaction and answer the given questions.

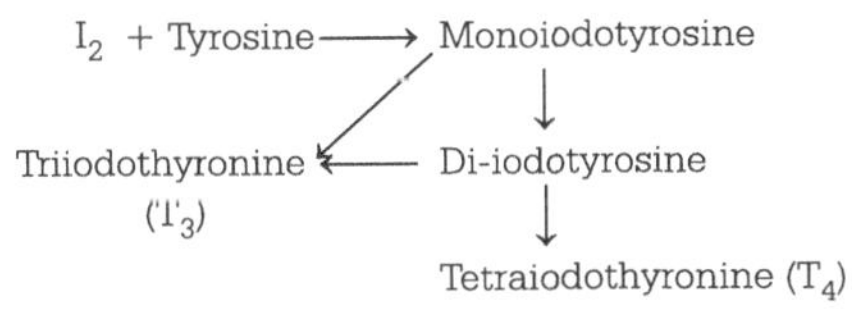

(i) This reaction takes place in which gland? ____________________

(ii) Which form of iodine is used in this reaction? ____________________

50. ______________ disorder is caused by insufficient thyroxine secretion or reduced iodine in our diet.

51. The enlargement of thyroid gland is commonly known as __________ (cretin/goitre).

52. Name the condition in which a pregnant lady gives birth of a baby with stunted growth, mental retardation, low IQ and abnormal skin.

____

53. Sometimes thyroid gland functions normally in childhood, but becomes atrophied during adulthood, this condition is known as ______________________ (Graves' disease/myxoedema).

54. The peculiar feature of myxoedema is that face and legs become swollen.

True ☐ False ☐

55. ______________________ is an auto-immune thyroiditis and occurs due to age factor or injury of thyroid gland.

56. ______________________ is a condition in which, the rate of synthesis and secretion of the thyroid hormones is increased to abnormal high levels.

57. Due to hyperthyroidism, enlargement of thyroid gland takes place, this disease is known as ______________________ (exophthalamic goitre/Hashimoto's disease).

58. Match the following columns.

	Column I (Diseases)		Column II (Conditions)		Column III (Symptoms)
(i)	Grave's disease	(a)	Hypothyroidism	1.	Lethargy, weight gain, loss of hair
(ii)	Hashimoto's disease	(b)	Hyperthyroidism	2.	Suicide of thyroid gland
(iii)	Myxoedema			3.	Protrusion of eyeballs, enlargement of thyroid gland
(iv)	Cretinism			4.	Stunted growth, delayed puberty

(i) ____________ (ii) ____________ (iii) ____________ (iv) ________

59. Which of the given hormone increase/decrease in simple goitre condition?

(i) Thyroxine ______________________

(ii) TRH ______________________

(iii) TSH ______________________

60. Consider the following flowchart of regulation of thyroid hormone, identify *A* ?

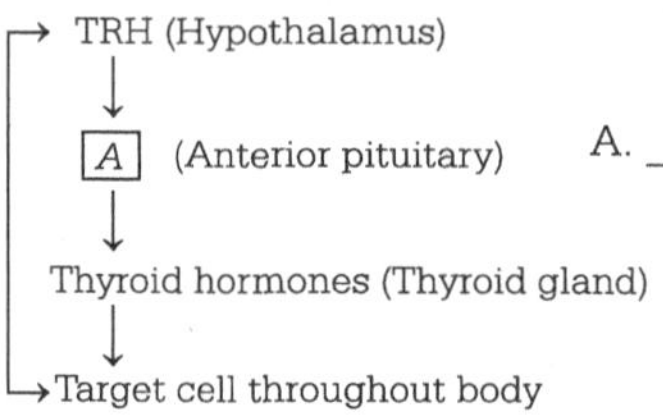

A. ______________________

61. Which condition occurs, when the part *A* from the above flowchart is inhibited?

____

62. Exposure to a cold environment triggers TRH secretion.

True ☐ False ☐

63. Thyroxine is essential for normal growth, skeletal maturation and mental development.

True ☐ False ☐

64. Given below are some physiological proceses of human body. What is the effect of thyroxine hormone on these processes? Write your answer by only mentioning increase or decrease.

(i) Basal metabolic rate ____________

(ii) Protein synthesis ____________

(iii) Glucose absorption ____________

65. Name a protein hormone secreted from thyroid gland.

____

66. Calcitonin is secreted by the C-cells in the ____________ (thyroid/pineal) gland.

67. Calcitonin reduces the reabsorption of calcium from bones and inhibits absorption of calcium by the renal tubules.
(*Rewrite the statement by changing the underlined word.*)

____

68. Thyrocalcitonin lowers the blood calcium levels in blood plasma.

True ☐ False ☐

69. Thyrocalcitonin opposed the effect of melatonin hormone.

(*Rewrite the statement by changing the underlined word.*)

____

70. Consider the following flowchart of regulation of Ca^{++} in human body. Identify *A*.

Calcium ions increased in body → T-cell stimulated $\xrightarrow{\text{Release}}$ *A* → Calcium is excreted through urine or accumulates in bone.

____

Topic 5 : Parathyroid Gland

71. By considering the given figure answer the following questions.
(i) Identify gland *A*.

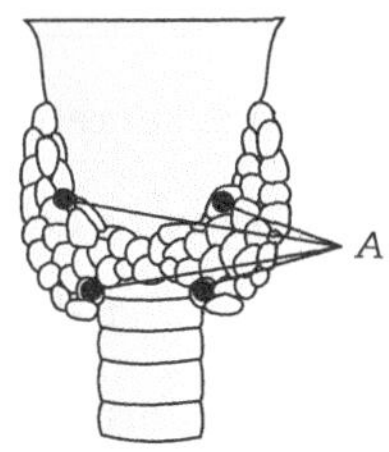

A. ____________

(ii) Which hormone is secreted from gland *A* ? ____________

(iii) Where is gland *A* situated? ____________

72. Complete the following analogy.

Parathyroid hormone : Hypercalcemic : : ____________ : Hypocalcemic

73. The secretion of parathyroid hormone is regulated by the circulating levels of <u>chlorine</u> ions.
(*Rewrite the statement by changing the underlined word.*)

____

74. Parathyroid hormone increases the Ca^{2+} levels in the ______________ .

75. PTH acts on bones and stimulates the process of bone deposition.

True ☐ False ☐

76. Parathyroid hormone stimulates ________ of Ca^{+2} by the renal tubules and ________ Ca^{2+} absorption from the digested food.

77. Consider the following flowchart of regulation of parathyroid hormone, identify *A*.

Plasma Ca^{+2} ↓ ⟶ Parathyroid gland ⟶ PTH↑ ⟶ Ⓐ

____

78. The parathyroid gland secretes excessive amounts of parathormone, this condition is known as ______________.

79. Hyperparathyroidism lead to calcium deposition in lungs causing dyspnoea.

True ☐ False ☐

80. Name the abnormalities in which tetany with symptoms like muscle spasms, convulsions and tremors arises. ______________________________

81. Differentiate between thyrocalcitonin and parathyroid hormone on the basis of action on calcium in human body.

____

Topic 6 : Thymus and Adrenal Glands

82. The <u>parathyroid</u> gland is a lobular structure located between lungs behind sternum on the dorsal side of aorta.
(*Rewrite the statement by changing the underlined word.*)

____

83. Thymosin is a ______________ hormones secreted by ______________.

84. Thymus is an endocrine gland as well as lymphatic gland.

True ☐ False ☐

85. Thymosin plays an important role in proliferation and maturation of B-lymphocytes.

True ☐ False ☐

86. Complete the following analogy.

T-lymphocytes : Cell-mediated immunity : : B-lymphocytes : ______________________.

87. The older people have low immunity. (*Give reason*)

____

88. How many pairs of adrenal gland are found in human body?

____

89. Adrenal glands are present at the _______________ part of each _______________ .

90. The centrally located tissue is called the adrenal cortex and outside this lies the adrenal medulla.

True ☐ False ☐

91. Complete the following flowchart.

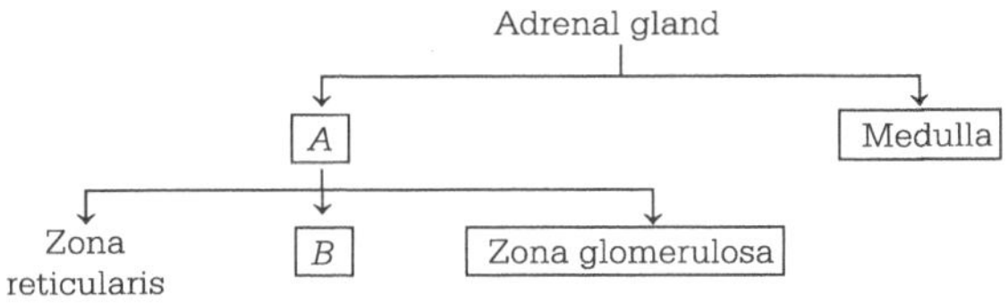

A. ________________________

B. ________________________

92. The <u>adipose</u> cells, present in medulla are the source of the hormone.
(*Rewrite the statement by changing the underlined word.*)

____

93. When adrenal medulla is removed, it does not affect the physiology of human. (*Give reason*)

____

94. Which hormones are secreted by part *A* in the given diagram of adrenal gland ?

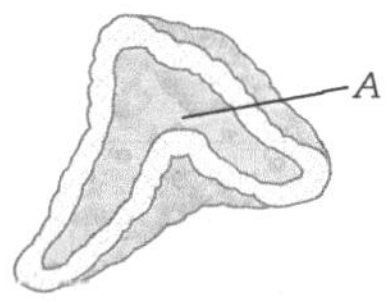

A. ________________________

95. _______________ and ____________________ are commonly called as catecholamines.

96. Medulla hormones are known as emergency hormones or <u>corticoids</u>.
(*Rewrite the statement by changing the underlined word.*)

____

97. Emergency hormones _____________ alertness, pupilary __________ and stimulate the breakdown of ____________ resulting in an increase concentration of glucose in blood.

98. Complete the following table with suitable words.

	Human systems	Effect of adrenaline	Effect of nor-adrenaline
(i)	Blood pressure	(A)	No effect
(ii)	Systolic	Increase	(B)
(iii)	Diastolic	(C)	Increase
(iv)	Respiration	(D)	Stimulated
(v)	Kidney	Renal vessel constrictor	(E)
(vi)	CNS	(F)	No effect

A. _______________ B. _______________ C. _______________

D. _______________ E._______________ F._______________

99. In the hypertension situation, ____________________ (nor-epinephrine/epinephrine) secreted in abnormal quantities.

____

100. The adrenal cortex produces three groups of hormones from cholesterol.

True ☐ False ☐

101. Give a collective term for the following chemicals.

Glucocorticoids, Mineralocorticoids, Androgens

____

102. Complete the following analogy.

Adrenal medulla : Adrenal cortex : : ______________ : Corticoids

103. Match the following columns.

	Column I (Hormones)		Column II (Functions)
A.	Glucocorticoids	1.	Regulate the balance of water and electrolyte
B.	Androgens	2.	Carbohydrate metabolism
C.	Mineralocorticoids	3.	Development of secondary sex characters

A. ________________ B. ________________ C. ________________

104. Glucocorticoids stimulate ________________ , ______________ and ______________

105. Glucocorticoids inhibit cellular uptake and utilisation of <u>ketones</u>.
(*Rewrite the statement by changing the underlined word.*)

____

106. Give a collective term for the following chemicals.

Cortisol, Cortisone, Hydrocortisone

____

107. Cortisol is involved in maintaining the cardio-vascular system and brain functions.

True ☐ False ☐

108. ____________________ (Cortisol/Aldosterone) produces anti-inflammatory reactions and suppresses the immune response.

109. Name the corticoids hormone, which acts in the given reaction.

Pyruvate $\rightarrow$ Glycerol $\rightarrow$ Glucose

____

110. The secretion of glucocorticoids is stimulated by ______________________________ from the anterior pituitary.

111. Some steps are incorrect in the following diagrammatic representation of regulation of glucocorticoids secretion. Redraw the flowchart with correct step.

Hypothalamus
↓
Corticotrophin Releasing Factor (CRF)
↓
Anterior lobe of pituitary gland
↓
ACTH
↓
Adrenal cortex
↓
Lower blood glucocorticoid level

____

112. ____________ is the chief mineralocorticoid hormone in our body.

113. Write the effect of aldosterone by mentioning Yes/No on following processes.

(i) Excretion of K^+ and phosphate ions. ____________

(ii) Reabsorption of Na^+. ____________

(iii) Maintenance of electrolyte. ____________

(iv) Reduce the number of lymphocytes. ____________

(v) Maintenance of osmotic pressure. ____________

(vi) Maintenance of blood pressure. ____________

114. What will happen to aldosterone when the sodium level falls in blood?

____

115. Name the synthetic aldosterone.

____

116. Androgenic steroids are secreted by the adrenal ____________ (medulla/cortex) which play a role in the growth of axial hair, pubic hair and facial hair during puberty.

117. Androgens are associated with deposition of protein in muscles and retention of nitrogen, especially in females.

True ☐ False ☐

118. The precursor of ____________ and ____________ synthesis is Dehydroxy Epiandrosterone (DHEA).

119. Complete the following analogy with respect to adrenal cortex secretion.

Hyposecretion : Hypersecretion : : ____________ : Cushing's syndrome

120. How many symptoms are related to Addison's disease.

> Acute weakness, Vomiting, Mental clouding, Hyperglycaemia, High BP, Acidosis, Renal failure, Fatigue

____

121. Name the disease in which symptoms like painful adiposity on face, excess catabolism, arrest of growth, osteoporosis, menstrual disturbances, psychotic depression are seen.

____

Topic 7 : Pancreas

122. Pancreas is a composite gland which acts as both ____________ and ____________ gland.

123. The endocrine part of pancreas consists of villi.
(*Rewrite the statement by changing the underlined word.*)

____

124. Complete the following flowchart.

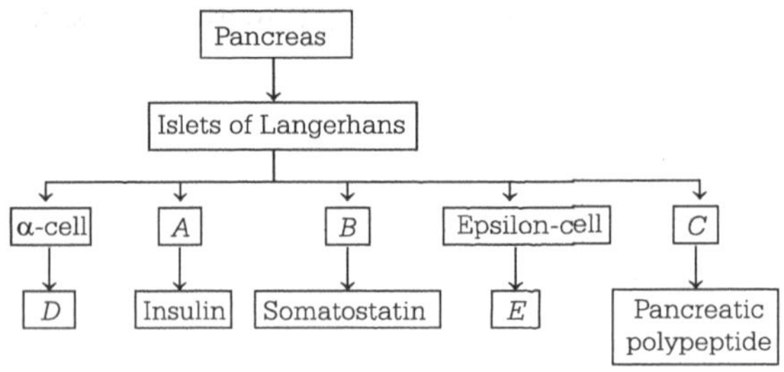

A. ____________________

B. ____________________

C. ____________________

D. ____________________

E. ____________________

125. Normal blood glucose level of body is maintain by a peptide hormone, i.e. ____________ .

126. Glucagon acts on the liver cells and stimulates glycolysis and glyconeogenesis.
(*Rewrite the statement by changing the underlined word.*)

____

127. Insulin acts mainly on __________________ and __________________ .

128. Complete the following analogy with respect to blood glucose level.

Insulin : ____________ : : Hypoglycemic : Hyperglycemic

129. Match the following columns.

	Column I		Column II
A.	Glucagon	1.	Increased blood sugar
B.	Insulin	2.	Decreased blood sugar
C.	Somatostatin	3.	Reduces cellular uptake of glucose
		4.	Inhibit the secretion of glucagon and insulin
		5.	Enhanced cellular uptake of glucose

A. ________________ B. ________________ C. ________________

130. Insulin stimulates conversion of glucose to glycogen, i.e. glycogenesis.

True ☐ False ☐

131. Name the hormone which acts antagonistic to insulin.

____

132. Insulin deficiency results in a disorder called diabetes insipidus.
(*Rewrite the statement by changing the underlined word.*)

____

133. ______________ is associated with loss of glucose through urine and formation of harmful compounds known as ______________ .

134. Diabetic patients are successfully treated with glucagon therapy.

True ☐ False ☐

135. Given below are some symptoms of diabetes mellitus. Give one word for the following symptoms.

(i) The excretion of glucose through urine. ______________

(ii) Excessive urination. ______________

(iii) Accumulation of the intermediate fat metabolite, acetyl coenzyme-A.

____

136. Write the types of diabetes mellitus.

____

Topic 8 : Testis and Ovary

137. The testes are situated inside the abdominal cavity within a pouch called scrotum.

True ☐ False ☐

138. Name the organ which acts as primary sex organ as well as an endocrine gland in male.

____

139. Testis is composed of ______________ and ______________

140. Androgens mainly testosterone are produced by Leydig cells or interstitial cells.

True ☐ False ☐

141. Which organs will not develop in a human, if he was undergone castrated in childhood?

____

142. Androgens play a major stimulatory role in the process of erythropoietin.
(*Rewrite the statement by changing the underlined word.*)

____

143. How many actions are stimulated by androgens from the given box?

Act on CNS, Catabolic effect on protein metabolism, Influence libido, Aggressiveness, Growth of facial hairs

____

144. Hyposecretion of androgens in man results into ________________ disorder.

145. Name primary female sex organ which produces one ovum during each menstrual cycle.

____

146. The ovarian steroids in females are ____________________ and ________________.

147. Consider the following flowchart, identify hormones *A* and *B*.

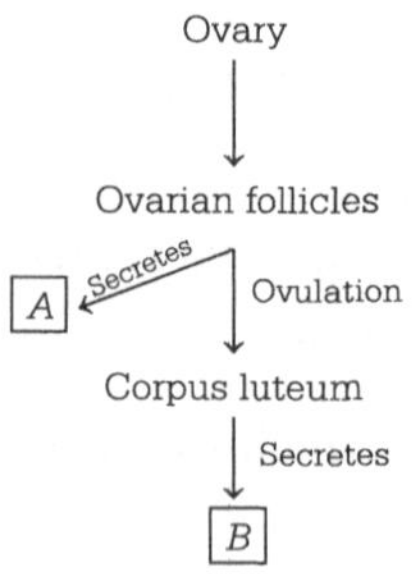

A. ______________________________

B. ______________________________

148. Identify the odd one out from the following words.

Relaxin, Oestrogen, Inhibin, Cortisol

____

149. Name the hormone which stimulates female secondary sex organs and helps in development of growing ovarian follicles and mammary gland.

____

150. Complete the following analogy considering 14th day of menstrual cycle.

FSH : LH : : Decrease : ________________ .

151. Pregnancy kit detect the presence or absence of hormone ______________________ from the urine of a female.

152. ____________________ (Progesterone/Oestrogen) is a luteal hormone which is synthesised by corpus luteum.

153. Progesterone is a ______________ hormone.

154. In the presence of progesterone, the thickening of the endometrium increases, it becomes secretory.

True ☐ False ☐

155. ____________________ (Progesterone/Relaxin) also acts on the mammary glands and stimulates the formation of alveoli and milk secretion.

156. Oestrogens and progesterone are ____________________ (decreased/increased) at the commencement of menstrual cycle.

157. Match the following columns.

	Column I (Female sex hormones)		Column II (Functions)
A.	Relaxin	1.	Its presence in urine indicates pregnancy
B.	Human chorionic gonadotropin	2.	It inhibits or activates gonadotropic activity of pituitary gland
C.	Inhibin/Activin	3.	It helps in relaxing of uterus

A. ________________ B. ________________ C. ________________

158. Precocious puberty occurs due to excessive production of ________________ .

159. The oestrogen and testosterone ratio disbalance in males is known as <u>hypogonadism</u>.
(*Rewrite the statement by changing the underlined word.*)

____

Topic 9 : Hormones of Heart, Kidney and Gastrointestinal Tract

160. The atrial wall of our heart secretes a peptide hormone called

____ .

161. Complete the following flowchart.

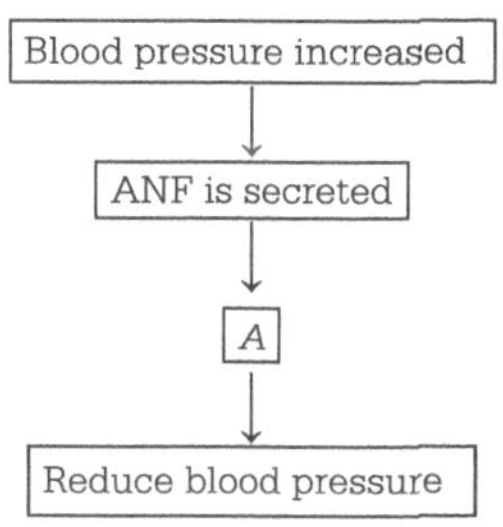

A. ________________

162. ANF inhibits renin secretion and initiates renin-angiotensin-aldosterone system.

True ☐ False ☐

163. Erythropoiesis stimulated by a peptide hormone called ________________ .

164. The <u>Henle's loop</u> cells of kidney produce erythropoietin.
(*Rewrite the statement by changing the underlined word.*)

____

165. The kidney play a significant role in regulation of blood calcium levels through the production of ________________ .

166. Name the hormones which is secreted from gastro-intestinal tract.

____

167. Gastrin stimulates the secretion of renin and amylase.
(*Rewrite the statement by changing the underlined words.*)

____

168. Secretin is a peptide hormone secreted by the small intestine.
True ☐ False ☐

169. Secretin stimulates secretion of ______________ and ______________.

170. GIP acts on both pancreas and gall bladder.
True ☐ False ☐

171. ______________ (CCK/GIP) inhibits gastric secretion and motility.

172. ______________ (Osteocalcin/Leptin) produced by osteoblasts, stimulates the pancreatic β-cells to increase insulin.

173. Match the following columns.

	Column I (Hormones)		Column II (Characteristics)
A.	Gastrin	1.	Stimulates secretion of pancreatic enzymes
B.	Secretin	2.	Acts on gastric gland
C.	CCK	3.	Acts on exocrine pancreas

A. ______________ B. ______________ C. ______________

174. Match the following columns.

	Column I (Local hormones)		Column II (Characteristics)
A.	Histamine	1.	Interleukins and interferons
B.	Pheromones	2.	Blood coagulation
C.	Cytokines	3.	Synthesised by the mast cell
D	Thromboxanes	4.	Communication among organisms

A. ______________ B. ______________ C. ______________ D. ______________

Topic 10 : Mechanism of Hormone Action

175. Hormones bind to the specific proteins, i.e. ______________ to produce their effects on target tissues.

176. Where are the hormone receptors present?

____

177. Membrane bound receptors are present on the cell membrane of the target cells.
True ☐ False ☐

178. ______________ receptors are present inside the target cell.

179. Hormone binds to its receptor leading to the formation of <u>peptide</u> receptor complex. (*Rewrite the statement by changing the underlined word.*)

____

180. Target tissues change their biochemical nature to form hormone-receptor complex.

True ☐ False ☐

181. Give a common biological terms for the following groups on the basis of chemical nature of hormones.

(i) Melatonin, Catecholamines, Adrenaline, Nor-adrenaline ________________

(ii) ADH, Oxytocin, ACTH, Calcitonin ________________

(iii) Insulin, Glucagon, Parathyroid, Relaxin ________________

182. Identify the odd one out from the following on the basis of their chemical nature.

Cortisol, Relaxin, Aldosterone, Testosterone, Oestrogen

____

183. Hormones which interact with membrane bound receptors, generate <u>enzymes</u>. (*Rewrite the statement by changing the underlined word.*)

____

184. Give a scientific collective term for the following.

Cyclic AMP, IP_3, Ca^{++}

____

185. Name the group of hormones which interact with intracellular receptors.

____

186. Consider the given diagram of mechanism of protein hormone action. Identify *A*, *B* and *C*.

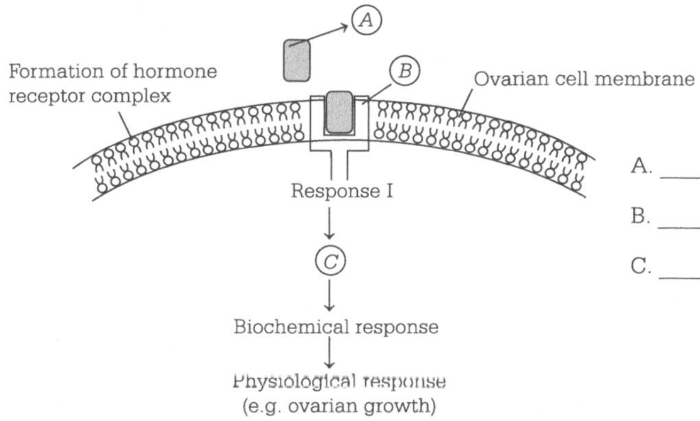

A. ________________

B. ________________

C. ________________

. FSH hormone receptor complex activates enzyme ______________ .

______________ and ______________ do not bind to the cell surface receptors.

Steroid and thyroid hormones are water soluble.

True ☐ False ☐

190. Consider the following figure of mechanism of steroid hormone action. Identify *A* and *B*.

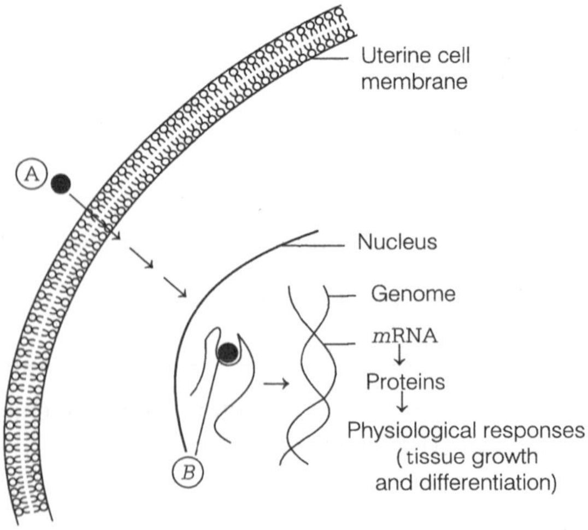

191. Steroid hormones have their receptor localised inside the mitochondria. (*Rewrite the statement by changing the underlined word.*)

____

192. ______________ , ______________ and ______________ hormones are water soluble, which cannot pass through the lipid membrane.

Answers

1. Hormones
2. Neural system, endocrine
3. Ductless
4. Hormone
5. Hormone
6. Endocrine glands, hormone
7. (i) Pituitary → Thymus → Pancreas → Adrenal → Ovary
   (ii) (a) *A*–Pituitary gland, *B*-Thymus, *C*–Pancreas, *D*-Adrenal, *E*-Testis, *F*–Ovary
   (b) Pancreas,
   (c) Ovary
8. Gastrointestinal tract, liver, kidney and heart.
9. Nuclei
10. False. The hormones released by hypothalamus regulate the synthesis and secretion of pituitary hormone.
11. (i) Releasing hormones and inhibiting hormones
    (ii) Releasing
    (iii) Inhibiting
    (iv) Anterior
12. Somatostatin
13. Somatostatin
14. Posterior pituitary
15. Sella tursica
16. Adenohypophysis, neurohypophysis
17. Pars distalis and pars intermedia.
18. Melanocyte stimulating hormone is odd one out. All hormones are released from pars distalis except melanocyte stimulating hormone which is secreted by pars intermedia.
19. 3(i, ii, iii)
20. (i) Gonadotropin Releasing Hormone,
    (ii) Growth Hormone,
    (iii) Prolactin,
    (iv) Thyroid Stimulating Hormone,
    (v) Adrenocorticotrophic Hormone,
    (vi) Luteinizing Hormone,
    (vii) Follicle Stimulating Hormone,
    (viii) Melanocyte Stimulating Hormone
21. Posterior pituitary
22. Posterior pituitary
23. Hypothalamus

24. A–Pars intermedia, B–Pars nervosa
25. (i) Acromegaly,
(ii) Hypopituitarism,
(iii) Hyperpituitarism
(iv) Cushing's syndrome
26. (i)-(b)-2, (ii)-(c)-(4), (iii)-(a)-(1), (iv)-(c)-(3)
27. LH and FSH
28. Androgens
29. FSH, androgens
30. (i) Graafian, (ii) Ovarian
31. Melanocytes
32. MSH
33. True
34. Oxytocin
35. Oxytocin
36. Distal tubules, diuresis
37. Vasopressin
38. (i) ADH is released.
39. (i) Diabetes insipidus,
(ii) Antidiuresis
40. Pineal gland
41. Diurnal, sleep-wake, temperature
42. Melatonin
43. Trachea
44. Follicles, stromal tissues
45. Tetraiodothyronine, thyrocalcitonin
46. (i) Triiodothyronine,
(ii) Tetraiodothyronine,
(iii) Anti-diuretic hormone,
(iv) Thyrocalcitonin
47. A–Thyroid gland, B–Trachea, C–Vocal cord
48. True
49. (i) Thyroid gland, (ii) Iodide
50. Hypothyroidism
51. Goitre
52. Hypothyroidism
53. Myxoedema
54. True
55. Hashimoto's disease
56. Hyperthyroidism
57. Exophthalmic goitre
58. (i)-(b)-(3), (ii)-(a)-(2), (iii)-(a)-(1), (iv)-(a)-(4)
59. (i) Decrease, (ii) Increase, (iii) Increase
60. A – TSH
61. Hypothyroidism
62. True
63. True
64. (i) Increase,
(ii) Increase,
(iii) Increase
65. Thyrocalcitonin (TCT)
66. Thyroid
67. Reabsorption
68. True
69. Parathyroid
70. A–Thyrocalcitonin
71. (i) A–Parathyroid gland,
(ii) Parathyroid hormone,
(iii) Parathyroid glands are present on the back side of the thyroid gland.
72. Thyrocalcitonin
73. Calcium
74. Blood
75. False. PTH acts on bones and stimulates the process of bone resorption.
76. Reabsorption increases
77. A–Increase plasma Ca^{+2}
78. Hyperparathyroidism
79. True
80. Hypoparathyroidism
81. Thyrocalcitonin decreases the calcium level in blood whereas parathyroid hormone increases the calcium level in blood.
82. Thymus
83. Peptide, thymus
84. True
85. False. Thymosin plays important role in proliferation and maturation of T-lymphocytes.
86. Humoral immunity.
87. Thymus is degenerated in old individuals resulting in a decreased production of thymosin. As a result, the immune responses of old persons become weak.
88. One pair
89. Anterior, kidney
90. False. The centrally located tissue is called the adrenal medulla and outside this lies the adrenal cortex.
91. A-Cortex, B-Zona fasciculata
92. Chromaffin
93. It does not affect the human physiology because hormone secreting cell, (i.e. chromaffin cell) are present throughout the body.
94. Adrenaline and nor-adrenaline
95. Adrenaline, nor-adrenaline
96. Hormones of fight or flight
97. Increase, dilation, glycogen
98. A-Increase, B-Increase, C-Decrease, D-Stimulated, E-Renal vessel constrictor, F-Increase mental anxiety.
99. Nor-epinephrine
100. True
101. Corticoids
102. Catecholamines
103. A-2, B-3, C-1
104. Gluconeogenesis, lipolysis, proteolysis
105. Amino acids
106. All of these are the examples of glucocorticoids.
107. False. Cortisol is involved in maintaining the cardio-vascular system and kidney functions.
108. Cortisol
109. Cortisol
110. Adrenocorticotrophic Hormone (ACTH)
111. Hypothalamus
↓
CRF
↓
Anterior lobe of pituitary gland
↓
ACTH
↓
Adrenal cortex
↓
Raised blood glucocorticoid level

rone

, (ii) Yes, (iii) Yes, (iv) No, es, (vi) Yes

ore aldosterone is secreted

Fludrocortisone

. Cortex

117. False. Androgens are associated with deposition of protein in muscles and retention of nitrogen, especially in males.

118. Testosterone, oestrogen.

119. Addison's disease

120. 5(Acute weakness, vomiting, mental clouding, acidosis, renal failure and fatigue).

121. Cushing's syndrome

122. Exocrine, endocrine

123. Islets of Langerhans

124. *A*–β-cell, *B*–δ-cell, *C*–F-cell, *D*–Glucagon, *E*–Acylghrelin

125. Glucagon

126. Glycogenolysis

127. Hepatocytes, adipocytes

128. Glucagon

129. A-1, 3, B-2, 5,C-4

130. True

131. Glucagon

132. Mellitus

133. Diabetes mellitus, Ketone bodies

134. False. Diabetic patients are successfully treated with insulin therapy.

135. (i) Glycosuria, (ii) Polyuria, (iii) Ketoacidosis

136. (i) Insulin dependent diabetes mellitus.
(ii) Insulin independent diabetes mellitus.

137. False. Testes are situated outside the abdominal cavity within pouch called scrotum.

138. Testis

139. Seminiferous tubules, Interstitial tissues

140. True

141. Epididymis, seminal vesicle, vas deferens and penis.

142. Spermatogenesis

143. 4(Act on CNS, influence libido, aggressiveness, growth of facial hairs).

144. Eunuchoidism

145. Ovary

146. Oestrogen, progesterone

147. *A*–Oestrogen, *B*–Progesterone

148. Cortisol is odd one out. All of the others are female sex hormone.

149. Oestrogen

150. Increase

151. HCG (Human Chronic Gonadotropin)

152. Progesterone

153. Pregnancy

154. False. In the presence of progesterone, the thickening of the endometrium stops it becomes secretory.

155. Progesterone

156. Decreased

157. A-3, B-1, C-2

158. Sex hormone

159. Gynaecomastia

160. Atrial Natriuretic Factor (ANF)

161. *A*–Dilation of blood vessels

162. True

163. Erythropoietin

164. Juxtaglomerular

165. Calcitriol

166. Gastrin, secretin, Cholecystokinin (CCK) and Gastric Inhibitory Peptide (GIP).

167. Hydrochloric acid, pepsinogen

168. True

169. Water, bicarbonate ions

170. False. CCK acts on both pancreas and gall bladder.

171. GIP

172. Osteocalcin

173. A-2, B-3, C-1

174. A-3, B-4, C-1, D-2

175. Hormone receptor

176. Target tissues

177. True

178. Intracellular

179. Hormone

180. True

181. (i) Amines, (ii) Peptides (iii) Proteins

182. Relaxin is odd one out. Rest all are steroids.

183. Second messengers

184. Secondary messengers

185. Steroid hormones and iodothyronines

186. *A*–Hormone (e.g. FSH), *B*–Receptor, *C*–Generation of second messenger, e.g. *c*AMP, Ca^{+2}

187. Adenylcyclase

188. Steroid hormone, thyroid hormone

189. False. Steroid and thyroid hormones are lipid soluble.

190. *A*–Hormone (e.g. oestrogen), *B*–Hormone-receptor complex.

191. Nucleus

192. Protein, polypeptide, amino acid derivatives.